Fast Start Calculus
for Integrated Physics

Daniel Ashlock

 Ashlock & McGuinness Consulting, Inc.

Ashlock & McGuinness Consulting, Inc.

Fast Start Calculus for Integrated Physics
Third Edition

ISBN-13: 978-1540651914

ISBN-10: 1540651916

About the cover

The cover of this text is a Newton's method fractal. In Chapter 6 Newton's method is developed for finding roots of a function. The picture shows a small part of the complex plane with corners $z = 0.6 - 1.38i$ and $w = 0.96 - 1.02i$. The colored areas are those points inside that subset of the complex plane that converge to three of the roots of a complex polynomial with six roots at $\pm 1.5, \pm 1.2i$ and $\pm 1.8i$. Each point in the subset is used as a starting guess for Newton's method and colors are selected by which root the guess eventually converges to. Guesses that converge to the three roots $\pm 1.5 + 1.2i$ and $1.8i$ are colored violet, cyan, and yellow while all other points are left transparent.

This fractal and many others appear on my Fractal Taxonomy Website.

http://eldar.mathstat.uoguelph.ca/dashlock/ftax/

Acknowledgments

This text was written for a course developed by a team including my co-developers Joanne M. O'Meara of the Department of Physics and Lori Jones and Dan Thomas of the Department of Chemistry. Andrew McEachern, Cameron McGuinness, and Jeremy Gilbert have served as head TAs and instructors for the course over the last five years and had a substantial impact on the development of both the course and this text. Martin Williams and Miranda Schmidt have been able partners on the physics side delivering the course and helping get the integration of the calculus and physics correct. I also owe five years of students thanks for serving as the test bed for the material. Many thanks to all these people for making it possible to decide what went into the text and what didn't. I also owe a great debt to Wendy Ashlock and Cameron McGuinness at Ashlock and McGuinness Consulting for removing a large number of errors and making numerous suggestions to enhance the clarity of the text. This is the third edition that corrects several mistakes. Errata sheets are available from the publisher for the second edition.

About the Author

Daniel Ashlock is a Professor of Mathematics at the University of Guelph. He has a Ph.D. in Mathematics from the California Institute of Technology, 1990, and holds degrees in Computer Science and Mathematics from the University of Kansas, 1984. Dr. Ashlock has taught mathematics at levels from 7th grade through graduate school for over three decades starting at the age of 17. Over this time Dr. Ashlock has developed a number of ideas about how to help students overcome both fear and deficient preparation. This text, covering the mathematics portion of an integrated mathematics and physics course, has proven to be one of the more effective methods of helping students learn mathematics with physics serving as an ongoing anchor and example.

Contents

Preface

This text covers single-variable calculus and selected topics from multi-variate calculus. The text was developed for a course that arose from a perennial complaint by the physics department at the University of Guelph that the introductory calculus courses covered topics roughly a year after they were needed. In an attempt to address this concern, a multi-disciplinary team created a two-semester integrated calculus and physics course. The idea is that the calculus will be delivered before it is needed, often just in time, and that the physics will serve as a giant collection of motivating examples that will anchor the student's understanding of the mathematics.

The course has run three times before this text was started, and it was used in draft form for the fourth offering of the course. Thus, there is a good deal of classroom experience and testing behind this text. There is also enough information to confirm our hypothesis that the course would help students. The combined drop and flunk rate for this course is consistently under 3% – where 20% is more typical for first-year university calculus. Co-instruction of calculus and physics works. It is important to note that we did not achieve these results by watering down the math. The topics covered, in two semesters, are about half again as many as are covered by a standard first-year calculus course. That's the big surprise: covering more topics faster increased the average grade and reduced the failure rate. Using physics as a knowledge anchor worked even better than we had hoped.

This text makes a number of innovations that have caused mathematical colleagues to raise objections. In mathematics it is traditional, even dogmatic, that math be taught in an order in which no thing is presented until the concepts on which it rests are already in hand. This is correct, useful dogma for mathematics students. It also leads to teaching difficult proofs to students who are still hungover from beginning-of-semester parties. This text neither emphasizes nor neglects theory, but it does move theory away from the beginning of the course in acknowledgement of the fact that this material is philosophically difficult and intellectually challenging. The course also presents a broad integrated picture as soon as possible. The fundamental theorem and its consequence that differentiation and integration are different points of view on the same thing is presented relatively early, for example. The text also emphasizes cleverness and computational efficiency. Remember that "mathematics is the art of avoiding calculation."

It is important to state what was sacrificed to make this course and this text work the way they do. This is not a good text for math majors, unless they get the theoretical parts of calculus later in a real analysis course. The text is relatively informal, almost entirely example driven, and application motivated. The author is a math professor with a CalTech Ph.D. and three decades of experience teaching math at all levels from 7th grade (as a volunteer) to graduate education including having supervised a dozen successful doctoral students. The author's calculus credits include calculus for math and engineering, calculus for biology, calculus for business, and multivariate and vector calculus. The author and development team welcome feedback – we are already working on an improved edition of this text.

Chapter 1

Math You Should Know and the Library of Functions

This book is a text on calculus, structured to prepare students for applying calculus to the physical sciences. The first chapter has no calculus in it at all; it is here because many students manage to get to the university or college level without adequate skill in algebra, trigonometry, or geometry. We assume you are already familiar with the concept of **variables** like x and y that denote numbers whose value is not known.

1.1. Solving Equations

An **equation** is an expression with an equals sign in it. For example:

$$x = 3$$

is a very simple equation. It tells us that the value of the variable x is the number 3. There are a number of rules you can use to manipulate equations. What these rules do is change an equation into another equation that has the same meaning but a different form. The things you can do to an equation without changing its meaning include:

- Add or subtract the same term from both sides. If that term is one that is already present in the equation, we may call this **moving the term to the other side**. When this happens, the term changes sign, from positive to negative or negative to positive. For example:

$$x - 4 = 5 \qquad \text{This is the original equation}$$
$$x = 5 + 4 \qquad \text{Add 4 to both sides (move 4 to the other side)}$$
$$x = 9 \qquad \text{Finish the arithmetic}$$

- Multiply or divide both sides by the same expression. For example:

$$3x - 4 = 5 \qquad \text{This is the original equation}$$
$$3x = 5 + 4 \qquad \text{Add 4 to both sides (move 4 to the other side)}$$
$$3x = 9 \qquad \text{Finish the addition}$$
$$\frac{3x}{3} = \frac{9}{3} \qquad \text{Divide both sides by 3}$$
$$x = 3 \qquad \text{Finish the arithmetic}$$

- Apply the same function or operation to both sides. For example:

$$\sqrt{x - 2} = 5 \qquad \text{This is the original equation}$$
$$\left(\sqrt{x - 2}\right)^2 = 5^2 \qquad \text{Square both sides}$$
$$x - 2 = 25 \qquad \text{Do the arithmetic}$$
$$x = 27 \qquad \text{Move 2 to the other side}$$

Knowledge Box 1.1.

The rules for solving equations include:

(a) Adding or subtracting the same thing from both sides.
(b) Moving a term to the other side; its sign changes.
(c) Multiplying or dividing both sides by the same thing.
(d) Performing the same operation to both sides, e.g. squaring or taking the square root.

These rules are illustrated by the examples in this section.

Solving an equation can get very hard at times and much of what we will do in this chapter is to give you tools for solving equations efficiently.

Example 1.1. Solve the equation $3x - 7 = x + 3$ for x.

Solution:

$$3x - 7 = x + 3 \qquad \text{This is the original equation}$$
$$2x = 10 \qquad \text{Move 7 and } x \text{ to the other side}$$
$$x = 5 \qquad \text{Divide both sides by 2}$$

$\Diamond$

All the examples so far have been solving for the variable x. You can solve for any symbol.

Example 1.2. Solve $PV = nRT$ for P.

Solution:

$$PV = nRT \qquad \text{This is the original equation}$$
$$P = \frac{nRT}{V} \qquad \text{Divide both sides by } V$$

$\Diamond$

When we solve an equation for a variable, we put it in **functional form**. We will learn more about functions in Section 1.4. When an equation is in functional form, the variable we solved for represents a value we are trying to find. In Example 1.2, which is the Ideal Gas Law, this would be the pressure of a gas. We call this the **dependent variable**. The variable (or variables) on the other side of the equation represent values that are changing. We call these the **independent variables**. In Example 1.2, these are n and T, the amount and temperature of the gas. (R is the universal gas constant – not a variable.)

Let's try a slightly harder example.

Example 1.3. Solve $y = \dfrac{2x - 1}{x + 2}$ for x.

Solution:

$$y = \frac{2x - 1}{x + 2} \qquad \text{This is the original equation}$$

$$y(x + 2) = 2x - 1 \qquad \text{Multiply both sides by } (x + 2)$$

$$xy + 2y = 2x - 1 \qquad \text{Distribute } y$$

$$xy + 2y + 1 = 2x \qquad \text{Move 1 to the other side}$$

$$2y + 1 = 2x - xy \qquad \text{Move } xy \text{ to the other side}$$

$$2y + 1 = x(2 - y) \qquad \text{Factor out } x \text{ on the right}$$

$$\frac{2y + 1}{2 - y} = x \qquad \text{Divide both sides by } (2 - y)$$

$$x = \frac{2y + 1}{2 - y} \qquad \text{Neaten up}$$

$\Diamond$

Example 1.3 used a strategy. First, clear the denominator; second, get everything with x in it on one side and everything else on the other side; third, factor to get a single x; and fourth, divide by whatever is multiplied by x in order to isolate x, obtaining the solution.

Example 1.4. Solve $x^2 + y^2 = 25$ for y.

Solution:

$$x^2 + y^2 = 25 \qquad \text{This is the original equation}$$

$$y^2 = 25 - x^2 \qquad \text{Move } x^2 \text{ to the other side}$$

$$y = \pm\sqrt{25 - x^2} \qquad \text{Take the square root of both sides}$$

The original equation in Example 1.4 graphs as a circle of radius 5 centered at the origin. When we take the square root of both sides, the fact that squaring a number makes it positive means that the plus and minus square roots are both correct. By convention, if we need a single expression, we take the positive square root.

Problems

Problem 1.1. *Solve the following equations for x.*

(a) $7x + 4 = -17$

(b) $8x - 6 = 66$

(c) $7x + 5 = 68$

(d) $5x + 3 = 43$

(e) $5x + 4 = 19$

(f) $2x - 5 = -9$

Problem 1.2. *Solve the following equations for x.*

(a) $\sqrt{6x + 3} = 15$

(b) $\sqrt{7x + 8} = 13$

(c) $\sqrt{32x + 4} = 6$

(d) $\sqrt{x - 1} = 10$

(e) $\sqrt{39x + 3} = 9$

(f) $\sqrt{2x + 3} = 19$

Problem 1.3. *Solve the following equations for y.*

(a) $x^2 + 14y^2 = 4$

(b) $2x^2 + 10y^2 = 3$

(c) $2x^2 + 10y^2 = 20$

(d) $8x^2 + 17y^2 = 16$

(e) $8x^2 + 18y^2 = 9$

(f) $15x^2 + 12y^2 = 8$

Problem 1.4. *Solve the following equations for x.*

(a) $y = \dfrac{7x + 6}{4x + 3}$

(b) $y = \dfrac{9x - 6}{x + 1}$

(c) $y = \dfrac{4x + 2}{3x + 3}$

(d) $y = \dfrac{4x - 3}{9x + 9}$

(e) $y = \dfrac{4x - 3}{4x + 4}$

(f) $y = \dfrac{2x - 1}{5x + 4}$

Problem 1.5. *Solve for i:*
$$Ni = Kq$$

Problem 1.6. *Solve for a:*
$$mua = n$$

Problem 1.7. *Solve for K:*
$$\frac{1}{o} = \frac{KO}{r}$$

Problem 1.8. *Solve for U:*
$$\frac{IL}{Fc} = U$$

Problem 1.9. *Solve for z:*
$$\frac{1}{D} = \frac{nM}{z}$$

Problem 1.10. *Solve for i:*
$$\frac{1}{ni} = \frac{XJ}{e}$$

Problem 1.11. *Solve for n:*
$$YV = \frac{1}{gn}$$

Problem 1.12. *Solve for U:*
$$\frac{F}{D} = \frac{1}{Un}$$

1.2. Lines

We will start with an example of a very simple line: $y = x + 1$ (Figure 1.1). For any value of x we add one to x and get y. We can draw the line by picking any two points on it, lining up a straight edge on the two points, and drawing along the straight edge.

The line $y = x + 1$ is made of all the points that look like $(x, x + 1)$. Five points like that have been plotted in Figure 1.1: (-2,-1), (-1,0), (0,1), (1,2), and (2,3).

Definition 1.1. A line can always be written in the form $y = mx + b$ where m is the **slope** of the line, and b is the **y-intercept** of the line. Both m and b are numbers.

There are different ways to interpret the parameters m and b.

- The number m is the amount the y-value increases for each increase of 1 in the x-value.

- If we pick two points on the line, then the **run** between those points is the change in the x-value, and the **rise** is the change in the y-value. The **slope**, m, is the rise divided by the run.

- The **y-intercept** b is where the line hits the y-axis. This means the point $(0, b)$ is on the line.

Let's look at what happens when we have lines with different slopes (Figure 1.2) or different intercepts (Figure 1.3).

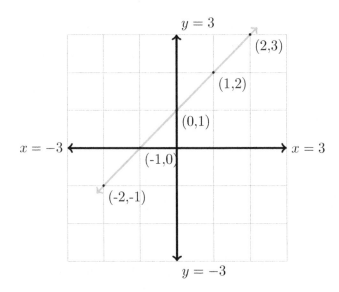

Figure 1.1. The line $y = x + 1$.

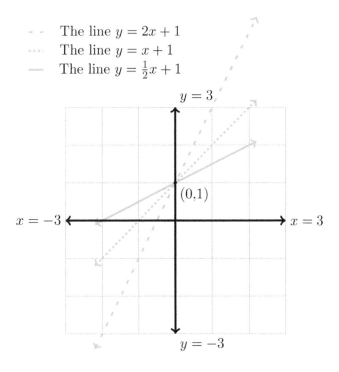

The line $y = 2x + 1$
The line $y = x + 1$
The line $y = \frac{1}{2}x + 1$

$y = 3$

$(0,1)$

$x = -3$

$x = 3$

$y = -3$

Figure 1.2. Changing slopes; same intercept.

1.2.1. Computing slopes, finding lines from points. If someone gives you the slope, m, and the intercept, b, for a line and asks you the equation of the line, then it is easy, $y = mx + b$. This form is called the **slope-intercept** form of a line. It is the standard form for reporting a line – if you're asked to "find" a line, then you find it in slope-intercept form unless the directions specifically request a different form.

Knowledge Box 1.2.

A line with slope m and y-intercept b has the equation:

$$y = mx + b$$

This is called **slope-intercept** form. It is the usual form for reporting a line.

A common state of affairs is to be given the slope of a line and a point that is on the line. This is one of the standard forms of a line and is easy to transform into slope intercept form. It has the very sensible name of **point-slope form**.

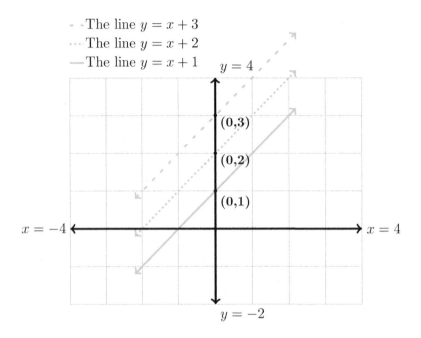

Figure 1.3. Changing intercepts; same slopes.

Knowledge Box 1.3.

A line with slope m containing the point (x_0, y_0) has the equation:

$$(y - y_0) = m(x - x_0)$$

This is called the **point-slope** form of a line.

Example 1.5. Find the line with slope $m = 2$ that passes through the point $(4, 4)$.

Solution:

$$
\begin{aligned}
y - y_0 &= m(x - x_0) && \text{Write the point-slope form} \\
y - 4 &= 2(x - 4) && \text{Put in the numbers} \\
y - 4 &= 2x - 8 && \text{Distribute the slope} \\
y &= 2x - 4 && \text{Add 4 to both sizes}
\end{aligned}
$$

So we see that the line (in slope-intercept form) is $y = 2x - 4$.

$\Diamond$

One of the things that you learn in elementary geometry is that **two points define a line**. This statement, while true, is a little short on details. Now that we have the point-slope form of a line, in order to find the equation of a line from two points, we need a way to find the slope from the two points. To do this, we need the fact that slope is rise over run. Given two points, you compute the rise and the run for those two points and then divide.

Knowledge Box 1.4.

Suppose that we have two points (x_0, y_0) and (x_1, y_1) and want to know the slope of the line that passes through both of them. If $x_0 \neq x_1$ then the slope of the line is

$$m = \frac{y_1 - y_0}{x_1 - x_0}$$

This is the **slope of a line through two points.**

Example 1.6. Find the slope of the line through the points $(1, 1)$ and $(3, 4)$.

Solution:

$$m = \frac{y_1 - y_0}{x_1 - x_0} = \frac{4 - 1}{3 - 1} = \frac{3}{2}$$

$$\Diamond$$

Now that we have a method for getting the slope of a line through two points, we can perform the slightly more complex task of finding the line through two points.

Example 1.7. Find the line through the points $(-2, 3)$ and $(2, 1)$.

Solution:

The first step is to find the slope of the line:

$$m = \frac{y_1 - y_0}{x_1 - x_0} = \frac{1 - 3}{2 - (-2)} = \frac{-2}{4} = \frac{-1}{2}$$

Now we pick either of the two points and apply the point-slope form. The point $(2,1)$ is slightly easier to work with so:

$$y - y_0 = m(x - x_0) \qquad \text{Write the point-slope form}$$

$$y - 1 = -\frac{1}{2}(x - 2) \qquad \text{Put in the numbers}$$

$$y - 1 = -\frac{1}{2}x + 1 \qquad \text{Distribute the slope}$$

$$y = -\frac{1}{2}x + 1 + 1 \qquad \text{Add 1 to both sides}$$

$$y = -\frac{1}{2}x + 2 \qquad \text{Simplify the equation}$$

So we see that the line (in slope-intercept form) is $y = -\frac{1}{2}x + 2$ whose graph looks like this:

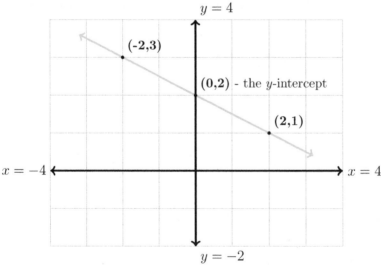

The line $y = -\frac{1}{2}x + 2$

The formula for a slope had a condition: $x_0 \neq x_1$. We now need to deal with this case. When the x-coordinates of the points are the same, we get a **vertical line**.

Knowledge Box 1.5.

The vertical line through the points (x_0, y_0) and (x_0, y_1) is the line

$$x = x_0$$

Unlike the other lines we have dealt with, it is *not* a function. In spite of that it is still a line. It contains the points: $(x_0, anything)$.

1.2.2. Parallel and orthogonal lines. Look at the lines in Figure 1.3. These three lines all have the same slope and they are parallel to one another. It turns out that this is always true.

Knowledge Box 1.6.

Parallel lines have the same slope.

Example 1.8. Find the line parallel to $y = 3x - 1$ that passes through the point (2,2).

Solution:

Since the line is parallel to $y = 3x - 1$, which has a slope of 3, we know that the slope of the new line is $m = 3$. Apply the point-slope formula:

$$y - 2 = 3(x - 2)$$
$$y - 2 = 3x - 6$$
$$y = 3x - 6 + 2$$
$$y = 3x - 4$$

and we have that the line parallel to $y = 3x - 1$ that passes through (2,2) is $y = 3x - 4$.

$\Diamond$

Figure 1.4 shows two lines that intersect at right angles. We say that lines that intersect at right angles are **orthogonal**.

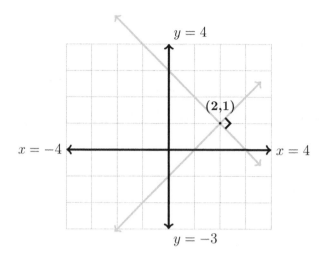

Figure 1.4. Two lines, $y = x - 1$ and $y = -x + 3$, that intersect at right angles.

There is a relationship between the slopes of any two orthogonal lines.

Knowledge Box 1.7.

Suppose that two lines intersect at right angles.
If the slope of the first line is m, then the slope of the second is:

$$-\frac{1}{m}$$

In words, orthogonal lines have negative reciprocal slopes.

Example 1.9. Find a line at right angles to the line $y = \frac{1}{2}x + 1$ that contains the point (2,2).

Solution:

Since we want the new line to be orthogonal to $y = \frac{1}{2}x + 1$, which has a slope of $\frac{1}{2}$, the new line will have a slope of $m = -\frac{1}{1/2} = -2$. Again, we use the point-slope formula to build the line.

$$y - 2 = -2(x - 2)$$
$$y - 2 = -2x + 4$$
$$y = -2x + 6$$

Make your own graph of these two lines to see if they are, actually, at right angles.

$\Diamond$

Example 1.10. Are the points (1,1), (3,3) and (6,0) the vertices of a right triangle?

Solution:

These three points can be grouped into three pairs of points. Each of these pairs of points define a line. If the three points are the vertices of a right triangle, then two of the lines have negative reciprocal slopes. Let's compute the three slopes of the three pairs of lines:

$$m_1 = \frac{3-1}{3-1} = \frac{2}{2} = 1$$
$$m_2 = \frac{0-3}{6-3} = \frac{-3}{3} = -1$$
$$m_3 = \frac{0-1}{6-1} = \frac{-1}{5}$$

Notice that $-\dfrac{1}{m_1} = -1 = m_2$, and so two of the lines are at right angles – which means the three points are the vertices of a right triangle.

Problems

Problem 1.13. *Graph the lines with the following slopes and intercepts.*

(a) m $= -\frac{7}{3}$, b $= 2$

(b) m $= -1$, b $= 5$

(c) m $= -\frac{7}{2}$, b $= -1$

(d) m $= -2$, b $= -2$

(e) m $= 2$, b $= -4$

(f) m $= -\frac{10}{3}$, b $= 0$

(g) m $= -2$, b $= 5$

(h) m $= -\frac{2}{3}$, b $= -1$

(i) m $= 1$, b $= -1$

(j) m $= -\frac{13}{5}$, b $= 0$

Problem 1.14. *For the following pairs of points, find the slope of the line through the points.*

(a) (0,0) and (2,2)

(b) (0,0) and (2,1)

(c) (0,3) and (3,0)

(d) (1,4) and (4,2)

(e) (-1,-1) and (4,2)

(f) (2,1) and (1,2)

(g) (2,2) and (-3,5)

(h) (5,0) and (2,2)

Problem 1.15. *True or false: it does not matter which of the points is the first or second point when you plug them into the formula for the slope of a line through two points. Write a paragraph explaining your answer.*

Problem 1.16. *For each of the pairs of points in Problem 1.14, find a formula for the line in point-slope form.*

Problem 1.17. *For each of the pairs of points in Problem 1.14, find a formula for the line in slope-intercept form.*

Problem 1.18. *For the lines defined in Problem 1.13 parts a-f, find a parallel line that passes through the point (1,2). Present your answer in slope-intercept form.*

Problem 1.19. *For each of the following triples of points, find a line parallel to the line through the first two points that contains the third.*

(a) (0,1) and (2,3) and (1,5)

(b) (-1,2) and (4,1) and (0,0)

(c) (4,0) and (2,7) and (2,1)

(d) (3,6) and (1,4) and (3,1)

(e) (2,2) and (8,3) and (4,-5)

(f) (5,-3) and (-2,4) and (-1,-1)

Problem 1.20. *For each of the following pairs of points, find a line at right angles to the line through the points that contains the first point.*

(a) (0,0) and (2,2)

(b) (1,3) and (3,1)

(c) (-1,1) and (4,-3)

(d) (2,2) and (0,0)

(e) $(\frac{1}{2},\frac{3}{2})$ and (1,1).

(f) (-8,4) and (3,2).

Problem 1.21. *Suppose that five roses cost $10.00 and seven roses cost $12.50. Assuming the cost is given by a linear function, determine the marginal cost per rose and the base cost for operating the flower shop. Use the resulting linear function to compute the cost of a dozen roses.*

Problem 1.22. *Suppose that a dozen eggs cost \$4.50 and eighteen eggs cost \$10.50. Assuming the cost is given by a linear function, determine the marginal cost per egg and the base cost for operating the grocery store. Use the resulting linear function to compute the cost of five dozen eggs.*

Problem 1.23. *A ball is thrown from a rooftop in a horizontal direction. Before it hits the ground, the distance from the building is measured as well as the height of the ball above the ground as a function of time. If we neglect the effects of air resistance, which of the two quantities follows a line? Explain in a few sentences.*

Problem 1.24. *Water comes out of an open spigot at the bottom of a cylindrical tank. The pressure of the water still in the tank is what drives the flow. True or false: the rate of water flow is a linear function of time. Explain in a few sentences.*

Problem 1.25. *Which of the following triples of points are the vertices of a right triangle? Remember that you must justify your conclusions with a sentence.*

(a) $A = (2,3)$, $B = (4,7)$, and $C = (6,1)$

(b) $A = (1,3)$, $B = (2,6)$, and $C = (4,2)$

(c) $A = (0,0)$, $B = (2,4)$, and $C = (7,1)$

(d) $A = (-2,1)$, $B = (0,6)$, and $C = (5,4)$

(e) $A = (-3,4)$, $B = (-1,8)$, and $C = (1,3)$

(f) $A = (4,4)$, $B = (1,1)$, and $C = (0,5)$

Problem 1.26. *Consider the points (1,2) and (3,4).*

(a) How many points (x,y) are there that form the vertices of a right triangle together with the given points?

(b) Find them.

Problem 1.27. *Prove or disprove that the points (1,3), (2,0), (4,4), and (5,1) are the vertices of a square.*

Problem 1.28. *Construct a procedure for testing four points to see if they are the vertices of a square and demonstrate it on (0,3), (2,0), (3,5), and (5,2) which do form the vertices of a square.*

Problem 1.29. *Find three points that, together with (0,0), are not the vertices of a rectangle. Give a reason your answer is correct.*

Problem 1.30. *True or false, and justify your answer.*

- Any three distinct points not on the same line are the vertices of a triangle.

- Any four distinct points, no three of which are on the same line, are the vertices of a quadrilateral.

Problem 1.31. *Suppose that t is time. Which of the following sets of points lie on a single line in the plane? In some cases plotting the points may help. You may want to consult Section 1.7 if you are unfamiliar with the sine and cosine functions.*

(a) Points: $(3t, t)$

(b) Points: $(\cos(t), \cos(t))$

(c) Points: $(\cos(t), \sin(t))$

(d) Points: $(t, t^2 - t + 1)$

(e) Points: $(t^2, 2t^2 + 1)$

Problem 1.32. *For the points (0,0), (1,2), (2,1), (3,5), and (2,4) there are ten possible lines that go through two of these points. What is the largest slope and the smallest slope of these lines?*

Problem 1.33. *Suppose we are tracking a point on a spinning disk over time. Find some computable, linear quantity that describes the point's position.*

1.3. Quadratic Equations

After a line, the simplest type of equation is a **quadratic equation**. Its name comes from the Latin word for "square," and it is characterized by always including a squared term. When graphed, it has a shape like that shown in Figure 1.5. This shape is called a **parabola**.

Knowledge Box 1.8.

A quadratic equation is an equation in the form
$$y = ax^2 + bx + c$$
where a, b, and c are constants and $a \neq 0$.

The usual goal, when we have a quadratic equation, is to find the values that make the equation zero or **satisfy** the equation. If we have the equation $y = x^2 - 3x + 2$ and if $x = 1$, we get $1 - 3 + 2 = 0$, and so $x = 1$ satisfies the equation. Similarly, if $x = 2$, then we get $4 - 6 + 2 = 0$, and so $x = 2$ also satisfies the equation. A number that satisfies an equation is a point where the graph of the equation crosses the x-axis. A solution to an equation is also called a **root** of the equation. How can a quadratic equation have two solutions?

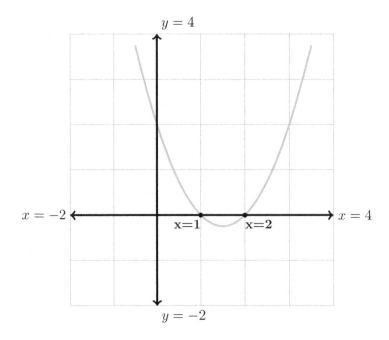

Figure 1.5. The equation $x^2 - 3x + 2 = 0$ has roots at $x = 1, 2$.

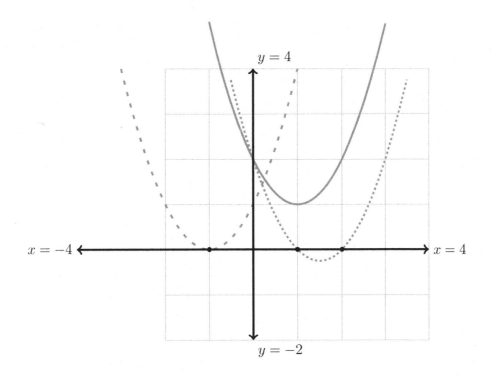

Figure 1.6. Quadratic equations with 0, 1, or 2 roots.

It turns out that a quadratic equation can have 0, 1, or 2 roots. Figure 1.6 shows how this is possible.

When the graph of the equation never crosses the x-axis, there are no roots; when it just touches the x-axis at one point, there is one root. When it crosses the x-axis twice, there are two roots. There is a formula we can use to tell how many roots a quadratic equation has. It is called the **discriminant**.

Knowledge Box 1.9.

The number of roots of $ax^2 + bx + c = 0$ is determined by looking at the **discriminant**:
$$D = b^2 - 4ac$$
If $D < 0$, then the equation has no roots; if $D = 0$, then the equation has one root; if $D > 0$, then the equation has two roots.

1.3.1. Factoring a quadratic. The simplest method of solving a quadratic equation is to factor it. If the equation has no roots, then it won't factor. A quadratic equation with one root is a perfect square of one term of the form $(x - a)$, possibly multiplied by a constant. When a quadratic has two roots, it factors into two different terms. Let's look at some examples of quadratics that factor:

Example 1.11.

$$x^2 - 3x + 2 = (x - 1)(x - 2)$$
$$x^2 - 9 = (x - 3)(x + 3)$$
$$x^2 - 8x + 15 = (x - 5)(x - 3)$$
$$x^2 + 4x + 4 = (x + 2)(x + 2)$$
$$x^2 - x - 2 = (x - 2)(x + 1)$$
$$2x^2 + 3x + 1 = (2x + 1)(x + 1)$$

$$(Ax + B)(Cx + D) = ACx^2 + (AD + BC)x + BD$$
$$\diamond$$

The last line of Example 1.11 is a **generic factorization**.

The rules for factors:

(a) The product of the two constants must be the constant term of the quadratic.

(b) Adding the results of multiplying the number in front of x in one term by the constant in the other term must be the number in front of x in the quadratic.

(c) The product of the numbers in front of the x's must be the number in front of x^2 in the quadratic.

So, for example, $x^2 - 3x + 2 = (x - 1)(x - 2)$ uses the facts that $(-1)(-2) = 2$ and $-1 - 2 = -3$ to tell us that the factors are $(x - 1)$ and $(x - 2)$, which makes the roots 1 and 2. If $(x - a)$ is a factor of a quadratic, then a is a root of the quadratic.

Example 1.12. Factor $x^2 - 7x + 12 = 0$, but first plug in $x = 3$.

Solution:

We plug in $x = 3$ and get that $9 - 21 + 12 = 0$. So plugging in 3 makes the quadratic zero. This means:

$$x^2 - 7x + 12 = (x - 3)(x-?)$$

But $12/3=4$ so $? = 4$ is a good guess. Let's check:

$$(x - 3)(x - 4) = x^2 - 3x - 4x + 12 = x^2 - 7x + 12,$$

and we have the correct factorization.

$$\diamond$$

The preceding example used a trick – plugging in 3 – to locate one of the factors of the quadratic. Let's make this a formal rule in Knowledge Box 1.10. In general you don't plug in 3, you plug in things that might work. Factors of the constant terms are usually good guesses – remember that they can be positive or negative.

Knowledge Box 1.10.

> Suppose that $f(x)$ is a quadratic equation and that $f(c) = 0$ for some number c. Then $(x - c)$ is a factor of $f(x)$.

Example 1.13. Factor $x^2 - 2x - 8 = 0$.

Solution:

The number 8 factors into 1×8 and 2×4. Since $4 - 2 = 2$ and the middle term is $2x$, the factorization is probably $(x \pm 2)(x \pm 4)$. Since we have -8, one needs to be positive, and the other needs to be negative. To get a $-2x$ we need to subtract $4x$ and add $2x$. This means the factorization must be:

$$x^2 - 2x - 8 = (x - 4)(x + 2)$$

$\Diamond$

1.3.2. Completing the square. You can't always find the roots of a quadratic by factoring. Another way to find the roots is to transform the quadratic from the form:

$$ax^2 + bx + c = 0$$

into the form:

$$(x - r)^2 = s$$

Then, you can find the roots by taking the square root of both sides and doing some algebra. This is a technique called **completing the square**. Completing the square consists of forcing a complete square in the form $(x - r)^2$ and then putting the leftovers on the other side of the equation. Notice that:

$$(x - a)^2 = x^2 - 2ax + a^2$$

To complete the square we follow these steps:

(1) From the equation above, we see that the constant term must be the square of half the number in front of x. We can force this by adding and subtracting the needed constant.

(2) Step 1 transforms the equation into a perfect square plus a constant. Take the constant to the other side.

(3) Solve by using the square root and normal algebra. Remember the $\pm$ on the square root and you will get two answers.

Example 1.14. Solve $x^2 + 2x - 1 = 0$ by completing the square.

Solution:

$$
\begin{array}{ll}
x^2 + 2x - 1 = 0 & \text{This is the original equation} \\
x^2 + 2x + (1 - 1) - 1 = 0 & \text{Add and subtract } \left(\frac{1}{2} \cdot 2\right)^2 \\
x^2 + 2x + 1 - 2 = 0 & \text{Regroup to get perfect square and constant} \\
(x + 1)^2 - 2 = 0 & \text{Make the perfect square explicit by factoring} \\
(x + 1)^2 = 2 & \text{Move 2 to the other side} \\
x + 1 = \pm\sqrt{2} & \text{Take the square root} \\
x = -1 \pm \sqrt{2} & \text{Move 1 to the other side}
\end{array}
$$

And we see the solutions to the quadratic equation are $x = -1 - \sqrt{2}$ and $x = -1 + \sqrt{2}$. This also means

$$x^2 + 2x - 1 = (x + 1 + \sqrt{2})(x + 1 - \sqrt{2}).$$

This factorization is not easy to see – which is why we need tools like completing the square.

$$\diamond$$

There is another thing that completing the square can do.

Example 1.15. Use completing the square to show $x^2 - 4x + 7 = 0$ does not have any solutions.

Solution:

$$
\begin{array}{ll}
x^2 - 4x + 7 = 0 & \text{This is the original equation} \\
x^2 - 4x + (4 - 4) + 7 = 0 & \text{Add and subtract } \left(\frac{1}{2} \cdot -4\right)^2 \\
x^2 - 4x + 4 + 3 = 0 & \text{Regroup to get perfect square and constant} \\
(x - 2)^2 + 3 = 0 & \text{Make the perfect square explicit by factoring} \\
(x - 2)^2 = -3 & \text{Move 3 to the other side} \\
x - 2 = \pm\sqrt{-3} & \text{Take the square root. D'oh! Negative!!}
\end{array}
$$

$$\diamond$$

Example 1.15 generated a root that was a square root of a negative. In fact, the right hand side of the equation, just before we take the square root, is exactly the discriminant $D = b^2 - 4ac$. Knowledge Box 1.9 told us that when the discriminant is negative, there are no roots.

Another nice thing about completing the square is that it gives you the geometrically important parts of the quadratic. Quadratic equations with a positive squared term, including all the ones we've seen so far in this section, graph as parabolas opening upward. Those with a negative squared term open downward. These show up in falling object equations. The root of the perfect square that shows up when we are completing the square is the point where the quadratic turns around and heads in the other direction. If we have a downward-opening quadratic, this tells us where the highest point on the quadratic is. The next example is a physics-like example, so we are going to use time t as the independent variable instead of x.

Example 1.16. Suppose that the height in meters of a thrown ball after t seconds is given by:

$$h = 6 + 4t - t^2 \text{ meters}$$

Complete the square to find when the ball is highest and at what time that greatest height occurs.

Solution:

We are going to complete the square with a small wrinkle: forcing the t^2 term positive to get the usual form for the quadratic.

$$
\begin{aligned}
6 + 4t - t^2 &= h && \text{This is the original equation} \\
t^2 - 4t - 6 &= -h && \text{Negate everything to get standard form} \\
t^2 - 4t + (4 - 4) - 6 &= -h && \text{Add and subtract } \left(\tfrac{1}{2}(-4)\right)^2 \\
t^2 - 4t + 4 - 10 &= -h && \text{Regroup to get perfect square and constant} \\
(t - 2)^2 - 10 &= -h && \text{Make the perfect square explicit by factoring} \\
10 - (t - 2)^2 &= h && \text{Negate again to get } h \text{ back} \\
h &= 10 - (t - 2)^2 && \text{Neaten up}
\end{aligned}
$$

We now have that h is 10 minus a squared quantity. That squared quantity is never negative and it is smallest (zero) when $t = 2$. This means the greatest height achieved by the ball is 10 meters at $t = 2$ seconds, and we have our answer.

$\Diamond$

1.3.3. The Quadratic Equation. It is possible to complete the square on the equation $ax^2 + bx + c = 0$ and get a general solution for any simple quadratic equation. This solution is called the **quadratic formula**.

Knowledge Box 1.11.

Quadratic Formula

If $ax^2 + bx + c = 0$ then

$$x = \frac{-b \pm \sqrt{b^2 - 4ac}}{2a}$$

Remember that if $D = b^2 - 4ac < 0$, then this requires you to take a square root of a negative, and so no roots exist.

Let's try applying the quadratic equation.

Example 1.17. Find the roots of $x^2 - x - 1 = 0$.

Solution:

Apply the quadratic equation. In this case $a = 1$, $b = -1$, $c = -1$ so we get:

$$x = \frac{1 \pm \sqrt{1 - 4(1)(-1)}}{2 \cdot 1} = \frac{1 \pm \sqrt{5}}{2},$$

and we have the solution.

The next example demonstrates that the quadratic equation results from completing the square in general. It's a bit over the top; don't worry if you can't follow it.

Example 1.18. Solve, by completing the square,

$$ax^2 + bx + c = 0.$$

Solution:

$ax^2 + bx + c = 0$	This is the original equation
$x^2 + \dfrac{b}{a}x + \dfrac{c}{a} = 0$	Divide by a to get a clean x^2
$x^2 + \dfrac{b}{a}x + \dfrac{b^2}{4a^2} - \dfrac{b^2}{4a^2} + \dfrac{c}{a} = 0$	Add and subtract $\dfrac{1}{2} \cdot \dfrac{b}{a}$ squared
$x^2 + \dfrac{b}{a}x + \dfrac{b^2}{4a^2} = \dfrac{b^2}{4a^2} - \dfrac{c}{a}$	Move constants not in the perfect square
$x^2 + \dfrac{b}{a}x + \dfrac{b^2}{4a^2} = \dfrac{b^2}{4a^2} - \dfrac{4ca}{4a^2}$	Get a common denominator
$x^2 + \dfrac{b}{a}x + \dfrac{b^2}{4a^2} = \dfrac{b^2 - 4ac}{4a^2}$	Simplify the fraction
$\left(x + \dfrac{b}{2a}\right)^2 = \dfrac{b^2 - 4ac}{4a^2}$	Factor the perfect square
$x + \dfrac{b}{2a} = \pm\dfrac{\sqrt{b^2 - 4ac}}{2a}$	Take the square root of both sides
$x = -\dfrac{b}{2a} \pm \dfrac{\sqrt{b^2 - 4ac}}{2a}$	Take $\dfrac{b}{2a}$ to the other side
$x = \dfrac{-b \pm \sqrt{b^2 - 4ac}}{2a}$	Simplify: that's the quadratic formula

$\diamondsuit$

Since the discriminant $D = b^2 - 4ac$ is the thing we take the square root of when applying the quadratic equation, it is built right into the equation. If you're trying to take the square root of a negative number, then there is no solution; if you're taking the square root of zero, the fact that zero is its own negative means that the $\pm$ doesn't really kick in, and you get one answer.

Let's conclude the section by doing several examples. Check the values given for the examples, and see if you can compute them.

Example 1.19.

Equations	Coefficients	Discriminant	Roots
$x^2 - 4x + 4 = 0$	$a = 1, b = -4, c = 4$	$D = 0$	$x = 2$
$x^2 - 4x + 3 = 0$	$a = 1, b = -4, c = 3$	$D = 4$	$x = 1, 3$
$x^2 + x + 1 = 0$	$a = 1, b = 1, c = 1$	$D = -3$	no roots
$2x^2 + 5x - 1 = 0$	$a = 2, b = 5, c = -1$	$D = 33$	$x = \dfrac{-5 \pm \sqrt{33}}{4}$
$x^2 - 7x + 4 = 0$	$a = 1, b = -7, c = 4$	$D = 33$	$x = \dfrac{7 \pm \sqrt{33}}{2}$
$x^2 + 5 = 0$	$a = 1, b = 0, c = 5$	$D = -20$	no roots

Problems

Problem 1.34. *Which of the following are quadratics? Demonstrate your answer is correct. Use the definition from Knowledge Box 1.8.*

(a) $y = 3x + 1$

(b) $y = 2x^2 + 3x - 5$

(c) $y = (x - 3)(2x - 1)$

(d) $y = (x + 3)^2 - 7$

(e) $y = (x - 1)(x - 2)(x - 3) - (x + 1)(x + 2)(x + 3)$

(f) $y = (x - 1.23)^3 + 1$

Problem 1.35. *Carefully graph each of the quadratic equations in Problem 1.34.*

Problem 1.36. *For each of the following quadratic equations, find the discriminant D of the equation and state the number of roots the equation has.*

(a) $y = x^2 - 5x + 6$

(b) $y = x^2 - 4x + 5$

(c) $y = x^2 + 6x + 9$

(d) $y = 2x^2 - x - 3$

(e) $y = 2x^2 - x + 2$

(f) $y = 5 - x^2$

(g) $y = x^2 + x + 1$

(h) $y = x^2 + x - 1$

Problem 1.37. *Factor each of the following quadratics. Not all of them factor without a bit of fiddling.*

(a) $y = x^2 - 5x + 6$

(b) $y = x^2 + 7x + 12$

(c) $y = 2x^2 + 6x + 4$

(d) $y = -4 + 4x - x^2$

(e) $y = (x - 1)(x + 1) - 3$

(f) $y = x^2 - 25$

(g) $y = x^2 - 3$

(h) $y = x^2 - x - 1$

Problem 1.38. *For each of the following quadratics, find all the roots or give a reason there are none.*

(a) $y = x^2 + 5x + 6$

(b) $y = x^2 + x - 6$

(c) $y = x^2 - 36$

(d) $y = x^2 - 20x + 91$

(e) $y = x^2 + x + 1$

(f) $y = x^2 + x - 1$

(g) $y = x^2 + 36$

(h) $y = 20x^2 + 9x + 1$

Problem 1.39. *For each of following quadratics, complete the square.*

(a) $y = x^2 - 2x - 2$

(b) $y = x^2 + 4x - 3$

(c) $y = 4x^2 + 4x - 2$

(d) $y = x^2 - x - \frac{3}{4}$

(e) $y = x^2 + 2x + 2$

(f) $y = x^2 + 6x + 10$

(g) $y = x^2 + x + 1$

(h) $y = x^2 + 3x + 1$

Problem 1.40. *If the height of a ball at time t in meters is given by*

$$h = 30 + 8t - 5t^2$$

find (without calculus) the time that ball attains its greatest height and the number of meters about the ground that the ball is at that time.

Problem 1.41. *If the height of a ball at time t in meters is given by*

$$h = 20 + 16t - 5t^2$$

find (without calculus) the time that ball attains its greatest height and the number of meters about the ground that the ball is at that time.

Problem 1.42. *If the height of a ball at time t in meters is given by*

$$h = 24 + 4t - 5t^2$$

find (without calculus) the time that ball attains its greatest height and the number of meters about the ground that the ball is at that time.

Problem 1.43. *Find a quadratic equation $y = ax^2 + bx + c$ with a, b, and c all whole numbers that has two roots, neither of which is a whole number divisor of c.*

Problem 1.44. *Find all values of q for which $y = x^2 - qx + 2$ has two roots.*

Problem 1.45. *Find all values of q for which $y = x^2 - 3x + q$ has two roots.*

Problem 1.46. *Find a quadratic $y = ax^2 + bx + c$ so that the equation has only one root at $x = 2$ and so that, when you plug in 3 for x, $y = 2$.*

Problem 1.47. *Find two quadratic equations whose graphs have no points in common. Explain why your solution is correct.*

Problem 1.48. *Find a quadratic equation that passes through the points (-1,0), (0,1), and (1,0).*

Problem 1.49. *Find a quadratic equation that passes through the points (0,3), (1,2), and (2,3).*

Problem 1.50. *Find a quadratic equation that passes through the points (2,-1), (3,4), and (4,9).*

Problem 1.51. *Suppose we have two points with distinct x coordinates. How many quadratic equations have graphs that include those two points?*

Problem 1.52. *Find and graph a quadratic equation that has roots at $x = 2, -3$.*

Problem 1.53. *Find a quadratic equation with a root at $x = -2$ that also passes through the point (1,1).*

Problem 1.54. *Find a quadratic equation with a root at $x = 3$ that also passes through the point (-1,2).*

Problem 1.55. *Find a quadratic equation with a root at $x = 4$ that also passes through the point (1,6).*

Problem 1.56. *Suppose we have three points on the same line. Can a quadratic equation pass through all three of those points?*

Problem 1.57. *Consider sets of three points that are not on the same line. What added conditions are needed to permit a quadratic equation to pass through all of the points?*

Problem 1.58. *Suppose that $y = ax^2 + bx + c$ is a quadratic equation with no roots and that, when you plug in 2 for x, $y = 4$. What can we deduce about a?*

Problem 1.59. *A quadratic equation is said to be a perfect square if it has the form $f(x) = (x - a)^2$ for some constant a. What is the discriminant of a perfect square?*

Problem 1.60. *Find three different quadratic equations with roots 2 and -2.*

Problem 1.61. *Suppose we have a quadratic equation with two roots. Is there another quadratic equation whose graph has exactly the same shape but that has a root only at zero. Either explain why not or find an example for $y = x^2 - 3x + 2$.*

Problem 1.62. *Suppose that the graph of two quadratic equations enclose a finite area. What can you deduce about the equations from this fact?*

1.4. Functions

A **function** is a way of assigning input numbers to output numbers so that each input number is assigned to one and only one output number. When we write (x, y) this means an **ordered pair** of numbers with x first and y second. Formally:

Knowledge Box 1.12.

A **function** is a set S of ordered pairs (x, y) in which no x value appears twice.

There is a problem with this definition – we haven't said what a set is yet. For now, a **set** is a collection of objects in which no objects appear twice.

When we are graphing a function there is a simpler way of capturing the notion of a function.

Knowledge Box 1.13.

The graph of a function intersects any vertical line at most once.

If you think about it, the vertical-line based definition agrees completely with the ordered-pair definition. Why have the ordered pair definition then? It will turn out to apply in all sorts of places where the vertical line definition doesn't even make sense; stay tuned and alert.

Example 1.20. Look at the two graphs below. The one on the left is the graph of $x^2 + y^2 = 1$, while the one on the right is the graph of $y = x^2 - 1$.

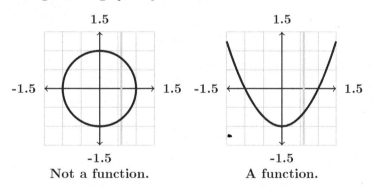

Not a function.   A function.

There are several places where a vertical line (we included one) hits the circle twice – it is clearly not a function. The parabola that is the graph of $y = x^2 - 1$, however, intersects a vertical line once and only once – even in the parts we cannot see.

$\Diamond$

Once we have the idea of functions, we adopt a new way of writing formulas when they happen to be functions.

Knowledge Box 1.14.

Notation for functions

When we have y equal to some formula, that may or may not give us a function. When it does, we use the name $f(x)$ instead of y. The symbol $f(x)$ is still a y-value, but writing it this way tells us that x is the input variable and also lets us know that this formula is a function.

Example 1.21. Where before we might have said

$$y = x^2 + 3x + 5,$$

the fact that any quadratic in this form is a function lets us say

$$f(x) = x^2 + 3x + 5.$$

$\Diamond$

Functions are pretty abstract right now, but they keep showing up later, so we wanted to get the basic concept in front of you now. As we develop new structures from now on, we will be careful to tell you which ones are or are not functions. We are a bit behind on that promise so:

Knowledge Box 1.15.

Any line $y = mx + b$ (not a vertical line) is a function. A quadratic equation $y = ax^2 + bx + c$ is also a function – each x is assigned to only one y by the formula.

One reason to use formulas to specify functions is that, as long as there is no ambiguity in the statement of the function, the formula maps each x to a single y or possibly to nothing at all if something impossible (e.g. dividing by zero) happens. The resulting collection of pairs (x, y) obeys the definition of function.

For now, circles are our big example of something that's not a function. This means that there is no way to write $f(x) = formula$ so that all the points on the circle show up as $(x, f(x))$ for some choice of x. No matter what the formula is, there will be points we cannot compute with it – unless the formula somehow violates the vertical line rule, e.g. by incorporating $\pm$, which isn't permitted.

1.4.1. Domain and Range. One thing we need to worry about is a formal way to define what numbers can go into or come out of a function. Negative numbers don't have a square root, for example.

Knowledge Box 1.16.

The **domain** of a function is the set of numbers that you can put into the function without a problem. Problems occur when dividing by zero or taking the square root of a negative number.

Example 1.22. What is the domain of the function

$$f(x) = \frac{1}{x-1}?$$

Solution:

We can take 1 divided by any number *except* zero. This means that any number can be plugged into this function except 1, which makes the thing we are dividing by zero. This means the answer is $x \neq 1$.

Knowledge Box 1.17.

The **range** of a function is the set of numbers that can come out of the function.

Example 1.23. What is the range of the function

$$f(x) = x^2?$$

Solution:

When we square a negative number the result is positive. Since any positive number has a square root every positive number is the result of squaring another number. We also have that $0^2 = 0$. That means any number that is not negative can come out of $f(x) = x^2$, and we see that the range is *all non-negative numbers.*

The domains and ranges of functions are usually intervals of numbers or individual numbers. At this point we want to introduce some notation that permits us to efficiently specify domains and ranges.

Definition 1.2. If we have two sets S and T, then the **union** of S and T is the set of all objects that are in either set. It is written:

$$S \bigcup T$$

Definition 1.3. If we have two sets S and T, then the **intersection** of S and T is the set of all objects that are in both sets. It is written:

$$S \bigcap T$$

Definition 1.4. An interval from the number a to the number b with $a < b$ has four forms. It can include both, neither, or one of a and b. We use square brackets [ ] to denote inclusion and parenthesis ( ) to denote exclusion.

Knowledge Box 1.18.

Standard and compact interval notation for intervals

$a < x < b \quad (a, b)$

$a < x \leq b \quad (a, b]$

$a \leq x < b \quad [a, b)$

$a \leq x \leq b \quad [a, b]$

Example 1.24. What are the domain and range of the function

$$f(x) = \sqrt{x^2 - 4}?$$

Solution:

To find the domain, we need to check for x values that are impossible. The only impossible thing that happens in this function is that, for some values of x, the values inside the square root function are negative. Solve:

$$x^2 - 4 = 0$$
$$(x - 2)(x + 2) = 0$$
$$x = \pm 2$$

We see that the function has roots at ± 2. Since the quadratic has a positive x^2 term, this means it opens upward, and so it is negative *between* the roots. This means the function is undefined there. So, the domain of the function is $-\infty < x \le -2$ together with $2 \le x < \infty$ or, in the more compact notation:

$$(-\infty, -2] \cup [2, \infty)$$

Now we need to find the range. No negative numbers come out of the square root, but zero does (at ± 2). The function $x^2 - 4$ goes to infinity. Since every positive number is the square root of some other positive number, we see all possible positive values will occur for some output of $x^2 - 4$. So, the range of $f(x)$ is $0 \le y < \infty$ or

$$[0, \infty)$$
$$\diamond$$

When we get to integration, the idea of odd and even functions will be handy.

Definition 1.5. A function is **even** if, for x where the function exists,

$$f(x) = f(-x).$$

A good example of an even function is $f(x) = x^2$. **Even functions forget signs.**

Definition 1.6. A function is **odd** if, for x where the function exists,

$$f(-x) = -f(x).$$

A good example of an even function is $f(x) = x^3$. **Odd functions remember signs.**

There is a lot more to say about domain and range, but we cannot say most of it until you learn more of the library of functions. A lot of the remaining functions in the library show up in the next three sections, so stay tuned. One important point is that some of the new functions will add to our list of impossible situations, e.g. you cannot take the logarithm of zero or negative numbers, and the tangent function does not exist for odd multiples of $\dfrac{\pi}{2}$. As we introduce the library of functions, we will state the domains and ranges of each function.

Problems

Problem 1.63. *Below are six graphs that may or may not be functions. Based only on the part of the graph shown, which of the following appear to be the graphs of functions?*

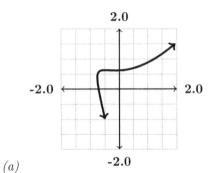

(a)

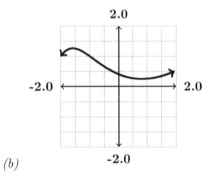

(b)

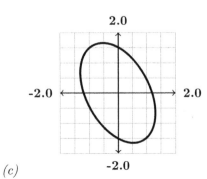

(c)

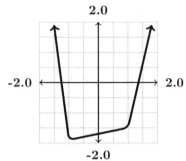

(d)

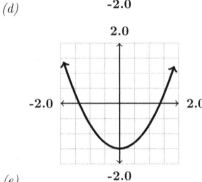

(e)

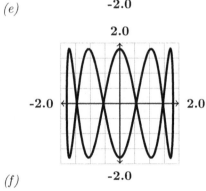

(f)

Problem 1.64. *Which of the following are functions? Give a reason for your answer.*

(a) $f(x) = 3x + 5$

(b) $g(x) = x^2 + 5x + 7$

(c) $h(x) = (2x - 5) + (x^2 + x + 1)$

(d) $r(x) = \dfrac{x}{x^2 + 1}$

(e) $s(x) = \dfrac{1 \pm \sqrt{x^2 + 3}}{x^2 + 1}$

(f) $q(x) = (x - 1)(x - 2)s(x)$ where $s(x)$ is the function above

Problem 1.65. *Find the domain of the following functions.*

(a) $f(x) = 4x - 2$

(b) $g(x) = \sqrt{4x - 2}$

(c) $h(x) = \dfrac{1}{4x - 2}$

(d) $r(x) = \dfrac{x^2 + 1}{x^2 - 4}$

(e) $s(x) = x^3 + x^2 + x + 1$

(f) $q(x) = \sqrt{9 - x^2}$

(g) $a(x) = (2x + 5)^3$

(h) $b(x) = \sqrt[3]{2x + 5}$

Problem 1.66. *Find the range of the following functions.*

(a) $f(x) = 5 - x$

(b) $g(x) = \sqrt{5 - x}$

(c) $h(x) = x^2 + 2x + 1$

(d) $r(x) = 1 - x^2$

(e) $s(x) = x^3$

(f) $q(x) = \sqrt{25 - x^2}$

(g) $a(x) = (1 - x)^3$

(h) $b(x) = \sqrt[3]{x - 1}$

Problem 1.67. *Which of the following sets are functions? Part (a) requires you be very careful with the definition of a function.*

(a) The set of points $(\cos(t), 20 \cdot \cos(t))$ where $-\infty < t < \infty$

(b) all pairs (x, y) where $x^2 = y^3$

(c) the points (1,2), (0,4), (-1,2), (2,5), and (-2,4)

(d) the points (2,1), (4,0), (2,-1), (5,2), and (4,-2)

(e) the set of points (x, y) where
$$x^2 + y^2 = 4$$
and $y \geq 0$

(f) the set of points (x, y) where
$$x^2 + (y - 1)^2 = 4$$
and $y \geq 0$

Problem 1.68. *For each of the functions in Problems 1.65 and 1.66 say if the functions are odd, even, or neither.*

Problem 1.69. *Describe carefully the largest subset(s) of a circle that are functions.*

Problem 1.70. *If $f(x)$ is an odd function show that $g(x) = x \cdot f(x)$ is an even function.*

Problem 1.71. *If $h(x)$ is an even function show that $r(x) = x \cdot h(x)$ is an odd function.*

Problem 1.72. *Construct an infinite set of points that have the property that the largest function that is a subset of the set contains one point.*

Problem 1.73. *Prove by logical argumentation that any subset of a function is a function. Remember that the definition of a function is that a function is a set of points with a special property; your argument should not touch on the graph of the function at all.*

Problem 1.74. *Can the graph of a function enclose a finite area? Explain your answer.*

Problem 1.75. *Using any method, compute the domain and range of the function*
$$f(x) = \frac{ax + b}{cx + d}$$
where $a, b, c,$ and d are real numbers. You may assume that the top and bottom of the fraction do not cancel out so as to remove the variables and that $a, b \neq 0$.

Problem 1.76. *Suppose that $f(x)$ and $g(x)$ are both functions whose domain is all real numbers: $(-\infty, \infty)$. If we define $h(x)$ to be equal to $f(x)$ when x is negative and to equal $g(x)$ when x is positive or zero, is $h(x)$ a function? Explain.*

1.5. Polynomials

We've already learned about lines, where the highest power of x is one, and quadratics, where the highest power of x is two. Polynomials are functions that let this highest power get as big as it wants to.

Knowledge Box 1.19.

A **polynomial** is a function that is a sum of constant multiples of powers of a variable.

Example 1.25. The following are examples of polynomials.

$$y = 3$$

$$f(x) = 3x + 4$$

$$g(x) = x^2 + 2x - 1$$

$$h(x) = x^3 + 4x - 1$$

$$k(x) = x^3 + 3x^2 + 3x + 1$$

$$y = x^4 + 2x^2 + 7$$

$$q(x) = x^6 - 2$$

$$\Diamond$$

Polynomials are very well-behaved functions. They have a domain of $(-\infty, \infty)$. The range of a polynomial is a bit trickier. As we've seen, a quadratic may have a range that doesn't include everything. We need some additional terminology before we go on.

Definition 1.7. The **standard form** of a polynomial is when it is written as a sum of constant multiples of a variable with the powers in descending order from left to right.

The polynomials in Example 1.25 are in standard form. A example of a polynomial not in standard form is:
$$f(x) = (x - 1)(x - 2)(x^2 + 4)$$
In standard form this polynomial would be:
$$f(x) = x^4 - 3x^3 + 6x^2 - 12x + 8$$

Definition 1.8. The **degree** of a polynomial is the highest power of x that appears in the polynomial when it is in standard form.

The polynomials in Example 1.25 have, from top to bottom, degrees as follows: 0, 1, 2, 3, 3, 4, and 6. The degree of a polynomial may not be obvious, as we see in the next example.

Example 1.26. What is the degree of

$$f(x) = (x-1)(x-2)(x-3)?$$

Solution:

To find the degree of this polynomial we have to multiply it out first.

$$\begin{aligned}
(x-1)(x-2)(x-3) &= (x-1)(x^2 - 5x + 6) \\
&= x(x^2 - 5x + 6) - 1(x^2 - 5x + 6) \\
&= x^3 - 5x^2 + 6x - x^2 + 5x - 6 \\
&= x^3 - 6x^2 + 11x - 6
\end{aligned}$$

Now, we can see that the degree of this polynomial is 3.

Actually, you don't have to multiply it out. It's possible to see that it will have a degree of three by imagining how the multiplication would come out. In fact:

Knowledge Box 1.20.

> If you multiply several polynomials, the degree of the product is the sum of the degrees of the polynomials you multiplied.

Definition 1.9. The **coefficients** of a polynomial are the constants multiplied by the powers of x.

Example 1.27. For the polynomial $f(x) = x^3 + 4x + 3$ the coefficients are 1, 4, and 3. We can be more specific: the coefficient of x^3 is 1; the coefficient of x is 4; and the constant coefficient is 3. Since there is no x^2 term, we can say that the coefficient of x^2 is 0.

Now that we know what degrees and coefficients are, we can say a little about the range of polynomial functions. These results are given here without explanation. We will revisit them in Chapter 6.

Knowledge Box 1.21.

The range of a polynomial of odd degree is $(-\infty, \infty)$.

Knowledge Box 1.22.

The range of a polynomial of positive even degree is:

- (C, ∞) for some constant C if the coefficient of its highest power is positive, and

- $(-\infty, D)$ for some constant D if the coefficient of its highest power is negative.

Normally "odd" and "even" cover all the possibilities, but zero is quite peculiar for an even number. $f(x) = cx^0$ is just a long way of saying $f(x) = c$, which has a very small range. Let's complete our Knowledge Box collection of possible ranges of polynomials with the following.

Knowledge Box 1.23.

The range of a polynomial of degree zero is a single constant c; the function has the form $f(x) = c$.

Example 1.28. The picture on the left is the graph of a second degree polynomial, while the one on the right is the graph of a third degree polynomial.

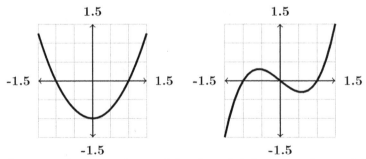

An even degree polynomial   An odd degree polynomial

Check these against the Knowledge Box rules for polynomials of even and odd degree and see if they agree.

◇

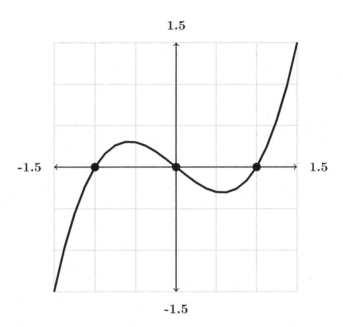

Figure 1.7. Roots of an odd degree polynomial.

The next definition comes up a lot when we are trying to solve problems. We have already seen that quadratic equations can have 0, 1, or 2 roots and, in Example 1.18, found a formula for those roots. We didn't formally define what a root was, so here you go.

Definition 1.10. A **root** of a polynomial function $f(x)$ is any number r for which $f(r) = 0$.

If you want to think about roots geometrically – they occur at places where the graph of a function crosses the x-axis. When we are solving problems, we can sometimes write a polynomial so that the places where the polynomial is zero *are* the solutions to the problem. This will come up frequently in Chapter 3 when we are doing optimization.

Figure 1.7 shows the odd degree polynomial from Example 1.28 with dots where its roots are.

The ranges of polynomials that we mentioned before arise from the way that a polynomial with positive degree heads toward infinity. This geometric behavior influences the roots as well. Polynomials can only change which direction they are going (up or down) a number of times that is one less than their degree. Notice a second degree polynomial changes from down to up or from up to down exactly once. All this has the following implications for roots of polynomials.

Knowledge Box 1.24.

A polynomial of odd degree n has from one to n roots. A polynomial of positive even degree n has from zero to n roots.

Polynomials have another interesting property: they form a **closed set** relative to addition, multiplication, and multiplication by constants. Adding or multiplying two polynomials or multiplying a polynomial by a constant results in another polynomial.

Knowledge Box 1.25.

Polynomials obey the following rules:

- A constant multiple of a polynomial is a polynomial.

- The sum of two polynomials is a polynomial.

- The product of two polynomials is a polynomial.

Next we note that something that was true of quadratics is also true for polynomials.

Knowledge Box 1.26.

Suppose that $f(x)$ is a polynomial and that $f(c) = 0$ for some number c. Then $(x - c)$ is a factor of $f(x)$.

This result is called the **Root-factor Theorem** for polynomials. If you're trying to factor a polynomial, one approach is to plug in numbers looking for a root (graphing the polynomial can narrow down the possibilities). Another way to look at this rule is the following restatement.

Knowledge Box 1.27.

Suppose that $f(x)$ is a polynomial and that $f(c) = 0$ for some number c. Then for some polynomial $g(x)$ we have that

$$f(x) = (x - c)g(x).$$

Chapter 6 goes into far more detail about the properties of polynomials – something we can do once we have the tools of calculus at our fingertips.

Problems

Problem 1.77. *For each of the following functions, determine if it is a polynomial. You may need to simplify the function to tell if it is polynomial.*

(a) $f(x) = 1 + (x+1) + (x^2 + x + 1) + (x^3 + x^2 + x + 1)$

(b) $g(x) = (x^2 + 1)^3 + (x^2 - x + 1)^2 + 7$

(c) $h(x) = \pi x + 7$

(d) $r(x) = 17$

(e) $s(x) = \dfrac{x}{x^2 + 1}$

(f) $q(x) = x^3 + 4.1x^2 - 3.2x^2 + 4.6x - 3.8$

(g) $a(x) = \dfrac{x^3 + x + 1}{x^2 + 1} - \dfrac{1}{x^2 + 1}$

(h) $b(x) = (x^2 + \sqrt{x})(x^2 - \sqrt{x})$

(i) $c(x) = (x^2 + \sqrt{x})^2$

(j) $d(x) = \dfrac{1}{x} - \dfrac{2}{x^2} + \dfrac{3x^3 - x + 2}{x^2}$

Problem 1.78. *Place each of the following polynomials into standard form.*

(a) $f(x) = 1 + (x+1) + (x+1)^2$

(b) $g(x) = (x^2 + x + 1)^2$

(c) $h(x) = (x^2 - 1)(x^2 + 1)(x+2)^2$

(d) $r(x) = x(x+1)(x+2)(x+4)$

(e) $s(x) = (x^2 + 1)(x+1) + (x^2 + 1)(x - 2) + (x^2 + 1)^2$

(f) $q(x) = x^3(x+1)^3(x-1)^3$

(g) $a(x) = (x+1)^3 - (x-1)^3$

(h) $b(x) = (x-2)^3 - (x^3 - 6x^2 + 12x)$

Problem 1.79. *Find the degree of each of the polynomials in Problem 1.78.*

Problem 1.80. *Give a simple rule for telling if a polynomial is an odd function, an even function, or neither.*

Problem 1.81. *Find the range of each of the following polynomials.*

(a) $y = 3$

(b) $y = 3x + 1$

(c) $y = x^2 + 6x + 12$

(d) $y = x^3 + 3x^2 - 7x + 8$

(e) $y = x^4 + 5x^2$

(f) $y = (x+1)(x+2)(x+3)(x^2 + 1)$

(g) $y = (x^2 + 5)^3$

Problem 1.82. *Argue convincingly that a positive whole number power of a polynomial is a polynomial.*

Problem 1.83. *Find a polynomial of degree 3 with one root at $x = 1$.*

Problem 1.84. *Find a polynomial of degree 4 with no roots at $x = 1$.*

Problem 1.85. *Find a polynomial of degree 3 with roots at $x = 0, \pm 3$.*

Problem 1.86. *Find a polynomial of degree 3 with roots at $x = 0, \pm 2$ so that $f(3) = 5$.*

Problem 1.87. *Prove that if two polynomials both have a root at $x = a$, then so does their sum.*

Problem 1.88. *Suppose that we have several polynomials. Fill in the box in the following sentence. The set of roots of the product of the polynomials is the ☐ of the sets of roots of each of the polynomials.*

Problem 1.89. *True or false (and explain): if we divide two polynomials the resulting function is a polynomial.*

Problem 1.90. *Demonstrate that the result of dividing two polynomials can be a polynomial.*

Problem 1.91. *Given that each of the following graphs is the graph of a polynomial, give as much information about the degree, coefficients, and number and value of roots as you can.*

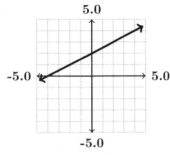

(a)

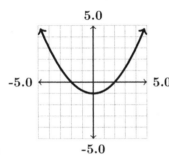

(b)

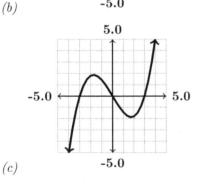

(c)

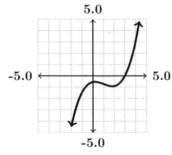

(d)

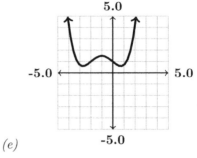

(e)

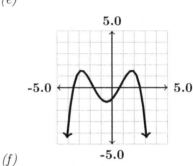

(f)

Problem 1.92. *Suppose that the largest number of points that any horizontal line intersects the graph of a polynomial $f(x)$ in is m. Prove that the degree of $f(x)$ is at least m.*

Problem 1.93. *Give an example of a polynomial of degree 8 that has the property that the maximum number of times it intersects any horizontal line is two.*

Problem 1.94. *Verify that the polynomial $f(x) = x^3 - 6x^2 + 11x - 6$ has the property that $f(1) = f(2) = f(3) = 0$. Use this information and the root-factor theorem to factor $f(x)$.*

Problem 1.95. *Use the root factor theorem and some cleverness to factor*

$$g(x) = x^4 - 625$$

Problem 1.96. *If*

$$h(x) = (x-2)(x^2+1)(x^2+x+1)$$

then how many roots does $h(x)$ have and what are they?

1.6. Powers, Logs and Exponentials

This section deals with a very important category of functions: logarithms and exponential functions, as well as the related algebra of powers. We link logs and exponentials together because each can undo what the other does, like square and square root. As with square root, there is also a concern with negative numbers when computing logs. Logarithms don't exist at zero, while square roots do, so there is a difference. We begin with the algebra of powers and their corresponding roots.

1.6.1. Powers and Roots. The simplest notion of a power is that of repeated multiplication. If you multiply a number a by itself m times, you get

$$a \times a \times \cdots \times a = a^m$$

We also adopt the convention that

$$\frac{1}{a} = a^{-1}$$

which means that the reciprocal of a number to a power is that number to the negative of the power. There are a number of rules about how powers interact, summarized in the following Knowledge Box.

Knowledge Box 1.28.

Algebraic Rules for Powers

- $a^{-n} = \dfrac{1}{a^n}$

- $a^n \times a^m = a^{n+m}$

- $\dfrac{a^n}{a^m} = a^{n-m}$

- $(a^n)^m = a^{n \times m}$

Example 1.29. This example showcases the rules for powers on a simple problem.

$$\left(2^3 \times 2^4\right)^5 = \left(2^{3+4}\right)^5$$
$$= \left(2^7\right)^5$$
$$= 2^{35}$$
$$= 34,359,738,368$$

The final step is probably not needed, as 2^{35} is the same number and much easier to read.

◇

Definition 1.11. For a positive whole number n we define

$$b = \sqrt[n]{a}$$

to be any number for which $b^n = a$. Notice that, when n is even, there are two possibilities, $\pm \sqrt[n]{a}$.

Because an even power of a number must be positive, even roots only exist for non-negative numbers. Odd roots exist for any number, positive or negative. Once we have the notion of roots, it becomes possible to see that roots are actually a type of power.

Definition 1.12. We define fractional powers in the following fashion.

$$\sqrt[n]{a} = a^{\frac{1}{n}}, \text{ and}$$

$$\sqrt[n]{a^m} = a^{\frac{m}{n}}$$

Example 1.30. Here are several equivalent ways of writing the fourth root of the third power of five:

$$\sqrt[4]{125} = \sqrt[4]{5^3} = \left(5^3\right)^{\frac{1}{4}} = 5^{3/4}$$

$\Diamond$

1.6.2. Exponentials and Logs. Exponential functions are functions involving powers in which the variables occur in the exponent. Logs are functions that undo exponential functions.

Knowledge Box 1.29.

> An **exponential function** with base **a>0** is any function of the form
> $$y = a^x$$
> This means to compute y we find the power x of a.

There are a number of problems with this definition of exponential functions. So far we only, strictly, know how to take whole or fractional powers of a constant a, but many numbers are not expressible as fractions. It's also sort of hard to understand what a^x means when a is negative. So, for now we are going to avoid the whole issue of negatives and only take powers of positive numbers.

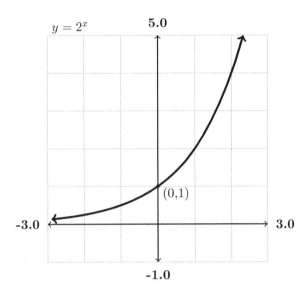

Figure 1.8. This is a graph of the function $y = 2^x$.

The graph in Figure 1.8 illustrates a number of properties of exponential functions of the form $y = a^x$.

Knowledge Box 1.30.

Properties of exponential functions

- The domain of $y = a^x$ is $(-\infty, \infty)$.
- The range of $y = a^x$ is $(0, \infty)$.
- The graph always contains the point (0,1).
- For $x < 0$, a^x is a positive number smaller than $y = 1$.
- For $x > 0$, a^x is a positive number bigger than $y = 1$.
- For $y = a^{-x}$, the last two facts are reversed.

The rules we just learned for powers apply to exponential functions. This means that, for example, $a^x \cdot a^y = a^{x+y}$.

Definition 1.13. If **c** is the **logarithm base b** of a number **a** we write

$$\log_b(a) = c$$

which is a different way of saying

$$b^c = a$$

The logarithm and exponential functions have the same relationship that the square and square roots functions do. The technical terms are that they are **inverses** of one another. Each reverses what the other does. The relationship between logs and exponentials is given in Knowledge Box 1.31.

Knowledge Box 1.31.

The relationship between logs and exponentials

- $b^{\log_b(c)} = c$ - $\log_b(b^a) = a$

Now that we have defined the logarithm function, we can list its algebraic properties. The fourth property is useful because it lets you compute the logarithm base *anything* once you can compute the logarithm base something.

Knowledge Box 1.32.

1 $\log_b(xy) = \log_b(x) + \log_b(y)$ **3** $\log_b(x^y) = y \cdot \log_b(x)$

2 $\log_b\left(\dfrac{x}{y}\right) = \log_b(x) - \log_b(y)$ **4** $\log_c(x) = \dfrac{\log_b(x)}{\log_b(c)}$

Let's look at the graph of a logarithm function (Figure 1.9).

Now that we can see the logarithm function, let's list its domain, range, and other properties.

Knowledge Box 1.33.

Properties of logarithmic functions

- The domain of $y = \log_b(x)$ is $(0, \infty)$.
- The range of $y = \log_b(x)$ is $(-\infty, \infty)$.
- The graph always contains the point $(1, 0)$.
- For $0 < x < 1$, $\log_b(x)$ is a negative number.
- For $x > 1$, $\log_b(x)$ is a positive number.

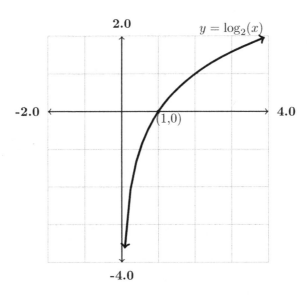

Figure 1.9. This is a graph of the function $y = \log_2(x)$.

The relationship between logs and exponentials means that we can use logs and exponentials to solve equations involving exponentials and logs. One of the rules for solving equations way back at the beginning of the chapter was "Performing the same operation to both sides, e.g. squaring or taking the square root." We add two new rules:

- You may take the logarithm of both sides of an equation.

- You may take a constant a to the power of each side of an equation.

Knowledge Box 1.34.

Taking the log of both sides of an equation

A handy implication of the algebraic properties of the logarithm function is the following:

$$\text{If } b^a = c, \text{ then } a = \log_b(c).$$

This can be used to simplify and solve equations.

Example 1.31. If $2^{x+1} = 7$ find x.

Solution:

$$2^{x+1} = 7$$
$$\log_2(2^{x+1}) = \log_2(7)$$
$$x + 1 = \log_2(7)$$
$$x = \log_2(7) - 1$$
$$x \cong 1.8073549$$

$\Diamond$

Knowledge Box 1.35.

Reversing a logarithm when solving an equation

The exponential result for solving equations with logs in them is that:
if $\log_b(c) = a$, then $c = b^a$.

The next example shows how to deal with a logarithm in an equation by taking a constant to the power of both sides. To get rid of the logarithm, the constant has to be the base of the logarithm.

Example 1.32. If $\log_3(x^2 + x + 1) = 1.6$, find x.

Solution:

$$\log_3(x^2 + x + 1) = 1.6$$
$$3^{\log_3(x^2+x+1)} = 3^{1.6}$$
$$x^2 + x + 1 = 3^{1.6}$$
$$x^2 + x + 1 - 3^{1.6} = 0$$

At this point we have converted the problem into a quadratic and apply the quadratic equation:

$$x = \frac{-1 \pm \sqrt{1^2 - 4 \cdot 1 \cdot (1 - 3^{1.6})}}{2 \cdot 1}$$

$$x \cong 1.7471195 \text{ or } -2.7471195$$

$\Diamond$

We conclude the section on logs and exponentials with the introduction of one of the odder features of these types of functions, the **natural logarithm** based on the number e. It takes some calculus to understand why we use a nutty number like e. For now, just accept that

$$\text{e} \cong 2.7182818.$$

For historical reasons there are two logarithm functions that are written without their base.

- $\ln(x)$ is shorthand for $\log_{\text{e}}(x)$, and

- $\log(x)$ is shorthand for $\log_{10}(x)$.

Most of the calculus of exponentials and logs is built around the twin functions $y = \ln(x)$ and $y = \text{e}^x$. These functions are mutual inverses and, in a sense we will understand later, they are the versions of the log and exponential function that arise naturally from the rest of mathematics.

Problems

Problem 1.97. *For each of the following expressions, find a simplified version of the the expression as a power of a single number, as in Example 1.29, where the number was 2.*

(a) $\left(3^6 \cdot 3^7\right)^2$

(b) $\dfrac{1}{\left(5^3 \cdot 5^4\right)^4}$

(c) $\left(2^4\right)^5 \cdot \left(4 \cdot 2^3\right)$

(d) $2^{\left(2^{\left(2^2\right)}\right)}$

(e) $\left(\left(2^2\right)^2\right)^2$

(f) $\dfrac{7 \cdot 7^2 \cdot 7^3 \cdot 7^4 \cdot 7^5}{7^6 \cdot 7^7}$

Problem 1.98. *Find the power of a specified by each of these expressions.*

(a) $\sqrt[5]{a^{15}}$

(b) $\sqrt[4]{a^3 \cdot a^5}$

(c) $\dfrac{\sqrt{a^5}}{a^2 \cdot \sqrt[3]{a}}$

(d) $\left(a^{1/3} \cdot a^{1/5}\right)^8$

(e) $a \cdot a^{1/2} \cdot a^{1/3} \cdot a^{1/4} \cdot a^{1/5} \cdot a^{1/6}$

(f) $a \cdot a^{-1/2} \cdot a^{1/3} \cdot a^{-1/4} \cdot a^{1/5}$

Problem 1.99. *For each of the following functions, find the domain and range of the function. For some of these, making a plot of the function may help you find the range.*

(a) $f(x) = \sqrt{x^2 + 1}$

(b) $g(x) = \sqrt{4 - x^2}$

(c) $h(x) = \left(\sqrt[3]{x} + 1\right)^3$

(d) $r(x) = \dfrac{\sqrt{1 + x}}{\sqrt{1 - x}}$

(e) $s(x) = \dfrac{1}{x}$

(f) $q(x) = \sqrt{x^3 - 6x^2 + 11x - 6}$

Problem 1.100. *Suppose $\log_b(u) = 2$, $\log_b(v) = -1$, and $\log_b(w) = 1.2$. Compute:*

(a) $\log_b(u \cdot v)$

(b) $\log_b(u^6 \cdot w)$

(c) $\log_b\left(\dfrac{u^2 \cdot v^3}{w^{1.2}}\right)$

(d) $\log_b(b^4 \cdot w^2)$

(e) $\log_b\left(\dfrac{b}{w} \cdot \dfrac{u}{v}\right)$

(f) $\log_b\left(\dfrac{b^4}{u^2}\right)$

Problem 1.101. *Solve the following for x.*

(a) $2^x = 14$

(b) $4^x - 5 \cdot 2^x + 6 = 0$

(c) $(2^x - 8)(3^x - 9) = 0$

(d) $2^{x^2 - 1} = 8$

(e) $9^x - 7 \cdot 3^x + 12 = 0$

(f) $5^{\sqrt{x^2 + 1}} = 625$

Problem 1.102. *For each of the following functions, find the domain and range of the function.*

(a) $f(x) = 2^{x^2 + 1}$

(b) $g(x) = 3^{1/x}$

(c) $h(x) = 1^{\sqrt{x + 1}}$

(d) $r(x) = \left(\dfrac{1}{2}\right)^{-x}$

(e) $s(x) = 2^x - 1$

(f) $q(x) = 2^x + 3^x$

Problem 1.103. *Solve the following for x.*

(a) $\log_5(x^2 - 6x + 8) = 1$

(b) $2\log_3(x) = \log_3(25)$

(c) $\log_2(1 - x) = 3$

(d) $\ln(e^x + 1) = 2x$

(e) $\log_3(x + 5) = 4$

(f) $\log_5(x^2 + 5x + 7) = 2$

Problem 1.104. *For each of the following functions, find the domain and range of the function. Part (f) may require you to complete a square.*

(a) $f(x) = \ln(x^2 + 1)$

(b) $g(x) = 5 - \log_2(x)$

(c) $h(x) = \log_3(3^x + 1)$

(d) $r(x) = \sqrt{\log_2(x)^2 + 1}$

(e) $s(x) = \ln(2 - x)$

(f) $q(x) = \log_5(x^2 + x + 1)$

Problem 1.105. *Show, using the algebraic rules for exponents, that*

$$\left(\frac{a}{b}\right)^n = \left(\frac{b}{a}\right)^{-n}$$

Problem 1.106. *Show that*

$$\sqrt[n]{\sqrt[m]{x}} = \sqrt[nm]{x}$$

Problem 1.107. *Look at the Knowledge Boxes for properties of exponential and logarithmic functions. How are the domain and range of these two types of functions related? Explain.*

Problem 1.108. *What is the geometric relationship between the graphs of the logarithm and exponential functions with the same base?*

Problem 1.109. *Explain in what sense the equation*

$$25^x - 4 \cdot 5^x + 3 = 0$$

is a quadratic. Having explained, solve it.

Problem 1.110. *Suppose that $y = C \cdot a^x$. If the points (0,5) and (2,20) are on the graph, then what are C and a?*

Problem 1.111. *Suppose that $y = C \cdot a^x$. If the points (1,6) and (3,54) are on the graph, then what are C and a?*

Problem 1.112. *Suppose $y = ax + b$ is a line with positive slope. Find the domain and range of the function*

$$f(x) = \sqrt{ax + b}$$

Your answer may be in terms of a and b.

Problem 1.113. *Suppose $y = ax + b$ is a line with positive slope. Find the domain and range of the function*

$$f(x) = \sqrt[3]{ax + b}$$

Your answer may be in terms of a and b.

Problem 1.114. *Suppose $y = ax + b$ is a line with positive slope. Find the domain and range of the function*

$$f(x) = \ln(ax + b)$$

Your answer may be in terms of a and b.

Problem 1.115. *Suppose $y = ax + b$ is a line with positive slope. Find the domain and range of the function*

$$f(x) = e^{ax+b}$$

Your answer may be in terms of a and b.

Problem 1.116. *In this section it is noted that even roots, like $\sqrt{x}$, only exist for positive numbers or zero, while odd roots, like $\sqrt[3]{x}$, exist for any number. In this context discuss in a few paragraphs, trying to understand $\sqrt[\sqrt{2}]{x}$, the following question as best you can: is $\sqrt{2}$ odd, even, or neither?*

Problem 1.117. *Suppose*

$$f(t) = A \cdot C^t$$

Compute and simplify $f(t+1)/f(t)$.

Problem 1.118. *Suppose that, for positive constants C and D, that $g(t) = A \cdot C^t$ and $h(t) = B \cdot D^t$, both exponential functions. Show that*

$$g(t)/h(t) = Q \cdot R^t$$

and so is also exponential. Give the conditions that determine if this function grows or shrinks as t increases.

1.7. Trigonometric Functions

This section does not cover trigonometry in detail – rather it gives the relationship between the basic trig functions, their domains and ranges, and how they relate to right triangles. A very important object for keeping track of trig functions is the **unit circle**, shown in Figure 1.10.

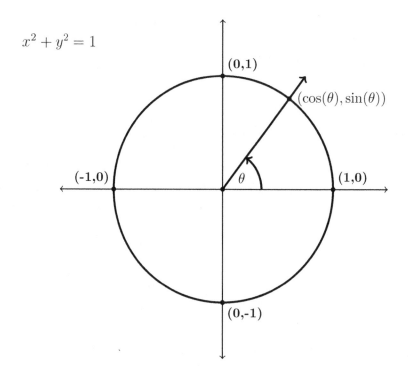

Figure 1.10. The unit circle.

The unit circle, a circle of radius one centered at the origin, has the property that a ray that makes an angle θ with the x-axis intersects the unit circle at the point $(\cos(\theta), \sin(\theta))$. There are a number of angles which have sines and cosines that work out fairly evenly. These are called the **special angles**, and they are displayed on a unit circle in Knowledge Box 1.36.

Knowledge Box 1.36.

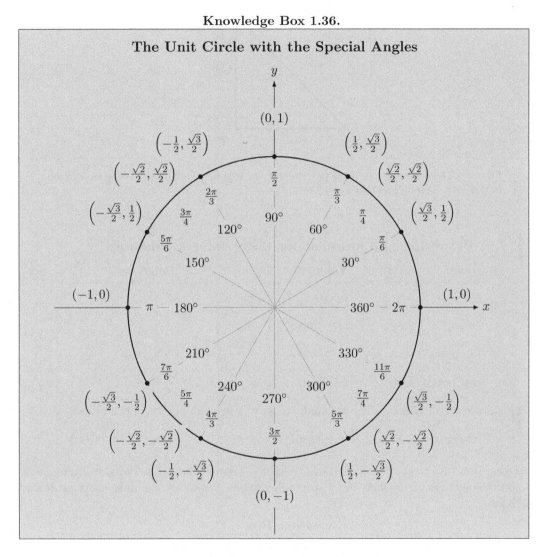

Knowledge Box 1.36 gives the angles in both degrees and radians. A full circle is 360° and 2π radians, making radians seem an odd choice. The reason for using radians – which we will do in the remainder of this text – is because they are the *natural* units of angle. A circle of radius 1 has a diameter of 2π. This means, if we use radians, then the length of an arc of a circle and the angle that subtends that arc have the same numerical value. Later on, as we develop trigonometric integrals, this will keep us from needing to perform unit conversions every time a trig function comes up in a solution.

There are six standard trigonometric functions, defined relative to the standard triangle shown in Figure 1.11. The number n in the following table represents any whole number. So $(2n + 1)$ is a way of saying "any odd whole number".

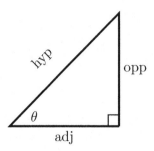

Figure 1.11. A standard right triangle showing the hypotenuse (hyp) and the sides adjacent to (adj) and opposite (opp) the angle θ.

Properties and formulae for trigonometric functions.

Name	Abbrev.	Value at θ	Domain	Range
sine	sin	opp/hyp	$(-\infty, \infty)$	[-1,1]
cosine	cos	adj/hyp	$(-\infty, \infty)$	[-1,1]
tangent	tan	opp/adj	$x \neq \dfrac{2n+1}{2}\pi$	$(-\infty, \infty)$
cotangent	cot	adj/opp	$x \neq n\pi$	$(-\infty, \infty)$
secant	sec	hyp/adj	$x \neq \dfrac{2n+1}{2}\pi$	$(-\infty, -1] \cup [1, \infty)$
cosecant	csc	hyp/opp	$x \neq n\pi$	$(-\infty, -1] \cup [1, \infty)$

Knowledge Box 1.37 shows relationships between the trigonometric functions, many of them are obvious consequences of the facts in the preceding table. These are examples of **trigonometric identities**.

Knowledge Box 1.37.

Some basic identities

For any angle θ, we have:

- $\tan(\theta) = \dfrac{\sin(\theta)}{\cos(\theta)}$

- $\tan(\theta) = \dfrac{1}{\cot(\theta)}$

- $\sec(\theta) = \dfrac{1}{\cos(\theta)}$

- $\cot(\theta) = \dfrac{\cos(\theta)}{\sin(\theta)}$

- $\csc(\theta) = \dfrac{1}{\sin(\theta)}$

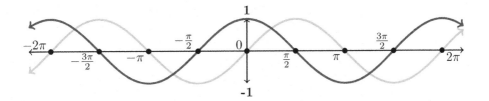

Figure 1.12. The sine (light) and cosine (dark) functions.

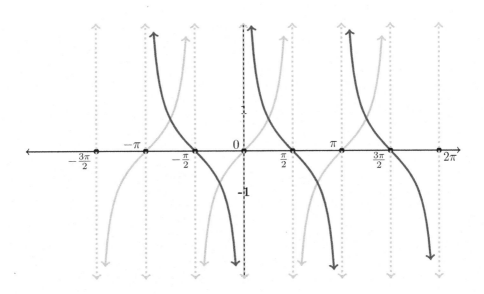

Figure 1.13. The tangent (light) and cotangent (dark) functions.

1.7.1. Graphs of the basic trig functions. Look at the graphs of the sine, cosine, tangent, cotangent, secant, and cosecant functions in Figures 1.12 to 1.14 and check them against the domains and ranges for the functions given earlier in this section.

The trigonometric functions are **periodic**, meaning that they repeat their values regularly. This is visible in their graphs. The sine, cosine, secant, and cosecant functions repeat every 2π, while the tangent and cotangent functions repeat every π units.

The functions and co-functions have simple relationships based on sliding the graph sideways. These relationships are shown in Knowledge Box 1.38.

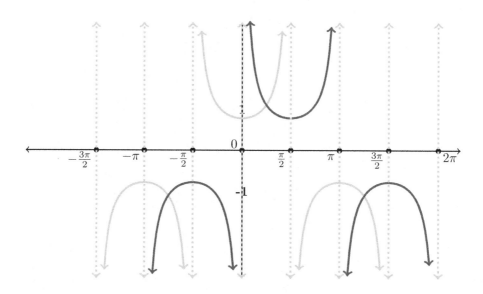

Figure 1.14. The secant (light) and cosecant (dark) functions.

Knowledge Box 1.38.

Periodicity identities

- $\sin(x + 2\pi) = \sin(x)$
- $\cos(x + 2\pi) = \cos(x)$
- $\sin(x) = \cos\left(x - \frac{\pi}{2}\right)$
- $\tan(x) = -\cot\left(x - \frac{\pi}{2}\right)$
- $\sec(x) = \csc\left(x + \frac{\pi}{2}\right)$
- $\cos(-x) = \cos(x)$

- $\sin(-x) = -\sin(x)$
- $\tan(x) = -\tan(x)$
- $\sin(x + \pi) = -\sin(x)$
- $\cos(x + \pi) = -\cos(x)$
- $\tan(x + \pi) = \tan(x)$

1.7.2. Theorems about triangles. The most basic fact about triangles is that the sum of the angles of a triangle in the plane is π radians. This means that if we know two of the angles, we can recover the third by taking π minus their sum.

If we have a right triangle with a hypotenuse of length c and legs of length a and b, then the Pythagorean theorem tells us that

$$a^2 + b^2 = c^2.$$

Homework problem 1.135 asks you to show that the fact

$$\sin^2(\theta) + \cos^2\theta = 1$$

is an instance of the Pythagorean theorem. In fact there are several useful relations between the trigonometric functions that arise from this fact.

The Pythagorean identities

For any angle θ, we have:
- $\sin^2(\theta) + \cos^2(\theta) = 1$
- $\tan^2(\theta) + 1 = \sec^2(\theta)$
- $1 + \cot^2(\theta) = \csc^2(\theta)$

There are some handy relationships that apply to the sides and angles of all triangles (not just right triangles). These are phrased in terms of the general triangle shown in Figure 1.15. Both of these are used to solve problems involving arbitrary triangles. They are called the **law of sines** and the **law of cosines**.

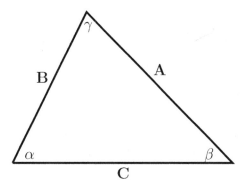

Figure 1.15. A general triangle with labelled side lengths and angles.

The law of sines

$$\frac{A}{\sin(\alpha)} = \frac{B}{\sin(\beta)} = \frac{C}{\sin(\gamma)}$$

The law of cosines

$$C^2 = A^2 + B^2 + 2AB \cdot \cos(\gamma)$$

Example 1.33. Suppose for the triangle shown in Figure 1.15 we have $A = 3$, $B = 5$, and $\beta = \dfrac{\pi}{7}$. What are α, γ, and C?

Solution:

Using the law of sines:

$$\frac{A}{\sin(\alpha)} = \frac{B}{\sin(\beta)}$$

$$\frac{3}{\sin(\alpha)} = \frac{5}{\sin(\pi/7)}$$

$$\sin(\alpha) = \frac{3\sin(\pi/7)}{5}$$

$$\alpha = \sin^{-1}\left(\frac{3\sin(\pi/7)}{5}\right)$$

$$\alpha \cong 0.273 \text{ rad}$$

For the moment, treat $\sin^{-1}$ as a button on your calculator. We will get to inverse trig functions in Section 2.3.3.

We know that $\alpha + \beta + \gamma = \pi$, so

$$\gamma = \pi - \frac{\pi}{7} - 0.273 \cong 2.42 \text{ rad}.$$

Now that we know γ, we can apply the law of sines again to get C:

$$C = \frac{\sin(\gamma) \cdot B}{\sin(\beta)} \cong 7.61$$

$\Diamond$

A **right triangle** (just in case you didn't already know) is one with a right angle in it, an angle with a value of $\pi/2$. An **isosceles triangle** is one where two of the sides have the same length. This forces it, via the law of sines, to have two equal angles as well. An **equilateral triangle** has all three sides the same and has all of its angles equal to $\dfrac{\pi}{3}$ radians. Equilateral triangles are also called **regular triangles**. Two triangles with the same three angles are **similar triangles** and, while they may be different sizes, have the same shape.

1.7.3. The sum, difference, double and half-angle identities. There are a large number of trig identities about the sine or cosine of the sum or difference of angles, and they are somewhat of a nightmare to memorize. From these it is not hard to derive the double and half angle identities (the last four). These last four will be very useful when we start doing integration.

Knowledge Box 1.42.

Sum and difference identities; double angle identities

- $\sin(\alpha + \beta) = \sin(\alpha)\cos(\beta) + \sin(\beta)\cos(\alpha)$
- $\cos(\alpha + \beta) = \cos(\alpha)\cos(\beta) - \sin(\alpha)\sin(\beta)$
- $\sin(\alpha - \beta) = \sin(\alpha)\cos(\beta) - \sin(\beta)\cos(\alpha)$
- $\cos(\alpha - \beta) = \sin(\alpha)\sin(\beta) + \cos(\alpha)\cos(\beta)$
- $\sin(2\theta) = 2\sin(\theta)\cos(\theta)$
- $\cos(2\theta) = \cos^2(\theta) - \sin^2(\theta)$
- $\cos^2(\theta/2) = \dfrac{1 + \cos(\theta)}{2}$
- $\sin^2(\theta/2) = \dfrac{1 - \cos(\theta)}{2}$

To reiterate – this section is not a complete treatment of trigonometry. Instead it summarizes the portions of trigonometry that come up in calculus. It might seem counter intuitive that trig identities will help with calculus, but they can be used to transform expressions from things that are impossible to deal with into things that are not too hard.

1.7.4. Euler's Identity. This section contains a computational trick for quickly recovering the various sum-of-angles, difference-of-angles, and the double angle identities. This section is an enrichment section; it contains a small amount of something we normally would not cover in a first-year calculus course.

Definition 1.14. $i = \sqrt{-1}$.

The problem with the above definition is that it defines something you have been told does not exist – probably ever since you first encountered square roots. In English i is the square root of negative one. The way you deal with this is that i is a number, albeit a funny one, with the added property that $i^2 = -1$. The material in this section makes sense just about when you finish Chapter 13. For the time being accept it as a notational shortcut. Once you have i available, one of the great truths of the universe, **Euler's Identity**, becomes possible to state.

Knowledge Box 1.43.

Euler's Identity

$$e^{i\theta} = i\sin(\theta) + \cos(\theta)$$

In order to make use of Euler's identity we need to know a little bit about **complex numbers**.

Definition 1.15. A **complex number** is a number of the form $a + bi$ where a and b are real numbers. If $a = 0$ we also call the number an **imaginary number**. If $b = 0$, then $a + bi = a$, and the number is a plain **real number**.

Here are the four basic arithmetic operations for complex numbers.

1 $(a + bi) + (c + di) = (a + b) + (c + d)i$

2 $(a + bi) - (c + di) = (a - b) + (c - d)i$

3 $(a + bi) \cdot (c + di) = (ac - bd) + (ad + bc)i$

4 $\dfrac{(a + bi)}{(c + di)} = \dfrac{ac + bd}{c^2 + d^2} + \dfrac{bc - ad}{c^2 + d^2}i$

One last fact is needed before we can reap the benefits of Euler's identity. If $a + bi = c + di$, then $a = c$ and $b = d$. In English, if two complex numbers are equal, then their real parts and their complex parts are also equal. On to harvest results!

Example 1.34. In this example we derive the two double angle identities as the real and imaginary parts of a single expression.

$$e^{2i\theta} = e^{i\theta + i\theta}$$
$$= e^{i\theta} \cdot e^{i\theta}$$
$$i\sin(2\theta) + \cos(2\theta) = (i\sin(\theta) + \cos(\theta)) \cdot (i\sin(\theta) + \cos(\theta))$$
$$= \cos^2(\theta) - \sin^2(\theta) + i\left(2\sin(\theta)\cos(\theta)\right)$$

Pulling out the real and imaginary parts we obtain the two double angle identities:

$$\cos(2\theta) = \cos^2(\theta) - \sin^2(\theta) \text{ (real)}$$
$$\sin(2\theta) = 2\sin(\theta)\cos(\theta) \text{ (imaginary)}$$

$\diamond$

All the sum and difference of angle identities can be derived in a similar fashion, something we leave for the homework.

Problems

Problem 1.119. *Remembering that sine and cosine are periodic with period 2π, find the exact values for the following, expressed with radicals rather than decimal numbers.*

(a) $\cos(5\pi/3)$

(b) $\sin(7\pi/6)$

(c) $\tan(\pi/3)$

(d) $\sec(3\pi/4)$

(e) $\cos(11\pi/4)$

(f) $\cot(27\pi/4)$

(g) $\sin(131\pi/4)$

(h) $\sin(-11\pi/6)$

Problem 1.120. *Find the set of all angles, in radians, that have a cosine of $\dfrac{\sqrt{2}}{2}$.*

Problem 1.121. *Find the domains of the following functions.*

(a) $f(x) = \sin\left(\dfrac{1}{x}\right)$

(b) $g(x) = \cot(\pi x)$

(c) $h(x) = \cos\left(\sqrt{x}\right)$

(d) $r(x) = \sec(\cos(x))$

(e) $s(x) = \csc(\sin(x))$

(f) $q(x) = \sqrt{\cos(x)}$

Problem 1.122. *Find the ranges of the following functions.*

(a) $f(x) = \tan\left(\dfrac{\pi \cdot \cos(x)}{2}\right)$

(b) $g(x) = \cos\left(x^2\right)$

(c) $h(x) = \csc\left(x^2\right)$

(d) $r(x) = \sin\left(\dfrac{\pi}{x^2 + 1}\right)$

(e) $s(x) = \tan\left(\dfrac{\pi}{x^2 + 1}\right)$

(f) $q(x) = \cos\left(\dfrac{1}{x}\right)$

Problem 1.123. *Give exact formulas, using radicals rather than decimals, for each of the following trig functions. Simplify your expressions as much as you can. Hint: the sum and difference of angle formulas may help.*

(a) $\cos\left(\dfrac{\pi}{12}\right)$

(b) $\sin\left(\dfrac{7\pi}{12}\right)$

(c) $\cos\left(\dfrac{5\pi}{12}\right)$

(d) $\tan\left(\dfrac{5\pi}{12}\right)$

(e) $\cot\left(\dfrac{\pi}{12}\right)$

(f) $\sec\left(\dfrac{7\pi}{12}\right)$

Problem 1.124. *Using the diagram from Figure 1.15: if $\alpha = \dfrac{\pi}{3}$, $A = 5$, and $B = 4$, what are C, β, and γ?*

Problem 1.125. *Using the diagram from Figure 1.15: if $\alpha = \dfrac{\pi}{3}$, $\beta = \dfrac{\pi}{4}$, and $A = 4$, what are B, C, and γ?*

Problem 1.126. *Using the diagram from Figure 1.15: if $\alpha = \dfrac{\pi}{5}$, $\beta = \dfrac{\pi}{5}$, and $A = 2$, what are B, C, and γ?*

Problem 1.127. *Show that the following are true, based on material given in this section.*

(a) $\sin(\theta) = \sqrt{1 - \cos^2(\theta)}$

(b) $\cos(\theta) = \sqrt{1 - \sin^2(\theta)}$

(c) $\tan(\theta + \tau) = \dfrac{\tan(\theta) + \tan(\tau)}{1 - \tan(\theta)\tan(\tau)}$

(d) $\tan(\theta - \tau) = \dfrac{\tan(\theta) - \tan(\tau)}{1 + \tan(\theta)\tan(\tau)}$

(e) $\tan(2\theta) = \dfrac{2\tan(\theta)}{1 - \tan^2(\theta)}$

(f) $\tan^2(\theta) = \dfrac{1 - \cos(2\theta)}{1 + \cos(2\theta)}$

Problem 1.128. *What happens to the law of cosines when the angle γ is a right angle? Refer to the definition of the law of cosines in the text.*

Problem 1.129. *Prove that a right isosceles triangle has, in addition to its right angle, two angles that are $\dfrac{\pi}{4}$ radians. Also show that the ratio of the side lengths of this triangle is $1 : 1 : \sqrt{2}$.*

Problem 1.130. *Prove that a right triangle with one of its other angles of size $\dfrac{\pi}{6}$ also has an angle of size $\dfrac{\pi}{3}$. Also show that the ratio of the side lengths of this triangle is $1 : \sqrt{3} : 2$.*

Problem 1.131. *Suppose that the repeated angle of an isosceles triangle has the value $\theta = \dfrac{\pi}{6}$, and that its longest side has length 2. Find the other angle and the lengths of the other sides.*

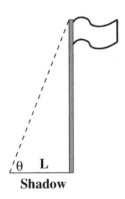

Shadow

Problem 1.132. *Refer to the diagram of the flag pole. If the shadow length is $L = 16m$ and the angle measures as $\theta = \dfrac{\pi}{4}$ rad, then how tall is the flagpole?*

Problem 1.133. *Refer to the diagram of the flag pole. If the shadow length is $L = 6m$ and the angle measures as $\theta = 1.1\ rad$, then how tall is the flagpole?*

Problem 1.134. *Refer to the diagram of the flag pole. If the shadow length is $L = 4m$ and the angle measures as $\theta = 1.4\ rad$, then how tall is the flagpole?*

Problem 1.135. *Explain for which right triangle $\cos^2(\theta) + \sin^2(\theta) = 1$ is an instance of the Pythagorean theorem.*

Problem 1.136. *Use Euler's identity to derive the sum-of-angles identities.*

Problem 1.137. *Use Euler's identity to derive the difference-of-angles identities.*

Problem 1.138. *Use Euler's identity to derive the double angle identities.*

Problem 1.139. *Use Euler's identity to derive the half angle identities.*

Problem 1.140. *Find an expression in terms of $\sin(\theta)$ and $\cos(\theta)$ for $\sin(3\theta)$.*

Problem 1.141. *Find an expression in terms of $\sin(\theta)$ and $\cos(\theta)$ for $\cos(3\theta)$.*

Problem 1.142. *A* Pythagorean triple *is a set of three whole numbers that could be the sides of a right triangle. The most famous is (3,4,5). Notice that*

$$3^2 + 4^2 = 9 + 16 = 25 = 5^2.$$

Find three additional Pythagorean triples for which the numbers do not have a common whole-number divisor bigger than 1. The would disallow (6,8,10) because these three numbers have a common factor of two.

Problem 1.143. *Find the angles, in radians to three decimals, for a right triangle with side lengths 3, 4, and 5.*

Problem 1.144. *Classify each of the six basic trig functions as odd, even, or neither.*

Chapter 2

Limits, Derivatives, Rules, and the Meaning of the Derivative

Traditional calculus courses begin with a detailed formal discussion of limits and continuity. This book departs from that tradition, with this chapter introducing limits only in an informal fashion so as to be able to get going with calculus. A formal discussion of limits and continuity appears in Chapter 8. The agenda for this chapter is to get you on board with a workable operational definition of limits; use this to give the formal definition of a derivative; develop the rules for taking derivatives; and end with a discussion of the physical meaning of the derivative.

2.1. Limits

Suppose you are given a function definition like:

$$f(x) = \frac{x^2 - 4}{x + 2}$$

Then, as long as $x \neq -2$, you can simplify as follows:

$$f(x) = \frac{x^2 - 4}{x + 2} = \frac{(x - 2)(x + 2)}{(x + 2)} = \frac{(x - 2)\cancel{(x + 2)}}{\cancel{(x + 2)}} = x - 2$$

So, this function is a line – as long as $x \neq -2$. What happens when $x = -2$? Technically, the function doesn't exist. This is where the notion of a **limit** comes in handy. If we come up with a whole string of x values and look where they are going as we approach -2, they all seem to be going toward minus 4. The key phrase here is *seem to be*, and the rigorous, precise definition of this vague phrase is the meat of Chapter 8. For now, let's examine a tabulation of the behavior of $f(x)$ near $x = -2$.

From below		From above	
x	$f(x)$	x	$f(x)$
-1	-3	-3	-5
-1.5	-3.5	-2.5	-4.5
-1.75	-3.75	2.25	-4.25
-1.8	-3.8	-2.2	-4.2
-1.9	-3.9	-2.1	-4.1
-1.95	-3.95	-2.05	-4.05
-1.99	-3.99	-2.01	-4.01
Heading for:			
-2	-4	-2	-4

Notice that this table approaches from below (numbers smaller than $x = -2$) and above (numbers larger than $x = -2$). In a well behaved function the approach from above and below are heading for the same place, but there are functions where they don't. We call these the **limit from above** and the **limit from below**. If they agree, their joint value is the **limit of the function**. In this case the limit of the function is -4 at $x = -2$.

Definition 2.1. We use the following symbols for the limits from above and below and the limit of a function $f(x)$ at a point $x = c$:

$$\lim_{x \to c^+} f(x) \qquad\qquad \lim_{x \to c^-} f(x) \qquad\qquad \lim_{x \to c} f(x)$$

Example 2.1. What the tabulated information about $f(x) = \dfrac{x^2 - 4}{x - 2}$ suggests is that:

$$\lim_{x \to 2^+} f(x) = -4 \qquad\qquad \lim_{x \to 2^-} f(x) = -4 \qquad\qquad \lim_{x \to 2} f(x) = -4$$

$$\Diamond$$

The problem with the example function $f(x) = \dfrac{x^2 - 4}{x - 2}$ is that it seems completely contrived. We will see shortly that functions with this sort of implausible structure arise naturally when we try to answer the simple question: **What is the tangent line to a function at a point?** This is the central question for this chapter.

Before we get there, we need the rules-of-thumb for taking limits. Suppose we are trying to compute L such that:

$$\lim_{x \to c} f(x) = L$$

We can follow these rules:

- If either of the limits from above or below don't exist, then L does not exist.

- If the limits from above and below exist but are not equal, then L does not exist.

- If the limits from above and below exist and are equal, then the limit of the function is the joint value of the upper and lower limits.

- If the function $f(x)$ is one that you can just plug into and is one that can be drawn without lifting your pencil (this is another Chapter 8 issue), then you can compute the limit by just plugging in.

- If the function can be turned into a well-behaved function by algebra that works everywhere but $x = c$ (like canceling $(x + 2)$ in our example), then plugging into the modified function computes the limit.

Example 2.2. Compute:

$$\lim_{x \to 2} x^3 + x^2 + x + 1$$

Solution:

Polynomials are the most well behaved functions possible; you can always take their limits by just plugging into them. So

$$\lim_{x \to 2} x^3 + x^2 + x + 1 = 8 + 4 + 2 + 1 = 15$$

$\Diamond$

Example 2.3. Compute:

$$\lim_{x \to 1} \frac{x\,e^x - e^x}{x - 1}$$

Solution:

This function is like our original example in the sense that

$$\frac{x\,e^x - e^x}{x - 1} = \frac{e^x(x - 1)}{x - 1} = \frac{e^x \cancel{(x - 1)}}{\cancel{x - 1}} = e^x,$$

which tells us that

$$\lim_{x \to 1} \frac{x\,e^x - e^x}{x - 1} = e^1 = e.$$

Example 2.4. Compute:

$$\lim_{x \to 2} \frac{x^3 - 6x^2 + 11x - 6}{x - 2}$$

Solution:

If we plug in $x = 2$ to the top of the fraction on a trial basis we get $8 - 24 + 22 - 6 = 0$. So, at $x = 2$ we have the forbidden configuration $\frac{0}{0}$. The root-factor theorem for polynomials tells us that $(x - 2)$ is a factor of $x^3 - 6x^2 + 11x - 6$. So, a little work gives us that

$$\frac{x^3 - 6x^2 + 11x - 6}{x - 2} = \frac{(x - 3)(x - 2)(x - 1)}{x - 2} = \frac{(x - 3)\cancel{(x - 2)}(x - 1)}{\cancel{x - 2}} = (x - 3)(x - 1)$$

So we have that:

$$\lim_{x \to 2} \frac{x^3 - 6x^2 + 11x - 6}{x - 2} = (2 - 3)(2 - 1) = -1$$

$\Diamond$

2.1.1. Split-rule functions. It is sometimes desirable to have functions that obey different rules for different values of x. Retailers often offer bulk pricing discounts, for example, with different costs per unit purchased for larger and smaller numbers of units. There is a notation for this kind of function, called **split rule notation**. Suppose that $s(x)$ is a function that squares negative numbers but adds one to positive numbers and zero. Then we would say:

$$f(x) = \begin{cases} x^2 & x < 0 \\ x + 1 & x \geq 0 \end{cases}$$

If we look at the graph of this function in Figure 2.1, we see that split-rule functions give us a lot of scope for creating functions that lack a limit at a point. Notice that the inequality at the change point is denoted in the graph by using a filled circle for the point that is part of the function and an empty circle for the point that is not part of the function.

Example 2.5. Examine the following function:

$$f(x) = \begin{cases} x^2 + 1 & x < 1 \\ 3x - 1 & x \geq 1 \end{cases}$$

What is the value of $\lim_{x \to 1} f(x)$?

Solution:

As we approach from below 1, the limit will be determined by the rule $x^2 + 1$, and so the limit at 1 is 2, from below. As we approach from above 1, the limit will be determined by the rule $3x - 1$, which will make the limit 2. Since the upper and lower limits both agree, the limit of the function at $x = 1$ is 2.

$\Diamond$

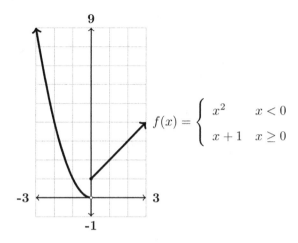

Figure 2.1. A split-rule function.

Example 2.6. Examine the following function: $g(x) = \dfrac{1}{x-2}$ What is the value of $\lim\limits_{x \to 2} f(x)$?

Solution:

For this function we need to look at the graph, at least until we learn more:

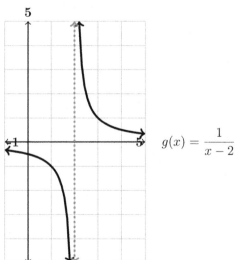

As we approach 2 from below, the value of $g(x)$ is negative. But, since we are dividing by numbers that are approaching more and more closely to zero, those numbers get bigger in absolute value, and the function shoots off toward $-\infty$. This means the limit doesn't exist. Similarly, the limit from above shoots off toward $+\infty$ and also fails to exist. This means the desired limit also does not exist.

◊

Problems

Problem 2.1. *For each of the following limits, give a reason the limit does not exist or compute its value.*

(a) $\lim\limits_{x \to 3} \dfrac{x^2 - 9}{x - 3}$

(b) $\lim\limits_{x \to 1} \dfrac{x^2 - 16}{x + 4}$

(c) $\lim\limits_{x \to -3} \dfrac{x^3 + 6x^2 + 11x + 6}{x + 3}$

(d) $\lim\limits_{x \to 0} \dfrac{e^x - 1}{e^x + 1}$

(e) $\lim\limits_{x \to 0} \dfrac{e^x + 1}{e^x - 1}$

(f) $\lim\limits_{x \to 1} \dfrac{2}{x^3 - 6x^2 + 11x - 6}$

(g) $\lim\limits_{x \to 2} \sqrt{x - 2}$

(h) $\lim\limits_{x \to 0} \ln(x)$

Problem 2.2. *For each of the following limits, give a reason the limit does not exist or compute its value. Use the functions $f(x)$, $g(x)$, and $h(x)$ that follow.*

$$f(x) = \begin{cases} x^3 & x \le 1 \\ -4x + 5 & x >= 1 \end{cases}$$

$$g(x) = \begin{cases} x^2 + 2 & x < -1 \\ 3 - x^2 & x >= -1 \end{cases}$$

$$h(x) = \begin{cases} 2x + 5 & x \le 2 \\ 9 - 2x & x >= 2 \end{cases}$$

(a) $\lim\limits_{x \to -1} f(x)$ (b) $\lim\limits_{x \to 1} f(x)$

(c) $\lim\limits_{x \to -1} g(x)$ (d) $\lim\limits_{x \to 1} g(x)$

(e) $\lim\limits_{x \to -2} h(x)$ (f) $\lim\limits_{x \to 2} h(x)$

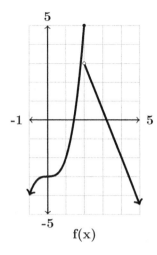

5

-1

5

-5

f(x)

Problem 2.3. *Using $f(x)$ shown in the graph above:*

(a) What is the value of the point $x = c$ where the rules change over?

(b) What is $\lim\limits_{x \to c^-} f(x)$?

(c) What is $\lim\limits_{x \to c^+} f(x)$?

(d) Does the limit of $f(x)$ at c exist?

Problem 2.4. *For which values of c does the function:*

$$f(x) = \begin{cases} x^2 - 1 & x < c \\ 2x + 5 & x \ge c \end{cases}$$

have a limit at $x = c$?

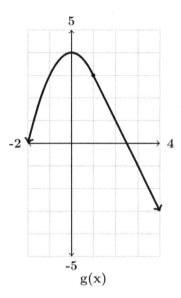

g(x)

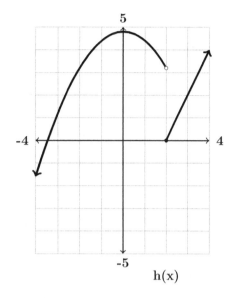

h(x)

Problem 2.5. *Using g(x) in the graph above:*

(a) What is the value of the point $x = c$ where the rules change over?

(b) What is $\lim\limits_{x \to c^-} g(x)$?

(c) What is $\lim\limits_{x \to c^+} g(x)$?

(d) Does the limit of $g(x)$ at c exist?

(e) What evidence is there that the function changes rules?

Problem 2.6. *For which values of c does the function:*

$$g(x) = \begin{cases} x^2 + 7x + 1 & x < c \\ x - 8 & x \geq c \end{cases}$$

have a limit at $x = c$?

Problem 2.7. *For which values of c does the function:*

$$h(x) = \begin{cases} x^2 + 4x & x < c \\ 3x - 2 & x \geq c \end{cases}$$

have a limit at $x = c$?

Problem 2.8. *Using h(x) in the graph above:*

(a) What is the value of the point $x = c$ where the rules change over?

(b) What is $\lim\limits_{x \to c^-} h(x)$?

(c) What is $\lim\limits_{x \to c^+} h(x)$?

(d) Does the limit of $g(x)$ at c exist?

Problem 2.9. *Using the notation that an empty circle indicates a missing point on a graph, graph the following functions on the indicated interval.*

(a) $f(x) = \dfrac{x^2 - 1}{x + 1}$ on $[-2, 2]$

(b) $g(x) = \dfrac{x^3 - 8}{x - 2}$ on $[-3, 3]$

(c) $h(x) = \dfrac{x^4 - 1}{x^2 - 1}$ on $[-2, 2]$

(d) $r(x) = \dfrac{x^2 - 25}{x + 5}$ on $[-8, 2]$

(e) $s(x) = \dfrac{x^2 - 3}{x - \sqrt{3}}$ on $[0, 2]$

(f) $q(x) = \dfrac{x^3 - 1}{x - 1}$ on $[-1, 3]$

2.2. Derivatives

We mentioned earlier that the central question of this chapter is: **What is the tangent line to a function at a point?** A derivative is the slope of that line. In order to compute the derivative, we need to use what we learned about limits in Section 2.1. So, what is a tangent line?

A **tangent line** is a line that touches a curve at exactly one point – at least near that point. The point is called the **point of tangency**. If the curve has a complex shape, then the tangent line may intersect the curve somewhere else as well. But, in a neighborhood of the point of tangency, it brushes the curve only once. The gray line in Figure 2.2 shows a line tangent to a curve.

A **secant line** is a line through two points on a curve. Figure 2.3 shows examples of several secant lines, all of which share one point – the point of tangency in the other picture.

This picture helps us to understand why we need limits to compute slopes of tangent lines. The slopes of the secant lines are all computed based on the two points they pass through. The slope of the tangent line is based on a single point – not possible using the slope formula for lines. If we think of the slope of the tangent line as the limit of the slopes of secant lines from a moving point to the point of tangency, then the limit as the moving point approaches the point of tangency will be the slope of the tangent line.

Suppose that the point of tangency is $(c, f(c))$, and that we examine the secant line through that point and a point "just a little" to the right – the distance to the right being h. Then, the second

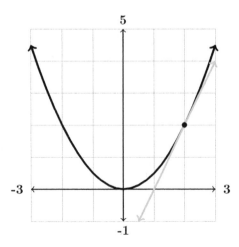

Figure 2.2. A function and a tangent line.

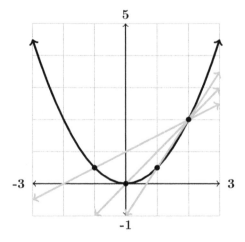

Figure 2.3. A function and several secant lines.

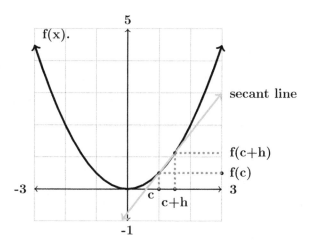

Figure 2.4. A function and a general secant line.

point on the secant line is $(c+h, f(c+h))$, giving the situation shown in the Figure 2.4. If we take the limit as $h \to 0$, then that limit should be the slope of the tangent line. Applying the formula for the slope of a line using two points, we get that the slope of the tangent line to $f(x)$ at $x = c$ is:

$$\lim_{h \to 0} \frac{f(c+h) - f(c)}{(c+h) - c} = \lim_{h \to 0} \frac{f(c+h) - f(c)}{h}$$

This formula is called the **definition of the derivative** and we have a special way of denoting it: $f'(c)$.

Knowledge Box 2.1.

The slope of the tangent line to $f(x)$ at the point $(c, f(c))$ is

$$f'(c) = \lim_{h \to 0} \frac{f(c+h) - f(c)}{h}$$

In the next example, we compute the slope of a tangent line and find the formula for that tangent line.

Example 2.7. Find the tangent line to $f(x) = x^2$ at the point (1,1).

Solution:

For this problem we have $c = 1$. To get the formula for the line we need a point and a slope. We have the point (1,1) on the tangent line, so all we need to calculate is the slope.

$$\lim_{h \to 0} \frac{f(1+h) - f(1)}{h} = \lim_{h \to 0} \frac{(1+h)^2 - 1^2}{h}$$

$$= \lim_{h \to 0} \frac{1 + 2h + h^2 - 1}{h}$$

$$= \lim_{h \to 0} \frac{2h + h^2}{h}$$

$$= \lim_{h \to 0} \frac{h(2 + h)}{h}$$

$$= \lim_{h \to 0} \frac{\cancel{h}(2 + h)}{\cancel{h}}$$

$$= \lim_{h \to 0} 2 + h = 2$$

Notice that this limit is one that requires algebraic manipulation to resolve. We could not just plug $h = 0$ into $\dfrac{2h + h^2}{h}$ because that yields $\dfrac{0}{0}$. All tangent-slope calculations yield limits that require algebraic manipulation – explaining the emphasis on this type of limit in the previous section. We now have the point $(1, 1)$ and a slope of $m = 2$. The line is thus $y - 1 = 2(x - 1)$ or $y = 2x - 1$.

Let's conclude by graphing the function and its tangent line at $c = 1$.

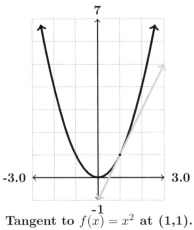

Tangent to $f(x) = x^2$ at (1,1).

$\Diamond$

We now know how to find the slopes of tangent lines at specific values $x = c$. It would be nice to have a general function for the derivative. We define the **general derivative** of a function as follows.

Knowledge Box 2.2.

The **general derivative** (or derivative) of $f(x)$ is
$$f'(x) = \lim_{h \to 0} \frac{f(x+h) - f(x)}{h}$$

Example 2.8. Compute the derivative of $f(x) = \dfrac{1}{x}$.

Solution:

$$
\begin{aligned}
f'(x) &= \lim_{h \to 0} \frac{f(x+h) - f(x)}{h} \\[2mm]
&= \lim_{h \to 0} \frac{\dfrac{1}{x+h} - \dfrac{1}{x}}{h} \\[2mm]
&= \lim_{h \to 0} \frac{\dfrac{x}{x(x+h)} - \dfrac{x+h}{x(x+h)}}{h} \\[2mm]
&= \lim_{h \to 0} \frac{\dfrac{x - (x+h)}{x(x+h)}}{h} \\[2mm]
&= \lim_{h \to 0} \frac{\dfrac{-h}{x(x+h)}}{h} \\[2mm]
&= \lim_{h \to 0} \frac{\dfrac{-\cancel{h}}{x(x+h)}}{\cancel{h}} \\[2mm]
&= \lim_{h \to 0} \frac{-1}{x(x+h)} \\[2mm]
&= \frac{-1}{x^2}
\end{aligned}
$$

So, for $f(x) = \dfrac{1}{x}$ we have that $f'(x) = \dfrac{-1}{x^2}$.

◊

The quantity

$$\frac{f(x+h) - f(x)}{h}$$

is called the **difference quotient** for $f(x)$. You can say that the derivative of a function is the limit of the difference quotient as $h \to 0$.

Example 2.9. Find the derivative of $f(x) = x^n$.

Solution:

$$\begin{aligned}
f'(x) &= \lim_{h \to 0} \frac{f(x+h) - f(x)}{h} \\
&= \lim_{h \to 0} \frac{(x+h)^n - x^n}{h} \\
&= \lim_{h \to 0} \frac{x^n + h \cdot n \cdot x^{n-1} + h^2 \cdot \text{stuff} - x^n}{h} \\
&= \lim_{h \to 0} \frac{h \cdot n \cdot x^{n-1} + h^2 \cdot \text{stuff}}{h} \\
&= \lim_{h \to 0} \frac{h\left(n \cdot x^{n-1} + h \cdot \text{stuff}\right)}{h} \\
&= \lim_{h \to 0} \frac{\not{h}\left(n \cdot x^{n-1} + h \cdot \text{stuff}\right)}{\not{h}} \\
&= \lim_{h \to 0} n \cdot x^{n-1} + h \cdot \text{stuff} \\
&= nx^{n-1}
\end{aligned}$$

$\Diamond$

This result is our first general purpose rule for derivatives, the **power rule**.

Knowledge Box 2.3.

The power rule for derivatives

If $f(x) = x^n$ then

$$f'(x) = nx^{n-1}$$

2.2.1. Derivatives of Sums and Constant Multiples. If $\lim_{x \to c} f(x) = L$ and $\lim_{x \to c} g(x) = M$ both exist, then

$$\lim_{x \to c} f(x) + g(x) = L + M.$$

Similarly, if a is a constant, then

$$\lim_{x \to c} a \cdot f(x) = a \cdot L.$$

Since derivatives are based on limits, we get two very handy rules from these facts.

Knowledge Box 2.4.

Two rules for derivatives

1 The derivative of a sum is the sum of the derivatives:

$$(f(x) + g(x))' = f'(x) + g'(x)$$

2 If a is a constant, then

$$(a \cdot f(x))' = a \cdot f'(x).$$

If we plug a constant value into the difference quotient, we get zero, and the limit as $h \to 0$ of 0 is just 0. This means that **the derivative of a constant is zero**.

We already have a derivative rule for powers of x. From Section 1.5 we know that a polynomial is a sum of constant multiples of powers of x. This means that our two new rules combine with the power rule to permit us to take the derivative of any polynomial.

Example 2.10. If $f(x) = x^3 + 5x^2 + 7x + 2$ find $f'(x)$.

Solution:

Using our new rules, the power rule, and remembering that the derivative of a constant is zero, we get the following:

$$\begin{aligned}
f'(x) &= \left(x^3 + 5x^2 + 7x + 2\right)' \\
&= \left(x^3\right)' + 5\left(x^2\right)' + 7(x^1)' + 2' \\
&= 3x^2 + 5 \cdot 2x + 7 \cdot 1x^0 + 0 \\
&= 3x^2 + 10x + 7
\end{aligned}$$

and we are done.

$\Diamond$

In general, to take the derivative of a polynomial in standard form, all you need to do is bring the power of each term out front, multiplying it by the existing coefficient, and subtract one from the power. So:

$$\left(x^6 + 7x^2 + 4x - 4\right)' = 6x^5 + 14x + 4$$

$$\left(3x^5 + 14x^3 - 8x^2 + 6x + 7\right)' = 15x^4 + 42x^2 - 16x + 6$$

$$\left(3x^9 - 9x^8 - 2x^7 + x^5 + 4x^4 - 7x^3 + 7x\right)' = 27x^8 - 72x^7 - 14x^6 + 5x^4 + 16x^3 - 21x^2 + 7$$

$$\left(5x^2 + 7\right)' = 10x$$

With a little practice this becomes a reflex.

Problems

Problem 2.10. *Using the definition of the derivative, find $f'(c)$ for each of the following pairs of functions and constants.*

(a) $f(x) = x^2$, $c = -1$

(b) $g(x) = x^3$, $c = 2$

(c) $h(x) = \dfrac{1}{x}$, $c = -2$

(d) $r(x) = x$, $c = 4$

(e) $s(x) = x(x+1)$, $c = 0$

(f) $q(x) = 3x + 7$, $c = 1$

Problem 2.11. *For each of the following functions, give the difference quotient. In the name of providence do not attempt to simplify!*

- $f(x) = x^3$

- $g(x) = \dfrac{1}{x^2}$

- $h(x) = \sqrt{x}$

- $r(x) = \cos(x)$

- $s(x) = e^x$

- $q(x) = \tan(x)$

Problem 2.12. *Using the definition of the derivative, find the general derivative of each of the following functions.*

- $f(x) = (x+1)^2$

- $g(x) = x^3$

- $h(x) = \dfrac{1}{x^2}$

- $r(x) = 17$

- $s(x) = \dfrac{1}{x+1}$

- $q(x) = x^2 + x$

Problem 2.13. *For the following functions and values $x = c$, find the tangent line to the function at $x = c$ in slope-intercept form.*

(a) $f(x) = x^2 - 1$, $c = 2$

(b) $g(x) = x^3 + x^2 + x + 1$, $c = -1$

(c) $h(x) = \dfrac{1}{x}$, $c = 1$

(d) $r(x) = \dfrac{3}{x^2}$, $c = 1$

(e) $s(x) = 2x^2 - 5x$, $c = 3$

(f) $q(x) = x^5 - 32$, $c = 2$

Problem 2.14. *For $f(x) = x^2 + 1$ find the tangent lines to $f(x)$ for each of the following x-values:* $\{-2, -1, 0, 1, 2\}$. *Graph the tangent lines and $f(x)$ on the same set of coordinate axes.*

Problem 2.15. *For each of the following functions, find the derivative by any method.*

(a) $a(x) = 2$

(b) $b(x) = 115x - 234$

(c) $f(x) = 3x^2 - 5x + 7$

(d) $g(x) = (x+1)(x+2)(x+3)$

(e) $h(x) = 2x^3 + 3x^2 + 7x - 11$

(f) $r(x) = 5 - x + x^2 - x^3 + x^4$

(g) $s(x) = (x+1)(x+2)(x+3)$

(h) $q(x) = (x+1)^3$

Problem 2.16. *Find the tangent line to $y = \sin(x)$ at $x = \dfrac{\pi}{2}$. Hint: this is a special case where you do not need the derivative to find the tangent line.*

Problem 2.17. *$f(x) = ax + b$ for constants a, b. Find the tangent line to $f(x)$ at $x = c$. No, you were not given an actual value for c.*

Problem 2.18. *For the function*

$$f(x) = \sqrt{x}$$

compute the derivative using the definition of the derivative. This requires an algebra trick, but it is not too hard.

Problem 2.19. *It is not too hard to compute the derivative of $f(x) = |x|$ for $x \neq 0$ because the function agrees with a line, one of $y = \pm x$, everywhere except at zero. Express and support an opinion: does the absolute value function have a derivative at $x = 0$?*

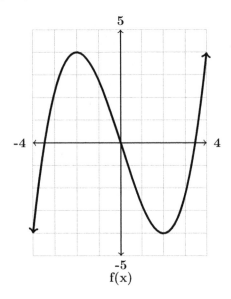

-5
f(x)

Problem 2.20. *For the function $f(x)$ in the picture above, answer the following questions.*

(a) Find $f'(-2)$,

(b) Find the tangent line to $f(x)$ at $c = 2$.

(c) Find the x-values shown where the tangent line has a negative slope.

(d) At which x-value does the slope of a tangent line have the largest negative value.

Problem 2.21. *Find the tangent line to $f(x) = x^2 + 1$ that is parallel to the line $y = 2x - 1$.*

Problem 2.22. *Find the tangent line to $f(x) = x^2 + 1$ that is at right angles to the line $y = 2x - 1$.*

Problem 2.23. *Find a quadratic function*

$$f(x) = ax^2 + bx + c$$

that has $y = 3x+2$ as a tangent line. Demonstrate that your answer is correct.

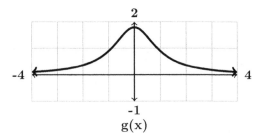

g(x)

Problem 2.24. *For the function $g(x)$ shown above, find the interval(s) on which tangent lines have negative slopes and the interval(s) on which tangent lines have positive slopes.*

Problem 2.25. *Find all tangent lines to the function $h(x) = x^4 - 4x$ that have slope $m = 1$.*

Problem 2.26. *Suppose that*

$$r(x) = ax^3 + bx^2 + cx + d.$$

Then what is the largest number of tangent lines that $r(x)$ can possess that have the same slope?

Problem 2.27. *Find a polynomial with no roots whose first derivative has three roots.*

Problem 2.28. *Suppose that $s(x)$ is a quadratic polynomial. What is the geometric interpretation of the point $(x, s(c))$ where $s'(c) = 0$.*

2.3. Derivatives of The Library of Functions

This section is a catalog of the derivatives of the library of functions. It also introduces three new functions: the inverses of the sine, tangent, and secant functions. We can already take the derivative of polynomial functions using combinations of powers of x. It turns out that the rule for powers of the form x^n also applies for powers that are not whole numbers.

Knowledge Box 2.5.

The general power rule for derivatives

If $f(x) = x^r$ for any real number r, then
$$f'(x) = rx^{r-1}$$

Example 2.11. Find the derivative of $f(x) = \sqrt{x}$.

Solution:
$$f(x) = \sqrt{x} = x^{1/2}$$

Apply the power rule and we get
$$f'(x) = \frac{1}{2}x^{1/2-1} = \frac{1}{2}x^{-1/2} = \frac{1}{2x^{1/2}} = \frac{1}{2\sqrt{x}}$$

$\Diamond$

At this point we just take off and give a whole bunch of derivative rules. It's hard to do good examples until we get to Section 2.4 where we get the rules for combining functions in various ways.

2.3.1. Logs and exponents. The rules for logarithm functions provide a sense of why $\ln(x)$ is called the natural logarithm. All the other logarithm functions have more complex derivative rules based on $\ln(x)$.

Knowledge Box 2.6.

Derivatives of log functions

- If $f(x) = \ln(x)$, then $f'(x) = \dfrac{1}{x}$.

- If $f(x) = \log_b(x)$, then $f'(x) = \dfrac{1}{x\ln(b)}$.

The exponential function $y = e^x$ has the simplest imaginable derivative.

Knowledge Box 2.7.

Derivatives of exponential functions

- If $f(x) = e^x$, then $f'(x) = e^x$.

- If $f(x) = a^x$, then $f'(x) = \ln(a) \cdot a^x$.

Example 2.12. Find the derivative of $f(x) = 3^x$.

Solution:

$$f'(x) = \ln(3) \cdot 3^x$$

$$\Diamond$$

2.3.2. The trigonometric functions. The derivatives of the trigonometric functions should be committed to memory. Some patterns that help with this:

- Only the derivatives of co-functions are negative.

- A function and its co-function have derivatives that are obtained by replacing functions with co-functions.

Knowledge Box 2.8.

Trigonometric derivatives

$f(x)$	$f'(x)$
$\sin(x)$	$\cos(x)$
$\cos(x)$	$-\sin(x)$
$\tan(x)$	$\sec^2(x)$
$\cot(x)$	$-\csc^2(x)$
$\sec(x)$	$\sec(x)\tan(x)$
$\csc(x)$	$-\csc(x)\cot(x)$

2.3.3. Inverse trigonometric functions. We are already somewhat familiar with inverse functions, like the log-exponential pair and the square-square root pair, but the time has come for a formal definition.

Knowledge Box 2.9.

> ### Definition of an inverse function
>
> A function $g(x)$ is the **inverse** of a function $f(x)$ on an interval $[a, b]$ if, for all x in $[a, b]$, we have
> $$f(g(x)) = g(f(x)) = x.$$
> The inverse of $f(x)$ is denoted $f^{-1}(x)$.

Example 2.13. Since $g(x) = \sqrt{x}$ only exists on the interval $[0, \infty)$, we have that $g(x) = \sqrt{x}$ is an inverse of $f(x) = x^2$ on the interval $[0, \infty)$.

When $g(x)$ is an inverse of $f(x)$ on some interval, we have a special name for it. The inverse of $f(x)$ is denoted:

$$f^{-1}(x)$$

which is read "the inverse of $f(x)$." This notation is traditional but problematic because it can be confused with the negative-first power of $f(x)$, i.e. its reciprocal. Usually the meaning of a negative first power is clear from context. If in doubt, ask.

Knowledge Box 2.10.

> ### Computing inverse functions
>
> Suppose that $y = f(x)$. If you can solve $x = f(y)$ for $y = g(x)$, then $g(x) = f^{-1}(x)$ on some interval.

Definition 2.2. A function has a **universal inverse** if there is a single function that is its inverse on its entire domain.

Example 2.14. Suppose that:

$$f(x) = \frac{x+3}{1-x}$$

Find $f^{-1}(x)$.

Solution:

Since we have $y = \dfrac{x+3}{1-x}$, we solve $x = \dfrac{y+3}{1-y}$ for y.

$$x = \frac{y+3}{1-y}$$
$$x(1-y) = y+3$$
$$x - xy = y+3$$
$$x - 3 = xy + y$$
$$x - 3 = y(x+1)$$
$$\frac{x-3}{x+1} = y$$
$$y = \frac{x-3}{x+1}$$

So, we have:

$$f^{-1}(x) = \frac{x-3}{x+1}$$

Now let's check that $f(f^{-1}(x)) = x$:

$$\frac{\dfrac{x-3}{x+1}+3}{1-\dfrac{x-3}{x+1}} = \frac{x-3+3(x+1)}{x+1-(x-3)}$$

$$= \frac{x-3+3x+3}{x-x+1+3}$$
$$= \frac{4x}{4}$$
$$= x$$

and we have verified that the inverse is correct.

$$\Diamond$$

We now have a firm enough grasp of inverse functions to go to work on the inverse trigonometric functions. There is an interesting feature of functions that permits them to have universal inverses

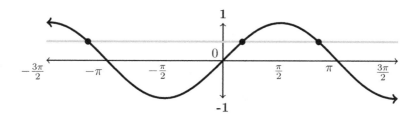

Figure 2.5. The sine function fails the horizontal line test.

– they must pass the **horizontal line test** – similar to the vertical line test for being a function. Any y-value where a horizontal line intersects the graph of a function in two places is a place where inverse values are ambiguous.

Figure 2.5 shows the horizontal line test applied to the sine function. Clearly, it fails the test. If you look back at the graphs of the other trigonometric functions in Section 1.7, you will see that all of them egregiously fail the horizontal line test. For that reason inverses are defined for only part of the domain of the trig functions. The following table gives the domain and range of each of the inverse trigonometric functions.

Properties and formulas for trigonometric functions

Name	Abbrev.	Domain	Range
inverse sine	$\sin^{-1}$	$[-1, 1]$	$\left[-\dfrac{\pi}{2}, \dfrac{\pi}{2}\right]$
inverse cosine	$\cos^{-1}$	$[-1, 1]$	$[0, \pi]$
inverse tangent	$\tan^{-1}$	$(-\infty, \infty)$	$\left(-\dfrac{\pi}{2}, \dfrac{\pi}{2}\right)$
inverse cotangent	$\cot^{-1}$	$(-\infty, \infty)$	$(0, \pi)$
inverse secant	$\sec^{-1}$	$(-\infty, -1] \cup [1, \infty)$	$\left[0, \dfrac{\pi}{2}\right) \cup \left(\dfrac{\pi}{2}, \pi\right]$
inverse cosecant	$\csc^{-1}$	$(-\infty, -1] \cup [1, \infty)$	$\left[-\dfrac{\pi}{2}, 0\right) \cup \left(0, \dfrac{\pi}{2}\right]$

Sometimes an alternate notion is used for inverse trig functions. The prefix "arc" is added to the function name instead of the exponent -1. So, for example, $\sin^{-1}$ is written arcsin.

Figures 2.6 to 2.8 show the graphs of the inverse trig functions, and Knowledge Box 2.11 gives the formulas for their derivatives.

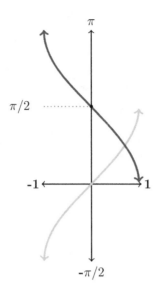

Figure 2.6. The inverse sine (light) and inverse cosine (dark) functions.

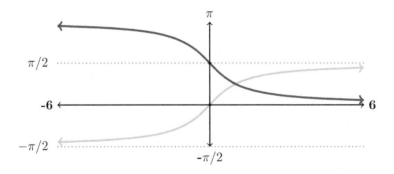

Figure 2.7. The inverse tangent (light) and inverse cotangent (dark) functions.

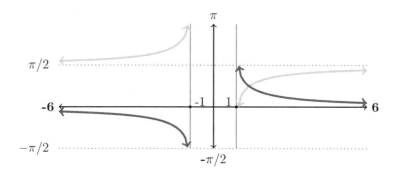

Figure 2.8. The inverse secant (light) and inverse cosecant (dark) functions.

Knowledge Box 2.11.

Inverse trigonometric derivatives

$f(x)$	$f'(x)$
$\sin^{-1}(x)$	$\dfrac{1}{\sqrt{1-x^2}}$
$\cos^{-1}(x)$	$\dfrac{-1}{\sqrt{1-x^2}}$
$\tan^{-1}(x)$	$\dfrac{1}{1+x^2}$
$\cot^{-1}(x)$	$\dfrac{-1}{1+x^2}$
$\sec^{-1}(x)$	$\dfrac{1}{|x|\sqrt{x^2-1}}$
$\csc^{-1}(x)$	$\dfrac{-1}{|x|\sqrt{x^2-1}}$

Problems

Problem 2.29. *Find the derivative of each of the following functions.*

- $f(x) = \sqrt[4]{x}$

- $g(x) = x^{3.1}$

- $h(x) = x^{\pi}$

- $r(x) = \sin(x) + \cos(x)$

- $s(x) = 4e^{x}$

- $q(x) = \tan^{-1}(x) + \pi/2$

Problem 2.30. *Find the tangent line to each of the following functions at the indicated point.*

- $f(x) = \sqrt{x}$ at $x = 1$

- $g(x) = \sin(x)$ at $x = \pi/3$

- $h(x) = \cos(x)$ at $x = \pi/4$

- $r(x) = \tan^{-1}(x)$ at $x = 0$

- $s(x) = \sin^{-1}(x)$ at $x = 0.5$

- $q(x) = \ln(x)$ at $x = \ln(2)$

Problem 2.31. *Find an inverse function for each of the following functions.*

- $f(x) = x^2 + 2x + 1$

- $g(x) = 13x - 27$

- $h(x) = \tan(3x + 2)$

- $r(x) = \dfrac{2x + 1}{x - 1}$

- $s(x) = e^{2x}$

- $q(x) = \dfrac{1}{x}$

Problem 2.32. *Find an inverse of the function $f(x) = x^2$ on the interval $(-\infty, 0\,]$.*

Problem 2.33. *Which of the following functions have universal inverses? Justify your answer.*

- $f(x) = \ln(x)$

- $g(x) = e^{x}$

- $h(x) = 2x + 5$

- $r(x) = x^2$

- $s(x) = x^3$

- $q(x) = (2x + 5)^3$

 Problem 2.34. *Find the universal inverse of*
$$f(x) = ax + b$$
when a and b are real numbers.

Problem 2.35. *For which values of x does*
$$y = \sin(x)$$
have a horizontal tangent line?

Problem 2.36. *Based on the information given in this section what is:*
$$\lim_{x \to \infty} \tan^{-1}(x)$$

Problem 2.37. *Do either of the functions $f(x) = \tan(x)$ or $g(x) = \tan^{-1}(x)$ have a universal inverse? Explain your answer.*

Problem 2.38. *Suppose you take the derivative of*
$$y = \sin(x)$$
104 times. What do you get?

Problem 2.39. *Inverses of functions exist over particular parts of their domain – a universal inverse exists everywhere in the domain.*
For what largest possible domains does
$$f(x) = x^2 + 4x + 4$$
have inverses. Hint: there are two answers. Find the inverses.

2.4. The product, quotient, reciprocal, and chain rules

In this section we learn the derivative rules that let us deal with functions built up out of other functions by both arithmetic and functional composition. Our first rule lets us take the derivative of a product of two functions. It is called the **product rule**.

Knowledge Box 2.12.

The product rule

$$(f(x) \cdot g(x))' = f(x)g'(x) + f'(x)g(x)$$

Example 2.15. Find the derivative of $h(x) = x \cdot \sin(x)$.

Solution:

Apply the product rule to the functions $f(x) = x$ and $g(x) = \sin(x)$:

$$\begin{aligned}
h'(x) &= x \cdot (\sin(x))' + (x)' \cdot \sin(x) \\
&= x \cdot \cos(x) + 1 \cdot \sin(x) \\
&= x \cdot \cos(x) + \sin(x).
\end{aligned}$$

$\Diamond$

Example 2.16. Find the derivative of $r(x) = \ln(x)\cos(x)$.

Solution:

Apply the product rule to the functions $f(x) = \ln(x)$ and $g(x) = \cos(x)$:

$$\begin{aligned}
r'(x) &= \ln(x) \cdot (\cos(x))' + (\ln(x))' \cdot \cos(x) \\
&= \ln(x) \cdot (-\sin(x)) + \frac{1}{x} \cdot \cos(x) \\
&= \frac{\cos(x)}{x} - \ln(x) \cdot \sin(x)
\end{aligned}$$

$\Diamond$

The next rule is the **quotient rule** which is used to deal with the ratio of two functions.

Knowledge Box 2.13.

The quotient rule

$$\left(\frac{f(x)}{g(x)}\right)' = \frac{g(x)f'(x) - f(x)g'(x)}{g^2(x)}$$

Example 2.17. Find the derivative of $h(x) = \dfrac{2x+1}{x+5}$.

Solution:

Apply the quotient rule to the functions $f(x) = 2x + 1$ and $g(x) = x + 5$.

$$\begin{aligned}
h'(x) &= \frac{(x+5)(2x+1)' - (2x+1)(x+5)'}{(x+5)^2} \\
&= \frac{(x+5)\cdot 2 - (2x+1)\cdot 1}{(x+5)^2} \\
&= \frac{2x + 10 - (2x+1)}{(x+5)^2} \\
&= \frac{9}{(x+5)^2}
\end{aligned}$$

In general we *do not* expand the denominator after using the quotient rule. It is often easier to deal with in factored form.

$$\diamond$$

Example 2.18. Find the derivative of $q(x) = \dfrac{x}{x^2+1}$.

Solution:

Apply the quotient rule to the functions $f(x) = x$ and $g(x) = x^2 + 1$.

$$\begin{aligned}
q'(x) &= \frac{\left(x^2+1\right)\cdot (x)' - x \cdot \left(x^2+1\right)'}{\left(x^2+1\right)^2} \\
&= \frac{\left(x^2+1\right)\cdot 1 - x \cdot (2x)}{\left(x^2+1\right)^2} \\
&= \frac{x^2 + 1 - 2x^2}{\left(x^2+1\right)^2} \\
&= \frac{1-x^2}{\left(x^2+1\right)^2}
\end{aligned}$$

$$\diamond$$

Example 2.19. Find the derivative of $r(x) = \dfrac{e^x}{\sin(x)}$.

Solution:

Apply the quotient rule to the functions $f(x) = e^x$ and $g(x) = \sin(x)$.

$$r'(x) = \frac{\sin(x)(e^x)' - e^x(\sin(x))'}{\sin^2(x)}$$
$$= \frac{\sin(x)e^x - e^x \cos(x)'}{\sin^2(x)}$$
$$= \frac{e^x(\sin(x) - \cos(x))}{\sin^2(x)}$$

$\Diamond$

When $f(x) = \dfrac{1}{g(x)}$, a simpler version of the quotient rule, called the **reciprocal rule**, may be used.

Knowledge Box 2.14.

The reciprocal rule

$$\left(\frac{1}{f(x)}\right)' = \frac{-f'(x)}{f^2(x)}$$

Example 2.20. Find the derivative of $h(x) = \dfrac{1}{x^2 + 1}$.

Solution:

Apply the reciprocal rule to the function for which the denominator is $f(x) = x^2 + 1$.

$$h'(x) = \frac{-\left(x^2 + 1\right)'}{\left(x^2 + 1\right)^2}$$
$$= \frac{-2x}{\left(x^2 + 1\right)^2}$$

$\Diamond$

Example 2.21. Find the derivative of $r(x) = \dfrac{1}{e^x + x^2}$.

Solution:

Apply the reciprocal rule to the function for which the denominator is $f(x) = e^x + x^2$.

$$q'(x) = \frac{-\left(e^x + x^2\right)'}{\left(e^x + x^2\right)^2}$$

$$= \frac{-\left(e^x + 2x\right)}{\left(e^x + x^2\right)^2}$$

$$= -\frac{e^x + 2x}{\left(e^x + x^2\right)^2}$$

$$\Diamond$$

2.4.1. Functional Composition and the Chain Rule. The **composition** of two functions results from applying one to the other. If the functions are $f(x)$ and $g(x)$, then their composition is written $f(g(x))$. Let's look at a few examples.

Example 2.22. If $f(x) = x + 7$ and $g(x) = x^2$, then

$$f(g(x)) = x^2 + 7$$

while

$$g(f(x)) = (x + 7)^2$$

$$\Diamond$$

Example 2.23. If $f(x) = \sin(x)$ and $g(x) = e^x$, then

$$f(g(x)) = \sin\left(e^x\right)$$

while

$$g(f(x)) = e^{\sin(x)}$$

$$\Diamond$$

The order in which two functions are composed matters a lot – the results are not symmetric. The **chain rule** is used to compute the derivative of a composition of functions.

Knowledge Box 2.15.

The chain rule

$$(f(g(x)))' = f'(g(x)) \cdot g'(x)$$

In the composition $f(g(x))$ we call $f(x)$ the **outer function** and $g(x)$ the **inner function**.

Example 2.24. Compute the derivative of $h(x) = e^{2x}$.

Solution:

Apply the chain rule to the functional composition for which the outer function is $f(x) = e^x$, and the inner function is $g(x) = 2x$. For these $f'(x) = e^x$ and $g'(x) = 2$. So:

$$h'(x) = e^{2x} \cdot 2$$
$$= 2e^{2x}$$

$\Diamond$

Example 2.25. Compute the derivative of $q(x) = \sin\left(x^2\right)$.

Solution:

Apply the chain rule to the functional composition for which the outer function is $f(x) = \sin(x)$, and the inner function is $g(x) = x^2$. For these $f'(x) = \cos(x)$ and $g'(x) = 2x$. So:

$$q'(x) = \cos\left(x^2\right) \cdot 2x$$
$$= 2x \cdot \cos\left(x^2\right)$$

$\Diamond$

The chain rule avoids a whole lot of multiplying out in some cases. Technically you could do the following example without the chain rule, but it would be purely awful.

Example 2.26. Compute the derivative of $r(x) = \left(x^2 + x + 1\right)^7$.

Solution:

Apply the chain rule to the functional composition for which the outer function is $f(x) = x^7$, and the inner function is $g(x) = x^2 + x + 1$. For these $f'(x) = 7x^6$ and $g'(x) = 2x + 1$. So:

$$r'(x) = 7\left(x^2 + x + 1\right)^6 \cdot (2x + 1)$$

Normally we don't multiply out answers like this.

$\Diamond$

Example 2.27. Compute the derivative of $a(x) = \sqrt{e^x + 2}$.

Solution:

Apply the chain rule to the functional composition for which the outer function is $f(x) = \sqrt{x}$, and the inner function is $g(x) = e^x + 2$. For these $f'(x) = \dfrac{1}{2\sqrt{x}}$ and $g'(x) = e^x$. So:

$$a'(x) = \frac{1}{2\sqrt{e^x + 2}} \cdot e^x$$
$$= \frac{e^x}{2\sqrt{e^x + 2}}$$

Example 2.28. Compute the derivative of $b(x) = \ln(\cos(x))$.

Solution:

Apply the chain rule to the functional composition for which the outer function is $f(x) = \ln(x)$, and the inner function is $g(x) = \cos(x)$. For these $f'(x) = \dfrac{1}{x}$ and $g'(x) = -\sin(x)$. So:

$$b'(x) = \frac{1}{\cos(x)} \cdot (-\sin(x))$$
$$= \frac{-\sin(x)}{\cos(x)}$$
$$= -\frac{\sin(x)}{\cos(x)}$$
$$= -\tan(x)$$

Notice how simplifying this answer requires one of the simpler trig identities.

With the product, quotient, reciprocal, and especially the chain rule, the variety of functions for which we can take the derivative is extended substantially. Practice is needed to develop a sense of which rule to use when.

Problems

Problem 2.40. *Find the derivatives of the following functions with the product rule. You will need the chain rule for some of these.*

(a) $f(x) = x \cdot \cos(x)$

(b) $g(x) = \sin(x) \cdot \cos(x)$

(c) $h(x) = x^5(x+1)^6$

(d) $q(x) = x^3 \cdot e^{2x}$

(e) $r(x) = \tan(x) \cdot \ln(x)$

(f) $s(x) = x^5 \cdot \tan^{-1}(x)$

(g) $a(x) = \cos(2x)e^{3x}$

(h) $b(x) = \cos(\pi x)\sin(ex)$

Problem 2.41. *Find the derivatives of the following functions with the quotient rule.*

(a) $f(x) = \dfrac{2x+1}{x^3+2}$

(b) $g(x) = \dfrac{\ln(x)}{1-x}$

(c) $h(x) = \dfrac{(x+1)^3}{(x-1)^3}$

(d) $q(x) = \dfrac{\tan(x)}{\cos(x)+1}$

(e) $r(x) = (e^x)/(1+e^x)$

(f) $s(x) = \dfrac{x^2+1}{x^2+x+1}$

(g) $a(x) = \dfrac{\sin(x)}{\cos(x)+1}$

(h) $b(x) = \tan^{-1}(x)/(x^2-1)$

Problem 2.42. *Find the derivatives of the following functions with the reciprocal rule.*

(a) $f(x) = \dfrac{1}{x^2+1}$

(b) $g(x) = \dfrac{1}{\cos(x)}$

(c) $h(x) = \dfrac{1}{\sqrt{2x+1}}$

(d) $q(x) = e^{-x}$

(e) $r(x) = \dfrac{1}{\sin(x)+\cos(x)}$

(f) $s(x) = \dfrac{1}{\ln(x)}$

Problem 2.43. *Find the derivatives of the following functions with the chain rule.*

(a) $f(x) = \left(x^2+1\right)^{11}$

(b) $g(x) = \cos\left(x^2+1\right)$

(c) $h(x) = (\cos(x)+1)^5$

(d) $q(x) = \sqrt{\cos(x)+1}$

(e) $r(x) = \left(\dfrac{1}{x}+1\right)^4$

(f) $s(x) = e^{x^2+x+1}$

(g) $a(x) = \ln\left(x^3+2x^2+7x+5\right)$

(h) $b(x) = e^{\tan(x)}$

(i) $c(x) = \ln(\tan(x))$

(j) $d(x) = \ln\left(e^x+1\right)$

Problem 2.44. *Find the derivatives of the following functions with the most appropriate technique(s).*

(a) $f(x) = \left(\dfrac{\sin(x)}{\cos(x)}\right)^5$

(b) $g(x) = x \cdot \ln(\cos^2(x)+1)$

(c) $h(x) = \dfrac{\left(x^2+1\right)^5}{x^7+1}$

(d) $q(x) = x^2 \cdot \ln(x^2+1)$

(e) $r(x) = x \cdot 5^{x^2+1}$

(f) $s(x) = \sin(x) \cdot \ln(x) \cdot \left(x^2+1\right)$

(g) $a(x) = \left(x^3+1\right)^{121} \cdot \dfrac{x}{x+4}$

(h) $b(x) = \ln\left(\dfrac{x^3+1}{x^2+2}\right)$

Problem 2.45. *Find the tangent lines to*

$$f(x) = \dfrac{1}{x^2+1}$$

where $x = \pm 1$. Give your answer in point-slope form.

Problem 2.46. *Find the tangent line to*
$$g(x) = x\,\mathrm{e}^{-x^2}$$
where $x = 0$. Give your answer in point-slope form.

Problem 2.47. *Find the tangent line to $h(x) = \tan^{-1}(x/2)$ where $x = \sqrt{3}$. Give your answer in point-slope form.*

Problem 2.48. *Remembering that*
$$\tan(x) = \frac{\sin(x)}{\cos(x)}$$
use the quotient rule to verify the rule for the derivative of $\tan(x)$.

Problem 2.49. *Remembering that*
$$\cot(x) = \frac{\cos(x)}{\sin(x)}$$
use the quotient rule to verify the rule for the derivative of $\cot(x)$.

Problem 2.50. *Remembering that*
$$\sec(x) = \frac{1}{\cos(x)}$$
use the reciprocal rule to verify the rule for the derivative of $\sec(x)$.

Problem 2.51. *Remembering that*
$$\csc(x) = \frac{1}{\sin(x)}$$
use the reciprocal rule to verify the rule for the derivative of $\csc(x)$.

Problem 2.52. *Derive the reciprocal rule by applying the quotient rule to*
$$g(x) = \frac{1}{f(x)}$$

Problem 2.53. *By using the product rule a couple times and simplifying, verify the triple product rule:*
$$(f(x)g(x)h(x))' = f(x)g(x)h'(x)$$
$$+ f(x)g'(x)h(x)$$
$$+ f'(x)g(x)h(x)$$

Problem 2.54. *Find the derivative of*
$$f(x) = x^5(x+1)^4(x+2)^3$$

Problem 2.55. *Would taking the logarithm of both sides first make Problem 2.54 any easier?*

Problem 2.56. *Find the derivative of*
$$g(x) = x^2 \cdot \sin(x) \cdot \mathrm{e}^x$$

Problem 2.57. *Find the derivative of*
$$h(x) = \sqrt{x+1} \cdot \tan(x) \cdot \ln(x)$$

Problem 2.58. *By using the chain rule as many times as needed, verify the triple chain rule:*
$$(f(g(h(x))))' =$$
$$f'(g(h(x))) \cdot g'(h(x)) \cdot h'(x)$$

Problem 2.59. *Find the derivative of*
$$f(x) = (\sin(\ln(x)))^5$$

Problem 2.60. *Find the derivative of*
$$g(x) = (\ln(\mathrm{e}^x + 1))^3$$

Problem 2.61. *Find the derivative of*
$$g(x) = \ln\left(\sin^4\left(x^4 + 1\right)\right)$$

Problem 2.62. *Prove that*
$$(f(ax+b))' = a \cdot f'(ax+b)$$
for any function $f(x)$ with a derivative.

Problem 2.63. *Find the derivative of $f(x) = \sin(x)\cos(x)$ with the chain rule instead of the product rule. Hint: use a trig identity.*

2.5. Physical Interpretation of Derivatives

So far we have gone from the limit-based definition of a derivative to knowing how to take the derivative of a pretty large number of different functions. The obvious missing piece here is the *meaning* of the derivative. The meaning of a derivative depends a lot on where you got it, of course, but there is a general principle that covers most of what a derivative means.

Knowledge Box 2.16.

What is a derivative?

If $f(x)$ measures a quantity, then $f'(x)$ is the rate at which that quantity is changing.

We have been using the geometric interpretation of the derivative: it is the slope of the tangent line to a curve at a point. A line $y = ax + b$ represents something that starts at b and adds a more per x-unit traversed; a line has a constant rate at which the quantity it is measuring changes. This explains why "$f'(x)$ is the rate of change of $f(x)$" and "$f'(x)$ is the slope of the tangent to the graph at x" are equivalent ideas.

At this point – to permit a number of innovative ways of using the derivative – we introduce a new notation that acknowledges that the derivative is a rate of change.

Knowledge Box 2.17.

Differential notation

Given that $y = f(x)$, another notation for the derivative is

$$\frac{dy}{dx} = f'(x)$$

This is spoken "the differential of y with respect to x." The new symbols dy and dx are the differential of y and of x respectively.

This notion of the derivative as a rate of change leads to a natural application in physics. We will need one more definition.

Definition 2.3. The derivative of the derivative of a function $y = f(x)$ is called the **second derivative** and is denoted by

$$\frac{d^2y}{dx^2} = f''(x)$$

The second derivative measures the rate at which the rate of change is changing. Ouch. It also measures the **curvature** of a graph. An example of this is shown in Figure 2.9; light hash marks show the degree of downward concavity or negative curvature; dark ones document upward concavity or positive curvature.

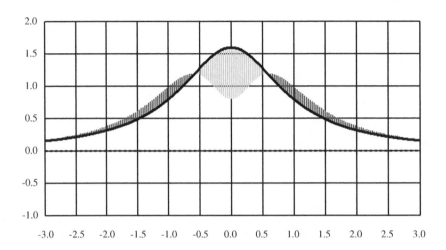

Figure 2.9. Graph of a function annotated to show curvature.

In addition to curvature, there is a straight-forward interpretation of the second derivative in physics – it is **acceleration**.

Knowledge Box 2.18.

Position, velocity, and acceleration

If $y = f(t)$ measures the **position** of an object at time t, then $\dfrac{dy}{dt} = f'(t)$ gives its **velocity**, while $\dfrac{d^2y}{dt^2} = f''(t)$ gives its **acceleration**.

If we can measure position, then the rate of change of position is velocity, and the rate at which velocity is changing is acceleration. If the unit of position is meters, then velocity is meters/second, and acceleration is meters/second/second. Each successive derivative increases the number of "per second" qualifiers by one. This is reflected in the differential notation:

- $y = f(t)$ (no denominator)

- $\dfrac{dy}{dt} = f'(t)$ (time in denominator)

- $\dfrac{d^2y}{dt^2} = f''(t)$ (time squared in denominator)

Example 2.29. Suppose that the distance traveled by a vehicle is given by

$$s(t) = 0.02t^2 + 0.5t + 2.$$

Find the velocity $v(t)$ and acceleration $a(t)$ of the vehicle.

Solution:

Take derivatives:

$$s(t) = 0.02t^2 + 0.5t + 2 \text{ meters}$$
$$v(t) = 0.04t + 0.5 \text{ meters/second}$$
$$a(t) = 0.04 \text{ meters/second}^2$$

$\diamond$

Velocity and acceleration are not the only rates of change we may be interested in. The next example shows how to compute the flow out of a tank from the volume of fluid in it.

Example 2.30. A cylindrical tank with an open spigot at the bottom has a volume of water that is

$$V(t) = 1216 - 72\sqrt{t} \text{ liters}$$

at a time t. Find the rate of flow of water from the tank.

Solution:

Take derivatives:

$$V(t) = 1216 - 72\sqrt{t} \text{ L}$$
$$V(t) = 1216 - 72t^{1/2}$$
$$V'(t) = -36t^{-1/2}$$
$$V'(t) = -\frac{36}{\sqrt{t}} \text{ L/sec}$$

Of course this formula only works while the tank has water in it, so we should also compute:

$$V(t) = 0$$
$$1216 - 72\sqrt{t} = 0$$
$$1216 = 72\sqrt{t}$$
$$1478656 = 5184t$$
$$t = 285.2$$

So the formula is only good for the first 285 seconds, after which the tank is empty.

2.5.1. Implicit Derivatives. In Section 1.4 we used the circle as an example of an interesting object that failed the vertical line test and so is not a function. On the other hand, the whole theory of derivatives we've built up works only for functions, at least so far. **Implicit derivatives** let us get around this problem. In order to use implicit derivatives we need a convenient fact about differentials: they cancel like variables.

Knowledge Box 2.19.

Algebraic properties of differentials

For any variables a, b, c,

$$\frac{da}{db} \cdot \frac{db}{dc} = \frac{da}{d\!\!\!/b} \cdot \frac{d\!\!\!/b}{dc} = \frac{da}{dc}$$

and similarly

$$\frac{da/db}{dc/db} = \frac{da}{dc}$$

The trick of implicit differentiation is that we can say "y is a function of x, but we are not explicitly solving for it." Then we take the derivative of everything, as if x and y are both "the variable." Since y is a function of x, we also pick up $\dfrac{dy}{dx}$ every time we take a derivative of y. This is required by the chain rule. Let's do an example.

Example 2.31. Suppose that

$$x^2 + y^2 = 25,$$

find the tangent line at the point (3,-4).

Solution:

There is an added step to finding a tangent line this way – you need to make sure the point of tangency is legitimate (on the curve). In this case

$$(3^2) + (-4)^2 = 9 + 16 = 25$$

so we are just fine. Now we take the derivative, normally but for both variables, sticking in a $\dfrac{dy}{dx}$ each time we take a derivative of y. This yields

$$2x + 2y\frac{dy}{dx} = 0$$

Plugging in the point gives us $2(3) + 2(-4)\dfrac{dy}{dx} = 0$, and we solve to get the slope of the tangent line, $\dfrac{dy}{dx} = \dfrac{3}{4}$. Plugging into the point-slope form, $y + 4 = \dfrac{3}{4}(x - 3)$ or

$$y = \frac{3}{4}x - \frac{25}{4}.$$

Let's look at a picture of the circle and the tangent line.

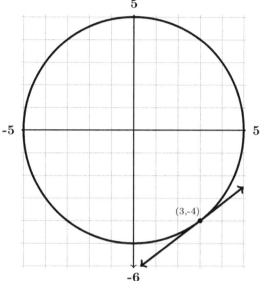

The circle and the tangent line.

Notice that we did not have to worry about whether the function we were using was the upper half of the circle or the lower half. We got a general expression that covered both halves of the circle. Which half we actually were working on was controlled by the choice of the point. (3,-4) is on the bottom half.

$$\Diamond$$

The next example permits us to use the algebraic properties of differentials.

Example 2.32. If $y = x^2$ and $x = t^2 + t + 3$, then what is $\dfrac{dy}{dt}$ when $t = 4$?

Solution:

When $t = 4$ we see that $x = 16 + 4 + 3 = 23$. Taking a couple of derivatives we see that $\dfrac{dy}{dx} = 2x$ and $\dfrac{dx}{dt} = 2t + 1$. When $t = 4$ we get that $\dfrac{dx}{dt} = 2 \cdot 4 + 1 = 9$. We can then compute that:

$$\frac{dy}{dt} = \frac{dy}{dx}\frac{dx}{dt} = 2x \cdot 9 = 2 \cdot 23 \cdot 9 = 414$$

$$\Diamond$$

One of the nice features of the implicit derivative is that it lets us avoid solving a function for y before we compute the tangent slope. The next example carefully avoids having to deal with the function

$$y = \sqrt[3]{17 - x^2}$$

Value the power!

Example 2.33. Given $x^2 + y^3 = 17$, find the tangent line to the curve at (3,2).

Solution:

First, again, check the point is on the curve. We get $3^2 + 2^3 = 9 + 8 = 17$. So, that's fine. Now compute the implicit derivative and get that:

$$2x + 3y^2 \frac{dy}{dx} = 0$$

Plug in (3,2), and we get that $6 + 12\frac{dy}{dx} = 0$. Solving for $\frac{dy}{dx}$ we get the slope of the line, which is $-\frac{1}{2}$. Using the point slope form we get that $y - 2 = -\frac{1}{2}(x - 3)$, which simplifies to $y = -\frac{1}{2}x + \frac{7}{2}$. Done!

$\Diamond$

Suppose we have a relationship that changes with time. Then we can take the (implicit) time derivative of everything. Any variable a gets a $\frac{da}{dt}$ tacked on, courtesy of the chain rule.

Example 2.34. Suppose a copper disk is being heated so that its radius is expanding at 0.02mm/minute, how fast is its surface area changing when the radius is 12mm?

Solution:

We start with the area formula for a disk:

$$A = \pi \cdot r^2$$

The variables are A and r. Taking the implicit time derivative we get:

$$\frac{dA}{dt} = 2\pi r \cdot \frac{dr}{dt}$$

The statement of the problem tells us the the rate of change of the radius with respect to time is $\frac{dr}{dt} = 0.02$ mm/min and that $r = 12$, so we plug in and get

$$\frac{dA}{dt} = 2\pi(12)(0.02) \cong 1.51 \text{ mm}^2/\text{min}$$

$\Diamond$

This example is a type of problem called a **related rate** problem. In general, if we have an equation that relates quantities that vary with time, we can use its implicit time derivative to get a relationship between the way the quantities are changing.

Example 2.35. Suppose 0.5 m^3 of hydrogen is being pumped into a spherical balloon each second, how fast is the radius of the balloon changing when it holds 1000 m^3 of hydrogen?

Solution:

This problem has a couple of moving parts. First of all the volume of a sphere, in terms of its radius, is:

$$V = \frac{4}{3}\pi r^3$$

So the (implicit) time derivative is:

$$\frac{dV}{dt} = 4\pi r^2 \frac{dr}{dt}$$

The problem asked for $\dfrac{dr}{dt}$ and gave us $\dfrac{dV}{dt} = 0.5$ m^3/sec. We also know we want the answer when $V = 1000$ m^3, **but** our formula is in terms of the radius. Solve:

$$1000 = \frac{4}{3}\pi r^3$$

for the radius and we get:

$$r = \sqrt[3]{\frac{3000}{4\pi}} \cong 6.2 \text{ m}$$

Now we plug in and get

$$0.5 = 4\pi(6.2)^2 \frac{dr}{dt}$$

or

$$\frac{dr}{dt} = \frac{1}{8\pi \cdot (6.2)^2} \cong 0.00104 \text{ m/sec}$$

Pretty slowly.

◊

Example 2.36. A car is approaching an intersection from the north at 100km/hr; a second is approaching from the west at 60km/hr. When the first car is 1km from the intersection, and the second is 1.2km from the intersection, how fast are the cars approaching one another?

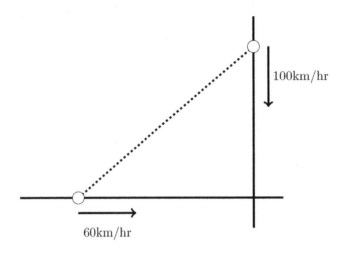

Solution:

The distance between the cars is the hypotenuse of a right triangle. If the distance of the first car from the intersection is y, and the distance of the second is x, then the distance D of the cars from one another is given by $D^2 = x^2 + y^2$. Taking the time derivative we get:

$$2D\frac{dD}{dt} = 2x\frac{dx}{dt} + 2y\frac{dy}{dt}$$

Divide through by two and compute $D = \sqrt{1^2 + 1.2^2} = \sqrt{2.44} \cong 1.56$ km. Plug in and get

$$2(1.56)\frac{dD}{dt} = 2(1.2)(60) + 2(1)(100)$$

or

$$\frac{dD}{dt} = \frac{344}{3.12} \cong 110.3 \text{ km/hr}$$

Notice that the answer is (and should be) plausible.

◇

2.5.2. Logarithmic differentiation. This section highlights another use for implicit differentiation that is so cool it gets its own name: **logarithmic differentiation**. The idea is this – if a function has a lot of products and/or powers, taking the log of both sides before you take the derivative might make life easier.

Recall that:

- $\ln(ab) = \ln(a) + \ln(b)$

- $\ln\left(a^b\right) = b \cdot \ln(a)$

Example 2.37. Suppose that $y = x^5(x+1)^7(x+3)^3$. Find the derivative $\dfrac{dy}{dx}$.

Solution:

Take the log of both sides and take the implicit derivative.

$$y = x^5(x+1)^7(x+3)^3$$

$$\ln(y) = 5\ln(x) + 7\ln(x+1) + 3\ln(x+3)$$

$$\frac{1}{y} \cdot \frac{dy}{dx} = 5\frac{1}{x} + 7\frac{1}{x+1} + 3\frac{1}{x+3}$$

$$\frac{dy}{dx} = y\left(\frac{5}{x} + \frac{7}{x+1} + \frac{3}{x+3}\right)$$

$$\frac{dy}{dx} = x^5(x+1)^7(x+3)^3\left(\frac{5}{x} + \frac{7}{x+1} + \frac{3}{x+3}\right)$$

Notice that the last step eliminates the y on the right side of the equation by substituting in the known value of y in terms of x. We usually do not simplify these expressions.

◇

Example 2.38. Given $y = \dfrac{x^6(x+2)^4}{(x^2+1)^5}$, find y'.

Solution:

Take the log of both sides and take the implicit derivative.

$$y = \frac{x^6(x+2)^4}{(x^2+1)^5}$$

$$\ln(y) = 6\ln(x) + 4\ln(x+2) - 5\ln(x^2+1)$$

$$\frac{1}{y} \cdot \frac{dy}{dx} = \frac{6}{x} + \frac{4}{x+2} + 5 \cdot \frac{2x}{x^2+1}$$

$$\frac{dy}{dx} = y\left(\frac{6}{x} + \frac{4}{x+2} + \frac{10x}{x^2+1}\right)$$

$$\frac{dy}{dx} = \frac{x^6(x+2)^4}{(x^2+1)^5}\left(\frac{6}{x} + \frac{4}{x+2} + \frac{10x}{x^2+1}\right)$$

$\Diamond$

Logarithmic differentiation also permits us to take the derivative of functions that we couldn't otherwise work with at all. The next example demonstrates this.

Example 2.39. Find the derivative of $y = x^x$.

Solution:

$$y = x^x$$

$$\ln(y) = x \cdot \ln(x)$$

$$\frac{1}{y} \cdot \frac{dy}{dx} = x \cdot \frac{1}{x} + 1 \cdot \ln(x)$$

$$\frac{dy}{dx} = y\left(1 + \ln(x)\right)$$

$$\frac{dy}{dx} = x^x\left(1 + \ln(x)\right)$$

$\Diamond$

Of course, once we have the ability to take derivatives, we can find tangent lines, do related relate problems, or anything else one can do with derivatives.

Problems

Problem 2.64. *Each of the functions below gives distance as a function of time. Compute velocity and acceleration.*

(a) $f(t) = 0.02t^3 + t^2$

(b) $g(t) = t \cdot \ln(t)$

(c) $h(t) = \dfrac{1}{t^2 + 1}$

(d) $r(t) = \sin(t)$

(e) $s(t) = \tan(t)$

(f) $q(t) = \left(\sqrt{t} + 1\right)^{3.6}$

Problem 2.65. *The height of a ball above the ground is given by the equation*

$$h(t) = 112 + 6t - 5t^2 \ m$$

(a) How high in the air is the ball at time $t = 0$?

(b) Is the ball moving up or down at $t = 0$?

(c) When does the ball hit the ground?

(d) Find an expression for the ball's velocity.

(e) What is the ball's acceleration?

Problem 2.66. *The amount of water in a tank at time t, in liters, is given by*

$$A(t) = 0.2t + \sqrt{t}$$

(a) At what rate is water flowing into the tank?

(b) Is the tank filling or emptying?

(c) At what point will there be 100 liters of water in the tank?

Problem 2.67. *Given a falling object that experiences constant, non-zero acceleration, explain why its position function is a quadratic function.*

Problem 2.68. *Does a ball falling from a great height above the earth experience constant acceleration?*

Problem 2.69. *Given the pairs of functions, find the indicated differential.*

(a) $y = 3x + 5$, $x = t^2 - 1$

Find $\dfrac{dy}{dt}$ at $t = 2$.

(b) $y = \ln(x)$, $x = 5t + 4$

Find $\dfrac{dy}{dt}$ at $t = 1$.

(c) $y = \dfrac{x + 1}{x - 1}$, $x = \sin(t)$

Find $\dfrac{dy}{dt}$ at $t = \dfrac{\pi}{3}$.

(d) $y = x(x + 1)$, $x = t^3 + t^2 + 2t + 1$

Find $\dfrac{dy}{dt}$ at $t = -2$.

(e) $y = e^x$, $x = t^2 + 51$

Find $\dfrac{dy}{dt}$ at $t = 1$.

(f) $y = 4x - 1$, $x = t \cdot e^{-t}$

Find $\dfrac{dy}{dt}$ at $t = 1$.

Problem 2.70. *Suppose* $\dfrac{dy}{dx} = 1.2$, $\dfrac{dx}{dt} = 3.4$, *and* $\dfrac{dt}{du} = -1.5$. *Find the following differentials.*

(a) $\dfrac{dy}{dt}$ (c) $\dfrac{dx}{du}$ (e) $\dfrac{du}{dy}$

(b) $\dfrac{dy}{du}$ (d) $\dfrac{dt}{dy}$ (f) $\dfrac{du}{dx}$

Problem 2.71. *Find the tangent line to each of the following curves at the specified point.*

(a) $x^2 + y^2 = 20$ at $(-4, 2)$

(b) $(xy + 1)^2 = 9$ at $(2, -2)$

(c) $x^2 + 3xy + y^2 = 5$ at $(1, 1)$

(d) $x \cdot \cos(y) = 1$ at $\left(\sqrt{2}, \dfrac{\pi}{4}\right)$

(e) $x^5 y + xy^5 + 1 = 3$ at $(-1, -1)$

(f) $x^3 + y^3 + 2xy + 4 = 17$ at $(1, 2)$

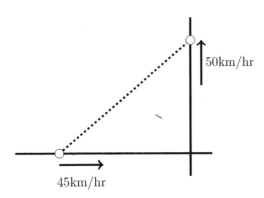

50km/hr

45km/hr

Problem 2.72. *A car is leaving an intersection northward at 50 km/hr; a second is coming from the west at 45km/hr. When the northward car is 100m from the intersection, and the eastward car is 60m from the intersection, what is the relative velocity of the cars?*

Problem 2.73. *A small metal disk is cooled with cold water so that its diameter is decreasing by 0.003 mm per sec. How fast is the surface area of one side decreasing when the diameter is 20mm?*

$h(t) = 4t + 0.13t^2$

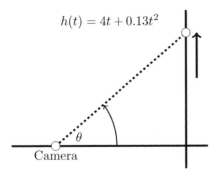

θ

Camera

Problem 2.74. *Suppose that the height of an object at time t is given by $h(t) = 4t + 0.13t^2$ m. The object is being tracked by a camera 10m away. Find an expression for the rate of rotation of the camera at time t.*

Problem 2.75. *The top of a 5m ladder is sliding down a wall at 1.4m per sec. How*

fast is the foot of the ladder sliding away from the wall when the top of the ladder is 4m off the floor?

Problem 2.76. *If three cubic meters of air per second are pumped into a spherical balloon, find the rate at which its diameter and surface area are changing when it holds 400 m^3 of air. Recall that for a sphere,*

$$V = \frac{4}{3}\pi r^3$$

and

$$A = 4\pi r^2$$

Problem 2.77. *Find $\dfrac{dy}{dx}$ for each of the following functions.*

(a) $y = (x^2 + 1)^4(x^2 + x + 1)^5$

(b) $y = \dfrac{x^6}{(x + 1)^{12}(x + 4)^5}$

(c) $y = e^x(x^2 + 1)^7(x^2 - 2)^5$

(d) $y = (x^2 + 1)^x$

(e) $y = x^{\tan(x)}$

(f) $y = (\sin(x) + 2)^x$

(g) $y = x^8 \cdot \tan^5(x) \cdot \sec^7(x)$

(h) $y = x(x+1)(x+2)(x+3)(x+4)(x+5)$

Problem 2.78. *Find the tangent line to*

$$y = \frac{x^6(x^2 + 1)}{(x^2 + x + 1)^5}$$

at the point (-1,2).

Problem 2.79. *Find the tangent line to*

$$y = x^{x^2+1}$$

at the point (2,32).

Problem 2.80. *Find the tangent line to*

$$y = (x + 1)^{x+3}$$

at the point (1,16).

Problem 2.81. *If two cars are approaching a right-angle intersection at the same speed S and are the same distance away from it, how fast are they approaching one another?*

Chapter 3

Applications of the derivative

This chapter covers two applications of the derivative:

- curve sketching and

- optimization

We will learn more about both topics in later chapters as our calculus grows more powerful, but the essentials of the applications are given here.

3.1. Limits at Infinity

We start curve sketching by using roots (places where a curve crosses the x-axis) and **vertical** and **horizontal** asymptotes. Finding horizontal asymptotes will require us to compute limits of a function as x grows infinitely far from zero – in either of the possible directions. The first section of the chapter is about **limits at infinity**.

The limit at positive infinity is the value a function gets close to as x gets larger and larger. The limit at negative infinity is the value a function gets close to as x gets smaller and smaller. The absolute value function gives us the ability to figure out if two things are "close". If $|f(x) - L|$ is small, then $f(x)$ is close to L. Traditionally, the Greek letter epsilon, ϵ, is used to represent small values. The situation is illustrated in Figure 3.1.

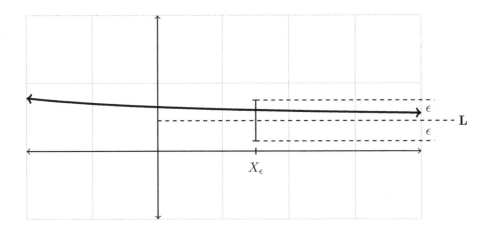

Figure 3.1. A limit at infinity: as long as $x > X_\epsilon$ we have that $f(x)$ is closer to L than ϵ.

This gives us the following formal definitions:

Definition 3.1. If we can find a number L so that for every number $\epsilon > 0$ we can find a value X_ϵ so that when $x \geq X_\epsilon$ we have that $|f(x) - L| < \epsilon$, then L is the limit of $f(x)$ as $x \to \infty$.

Definition 3.2. If we can find a number L so that for every number $\epsilon > 0$ we can find a value X_ϵ so that when $x \leq X_\epsilon$ we have that $|f(x) - L| < \epsilon$, then L is the limit of $f(x)$ as $x \to -\infty$.

There are a number of cases where we can compute the limit using algebra. It is entirely possible that no such number as L exists. Examine the sine function. As $x \to \infty$ it continues to alternate back and forth, settling down to no value L.

A simpler way for the limit to fail to exist is for it to **diverge**. This means that as x gets bigger and bigger $f(x)$ does as well. Or, more formally:

Definition 3.3. If for any constant c, we can find X_c so that $x > X_c$ means $f(x) \geq c$, then

$$\lim_{x \to \infty} f(x)$$

diverges to infinity.

We will begin by doing examples where we compute limits using the definition.

Example 3.1. Using the definition of the limit (Definition 3.1), show that:

$$\lim_{x \to \infty} \frac{1}{1 + x^2} = 0$$

Solution:

For each $\epsilon > 0$ we need to find X_ϵ. Assume we have fixed ϵ and compute X_ϵ. Remember that when you take the reciprocal of both sides of an inequality, it reverses.

$$\left| \frac{1}{1 + X_\epsilon^2} - 0 \right| < \epsilon$$

$$\frac{1}{1 + X_\epsilon^2} < \epsilon$$

$$1 + X_\epsilon^2 > \frac{1}{\epsilon}$$

$$X_\epsilon^2 > \frac{1}{\epsilon} - 1$$

$$X_\epsilon > \sqrt{\frac{1}{\epsilon} - 1}$$

So the value $X_\epsilon = \sqrt{\frac{1}{\epsilon} - 1}$ meets the requirements of the definition for $\epsilon \leq 1$. Since the function has a maximum value of 1 (at zero) satisfying the inequality for larger values of ϵ is easy.

$\Diamond$

Example 3.2. Show, with the definition of divergence, (Definition 3.3) that $\lim_{x \to \infty} x^3$ diverges to ∞.

Solution:

For a constant c, set $X_c = \sqrt[3]{c}$. If

$$x \geq X_c$$
$$x \geq \sqrt[3]{c}$$
$$x^3 \geq c$$
$$f(x) \geq c$$

and we have the desired property for the given X_c.

$$\Diamond$$

Now let's solve the limit from Example 3.1 algebraically. To do this we need to accept the fact that a constant quantity divided by a quantity that grows without limit is, itself, going to zero. In this case:

Knowledge Box 3.1.

Limit of a constant over a growing power of x

$$\lim_{x \to \infty} \frac{1}{x^\alpha} = 0 \text{ for } \alpha > 0$$

Example 3.3. Compute

$$\lim_{x \to \infty} \frac{1}{x^2 + 1}$$

using the fact in Knowledge Box 3.1.

Solution:

$$\lim_{x \to \infty} \frac{1}{x^2 + 1} = \lim_{x \to \infty} \left(\frac{\frac{1}{x^2}}{\frac{1}{x^2}} \right) \frac{1}{x^2 + 1}$$

$$= \lim_{x \to \infty} \frac{\frac{1}{x^2}}{\frac{x^2}{x^2} + \frac{1}{x^2}}$$

$$= \lim_{x \to \infty} \frac{\frac{1}{x^2}}{1 + \frac{1}{x^2}}$$

$$= \frac{0}{1 + 0}$$

$$= 0$$

When all the x's vanish we are left with a constant that does not depend on x. The limit of a constant is the constant itself.

$$\Diamond$$

More generally, this trick (divide top and bottom by the highest power of x) works whenever we have a polynomial divided by a polynomial. Let's do another example.

Example 3.4. Find:

$$\lim_{x \to \infty} \frac{2x^3}{x^3 + 3x + 1}$$

Solution:

$$\lim_{x \to \infty} \frac{2x^3}{x^3 + 3x + 1} = \lim_{x \to \infty} \frac{1/x^3}{1/x^3} \cdot \frac{2x^3}{x^3 + 3x + 1}$$

$$= \lim_{x \to \infty} \frac{2}{1 + \frac{3}{x^2} + \frac{1}{x^3}}$$

$$= \frac{2}{1 + 0 + 0}$$

$$= 2$$

$$\Diamond$$

This leads to a simple rule for the ratio of two polynomials.

Knowledge Box 3.2.

Limit of the ratio of polynomials

Suppose that $f(x)$ is a polynomial of degree n and $g(x)$ is a polynomial of degree m. Then

$$\lim_{x \to \infty} \frac{f(x)}{g(x)} \text{ is:}$$

- 0 if $m > n$

- ∞ if $n > m$

- Equal to the ratio of the coefficients of the highest degree terms on the top and bottom if $n = m$

Example 3.5.

$$\lim_{x \to \infty} \frac{3x^2 + 5x + 1}{x^2 + 2x + 7} = 3$$

Notice how the rule in Knowledge Box 3.2 makes this sort of problem have a one-step immediate solution?

$\Diamond$

This technique for resolving limits at infinity can be used more generally by making the rule about a constant over something growing more general.

Knowledge Box 3.3.

Limit of a constant over a growing quantity

Suppose that $\lim_{x \to \infty} f(x) = \infty$ and that C is a constant. Then

$$\lim_{x \to \infty} \frac{C}{f(x)} = 0.$$

With this new fact, the rule changes from "divide by the highest power of x" to "divide by the fastest growing quantity".

Example 3.6. Compute:

$$\lim_{x \to \infty} \frac{e^x}{1 + 2e^x}$$

Solution:

$$
\begin{aligned}
\lim_{x \to \infty} \frac{e^x}{1 + 2e^x} &= \lim_{x \to \infty} \frac{e^{-x}}{e^{-x}} \cdot \frac{e^x}{1 + 2e^x} \\
&= \lim_{x \to \infty} \frac{e^{-x} \cdot e^x}{e^{-x} + e^{-x} \cdot 2e^x} \\
&= \lim_{x \to \infty} \frac{1}{e^{-x} + 2} \\
&= \lim_{x \to \infty} \frac{1}{\dfrac{1}{e^x} + 2} \\
&= \frac{1}{0 + 2} \\
&= \frac{1}{2}
\end{aligned}
$$

$$\diamond$$

Since there is only one fast growing quantity in this example – e^x – it's not hard to spot the fastest growing quantity. The next Knowledge Box has information about which types of functions grow faster. Except for "constants" all the functions in Knowledge Box 3.4 grow to infinity as x does.

Knowledge Box 3.4.

Which functions grow faster?

- Logarithms grow faster than constants

- Positive powers of x grow faster than logarithms

- Larger positive powers of x grow faster than smaller positive powers of x

- Exponentials (with positive exponents) grow faster than positive powers of x

- Exponentials with larger exponents grow faster than those with smaller exponents

Let's do some one-step examples with these new rules.

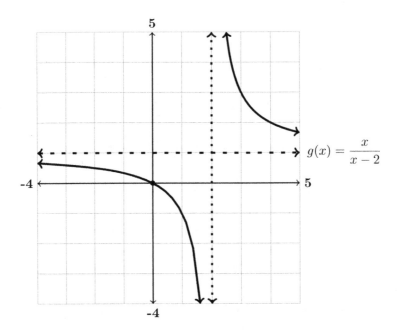

Figure 3.2. A function with a vertical and horizontal asymptote.

Example 3.7.

$$\lim_{x \to \infty} \frac{3}{\ln(x) + 1} = 0$$

$$\lim_{x \to \infty} \frac{x^5}{e^x + 1} = 0$$

$$\lim_{x \to \infty} \frac{e^{2x}}{1 + e^x} = \infty$$

$$\Diamond$$

Aside from keeping track of signs, limits as

$$x \to -\infty$$

obey the same rules as those going to positive infinity. With limits at infinity under some degree of control, we can now discuss asymptotes.

Definition 3.4. An **asymptote** is a horizontal or vertical line that the graph of a function approaches arbitrarily close to but does not touch.

The function shown in Figure 3.2 has a horizontal asymptote (dashed line) and a vertical one, dotted line, as well as a **root** at (0,0).

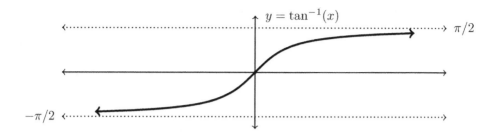

Figure 3.3. The inverse tangent function, two horizontal asymptotes.

Knowledge Box 3.5.

Rules for finding asymptotes

The **horizontal asymptotes** to the graph of $f(x)$ are the y values:

$$y = \lim_{x \to \pm\infty} f(x)$$

There are at most two horizontal asymptotes. **Vertical asymptotes** occur whenever there is a point $x = v$ where $f(v)$ fails to exist because of a division of a non-zero quantity by zero.

Notice that the example function has a horizontal asymptote at $y = 1$, because

$$\lim_{x \to \pm\infty} \frac{x}{x-2} = 1$$

and a vertical asymptote at $x - 2 = 0$ or $x = 2$ where a divide-by-zero occurs. It also has a root at $x = 0$.

Figure 3.3 shows an example of a function with two horizontal asymptotes: $g(x) = \tan^{-1}(x)$. The function approaches $\pi/2$ as x goes to positive infinity and $-\pi/2$ as x goes to $-\infty$. The function also has a single root at (0,0).

When sketching the graph of a function the first three things we list are the roots, the horizontal asymptotes, and the vertical asymptotes.

Example 3.8. Find the roots and asymptotes of

$$f(x) = \frac{3 - x^2}{x^2 - 4}.$$

Solution:

First of all, we need to find the roots by solving for those x-values that yield a y value of zero. An important part of this is to remember that a fraction is zero only when its numerator is zero at places where its denominator is not zero.

$$3 - x^2 = 0$$
$$3 = x^2$$
$$x = \pm\sqrt{3}$$

Since neither of these are values that make $x^2 - 4$ zero, our roots are at $x = \pm\sqrt{3}$.

Next we need the vertical asymptotes, which will be the points where the denominator is zero; $x^2 - 4 = (x + 2)(x - 2) = 0$, so there are vertical asymptotes at $x = \pm 2$.

The horizontal asymptotes can be found with the rule for ratios of polynomials:

$$\lim_{x \to \infty} \frac{3 - x^2}{x^2 - 4} = \frac{-1}{1} = -1$$

So we have a horizontal asymptote at $y = -1$. For now, plot a few points near a vertical asymptote to figure out which way the curve is going as you approach the asymptote from either side. The next section will give us another tool to figure this out. Now, we can sketch the graph of $f(x)$. Notice that roots and both sorts of asymptotes are shown.

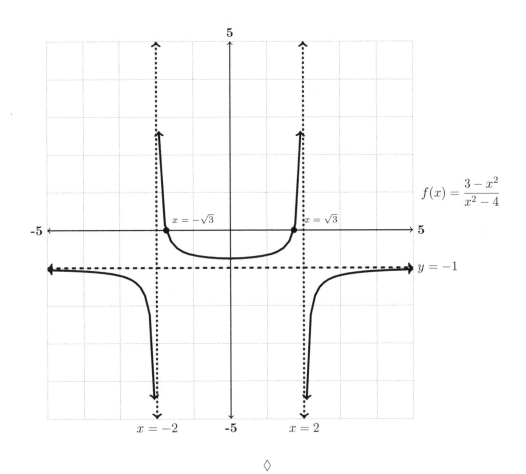

$$f(x) = \frac{3 - x^2}{x^2 - 4}$$

$\Diamond$

Problems

Problem 3.1. *Using the definition of a limit at infinity show that*

$$\lim_{x \to \infty} \frac{2}{x^2 + 4} = 0.$$

Problem 3.2. *Using the definition of a limit at infinity show that*

$$\lim_{x \to \infty} \frac{x + 1}{x - 1} = 1.$$

Problem 3.3. *Compute each of the following limits using explicit algebra.*

(a) $\displaystyle\lim_{x \to \infty} \frac{x}{2x + 1}$

(b) $\displaystyle\lim_{x \to \infty} \frac{x^2}{1 + 5x}$

(c) $\displaystyle\lim_{x \to \infty} \frac{2x^2 + 3x + 5}{x^2 - 4x + 6}$

(d) $\displaystyle\lim_{x \to \infty} \frac{x^3 + 3x + 2}{x^3 + 3x^2 + 2}$

(e) $\displaystyle\lim_{x \to \infty} \frac{3x(x + 1)(x - 1)}{x^3 + 4}$

(f) $\displaystyle\lim_{x \to \infty} \frac{5x^2}{4x^2 + 1}$

Problem 3.4. *Compute each of the following limits by any method.*

(a) $\displaystyle\lim_{x \to \infty} \frac{3e^x}{1 + e^x}$

(b) $\displaystyle\lim_{x \to \infty} \frac{5e^x}{1 + e^{2x}}$

(c) $\displaystyle\lim_{x \to \infty} \frac{x^2}{x \cdot \ln(x)}$

(d) $\displaystyle\lim_{x \to \infty} \frac{\ln(x)}{\sqrt{x} + 1}$

(e) $\displaystyle\lim_{x \to \infty} \frac{\sqrt{x} + 2}{\sqrt[3]{x} + 1}$

(f) $\displaystyle\lim_{x \to \infty} \frac{e^x}{\ln(x^5)}$

Problem 3.5. *Find the root and asymptotes (vertical and horizontal) for each of the following functions.*

(a) $f(x) = \dfrac{x^2 - 1}{x}$

(b) $g(x) = \dfrac{x}{x^2 - 1}$

(c) $h(x) = \ln(1/x)$

(d) $r(x) = \dfrac{e^x - 1}{e^x + 1}$

(e) $s(x) = \dfrac{2x^2 - 8}{x^2 - 9}$

(f) $q(x) = \dfrac{1}{x} + \dfrac{1}{x - 2}$

(g) $a(x) = \tan^{-1}(x^2)$

(h) $b(x) = \dfrac{x^3 + 6x^2 + 11x + 6}{x^3 - 6x^2 + 11x - 6}$

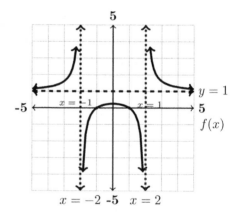

Problem 3.6. *For the graph above, find the roots and the vertical and horizontal asymptotes.*

Problem 3.7. *Using the definition of the limit at infinity, show $\displaystyle\lim_{x \to \infty} x^2$ diverges to ∞.*

Problem 3.8. *Using the definition of the limit at infinity, show*

$$\lim_{x \to \infty} \frac{x^2 - 1}{x}$$

diverges to ∞.

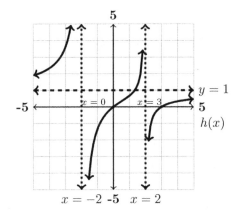

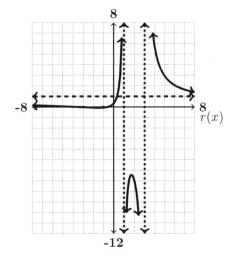

Problem 3.9. *For the graph shown above, assume that the function plotted is the ratio of two quadratic equations.*

(a) List the roots

(b) List the vertical asymptotes

(c) List the horizontal asymptotes

(d) Give a sensible formula for $h(x)$

Problem 3.10. *For each of the following functions, sketch the graph to the best of your ability, showing and labeling the roots and asymptotes.*

(a) $\lim\limits_{x \to \infty} \dfrac{2x}{x+1}$

(b) $\lim\limits_{x \to \infty} \dfrac{x^2}{1 + 5x^2}$

(c) $\lim\limits_{x \to \infty} \dfrac{x^2 - 9}{x^2 - 1}$

(d) $\lim\limits_{x \to \infty} \dfrac{x^2 + 3x}{x^2 - 1}$

(e) $\lim\limits_{x \to \infty} \dfrac{x^4}{x^4 - 1}$

(f) $\lim\limits_{x \to \infty} \dfrac{5x^2}{x^2 - 4}$

Problem 3.11. *For the graph shown above, assume that the function plotted is the ratio of two quadratic equations.*

(a) List the roots

(b) List the vertical asymptotes

(c) List the horizontal asymptotes

(d) Give a sensible formula for $r(x)$

Problem 3.12. *Describe the roots and asymptotes of*

$$f(x) = \frac{1}{x^n}$$

where n is a positive whole number including the way the function approaches the asymptotes. Hint: there are two outcomes.

Problem 3.13. *If $f(x)$ and $g(x)$ are both polynomials, what is the largest number of horizontal asymptotes that*

$$h(x) = \frac{f(x)}{g(x)}$$

can have? Explain your answer carefully.

Problem 3.14. *Find the number of horizontal asymptotes of*

$$f(x) = \frac{\left|x^3\right|}{x^3 + 1}$$

3.2. Information from the Derivative

There are two sorts of useful information for sketching a curve that we can pull out of the derivatives of a function. We can compute where it is **increasing** and **decreasing** and we can compute where it is curved up (**concave up**) or curved down (**concave down**).

3.2.1. Increasing and decreasing ranges. Remember that a derivative is a rate of change. This means that when $f'(x) > 0$ in a range, the function is increasing in that range, and when $f'(x) < 0$ the function is decreasing in that range. Let's nail down exactly what it means to be increasing or decreasing on a range.

Definition 3.5. A function is **increasing** on an interval if, for each $u < v$ in the interval $f(u) < f(v)$.

Definition 3.6. A function is **decreasing** on an interval if, for each $u < v$ in the interval $f(u) > f(v)$.

Now we are ready for the derivative-based rules on when a function is increasing or decreasing.

Knowledge Box 3.6.

Derivative-based function rules

- A function $f(x)$ is **increasing** where $f'(x) > 0$
- A function $f(x)$ is **decreasing** where $f'(x) < 0$
- Those $x = c$ where $f'(c) = 0$ are called **critical values**
- The points $(c, f(c))$ where $f'(c) = 0$ are called **critical points**

Example 3.9. Find the critical point(s) and increasing and decreasing ranges for

$$f(x) = x^2 - 4.$$

Solution:

We see $f'(x) = 2x$. Solving $2x = 0$ we get that $x = 0$ is the only critical value. So the critical point is $(0, -4)$. Solving $2x < 0$ we see $f'(x) < 0$ on $(-\infty, 0)$; similarly $f'(x) > 0$ on $(0, \infty)$. The following graph permits you to check all this against the actual behavior of the function.

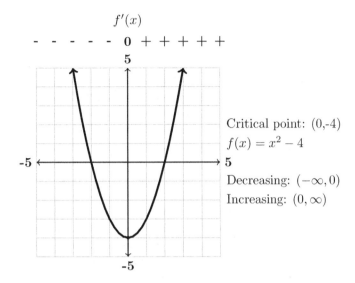

$f'(x)$

$- \quad - \quad - \quad - \quad - \quad - \quad \mathbf{0} \quad + \quad + \quad + \quad + \quad +$

Critical point: (0,-4)

$f(x) = x^2 - 4$

Decreasing: $(-\infty, 0)$

Increasing: $(0, \infty)$

Notice that we have a row of +, -, 0 symbols across the top of the graph: these show the sign of $f'(x)$ and are handy for analysis of increasing and decreasing ranges, as we will see in a minute.

$\Diamond$

Example 3.10. Find the critical point(s) and increasing and decreasing ranges for

$$f(x) = x^3 - 4x.$$

Solution:

Finding the increasing and decreasing ranges requires that we first find the critical values, where $f'(c) = 0$.

$$f'(x) = 3x^2 - 4$$
$$3x^2 - 4 = 0$$
$$3x^2 = 4$$
$$x^2 = \frac{4}{3}$$
$$x = \pm \frac{2}{\sqrt{3}}$$

The derivative can only change between positive and negative at $c = \pm \frac{2}{\sqrt{3}}$, so we plug in values in each of the resulting ranges. The value $c = \frac{2}{\sqrt{3}}$ is a little larger than one, so let's look at -2 and 2. We can make a table of values.

x	$f'(x)$	$\pm$
$-\infty$	na	na
-2	$8 > 0$	$+$
$-\dfrac{2}{\sqrt{3}}$	0	0
0	$-4 < 0$	$-$
$\dfrac{2}{\sqrt{3}}$	0	0
2	$8 > 0$	$+$
∞	na	na

So, the function is:

Increasing on: $(-\infty, -\dfrac{2}{\sqrt{3}}) \cup (\dfrac{2}{\sqrt{3}}, \infty)$

Decreasing on: $(-\dfrac{2}{\sqrt{3}}, \dfrac{2}{\sqrt{3}})$

with critical points at $\left(\pm\dfrac{2}{\sqrt{3}}, \mp\dfrac{16}{3\sqrt{3}}\right)$

Notice that the use of $\mp$ means that the sign of the second coordinate of the critical points is the opposite of the sign of the first. We also usually use a much more compact form for the table of signs given above:

$$f'(x): \ (-\infty) + + + \left(-\dfrac{2}{\sqrt{3}}\right) - - - \left(\dfrac{2}{\sqrt{3}}\right) + + + (\infty)$$

Numbers inserted into the chain of "+" and "-" symbols represent critical points. This device is called a **sign chart** for increasing and decreasing ranges. Below is a picture of the function with the features we just located shown. The critical points are plotted.

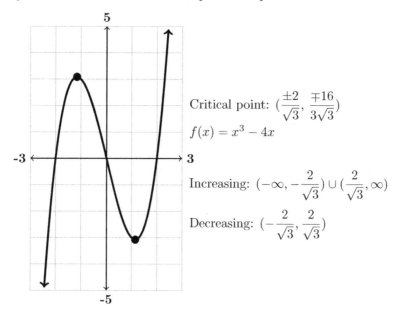

Critical point: $(\dfrac{\pm 2}{\sqrt{3}}, \dfrac{\mp 16}{3\sqrt{3}})$

$f(x) = x^3 - 4x$

Increasing: $(-\infty, -\dfrac{2}{\sqrt{3}}) \cup (\dfrac{2}{\sqrt{3}}, \infty)$

Decreasing: $(-\dfrac{2}{\sqrt{3}}, \dfrac{2}{\sqrt{3}})$

$\Diamond$

An alert reader will have noticed that we carefully avoided, so far in this section, examples involving asymptotes. The reason for this is that they can influence the increasing/decreasing ranges as well.

Knowledge Box 3.7.

A continuous, differentiable function can only change between increasing and decreasing at a critical value or at a vertical asymptote.

This means that we include the position of vertical asymptotes along with critical values on the sign chart for finding increasing and decreasing ranges for a function.

Example 3.11. Find the critical points, vertical asymptotes, and increasing and decreasing ranges for

$$f(x) = \frac{x^2 - 1}{x^2 - 4}$$

Solution:

The vertical asymptotes are easy: $x^2 - 4 = 0$ at $x = \pm 2$. The critical points require us to solve $f'(x) = 0$, which gives us:

$$\frac{(x^2 - 1)(2x) - (x^2 - 4)(2x)}{(x^2 - 4)^2} = 0$$
$$(x^2 - 1)(2x) - (x^2 - 4)(2x) = 0$$
$$2x^3 - 2x - 2x^3 + 8x = 0$$
$$6x = 0$$
$$x = 0$$

So there is a critical value at $x = 0$ and a critical point at $(0, \frac{1}{4})$. Remember that a fraction is zero only where its numerator is zero. Let's make the sign chart with the critical values and vertical asymptotes. We can plug in any value in an interval to get the $\pm$ value for $f'(x)$:

$$(-\infty) + + + (-2) + + + (0) - - - (2) - - - (\infty)$$

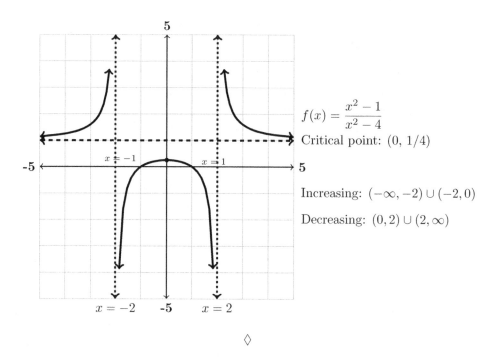

$$f(x) = \frac{x^2 - 1}{x^2 - 4}$$

Critical point: $(0, 1/4)$

Increasing: $(-\infty, -2) \cup (-2, 0)$

Decreasing: $(0, 2) \cup (2, \infty)$

$\Diamond$

Notice that the function can increase both up to and after the asymptote; but at the asymptote the function goes to $+\infty$ or comes from $-\infty$. It does not increase through the asymptote. This means that these points are **gaps** in the sign chart and the increasing/decreasing ranges.

3.2.2. Concavity. This section introduces the last quantity we will be using for sketching the graphs of functions: concavity. Concavity – as we learned in Chapter 2 – depends on the sign of the second derivative. We will use sign charts for concavity, just as for increasing and decreasing values.

Knowledge Box 3.8.

Concavity rules

- Where $f''(x) > 0$ we say a function is **concave up**
- Where $f''(x) < 0$ we say a function is **concave down**
- Values $x = c$ where $f''(c) = 0$ are **inflection values**
- Points $(c, f(c))$ there $f''(c) = 0$ are **inflection points**

Example 3.12. Find the inflection points and concave up and down ranges of

$$f(x) = x^3 - 4x.$$

Solution:

We need the second derivative:

$$f'(x) = 3x^2 - 4$$
$$f''(x) = 6x$$

Solve $6x = 0$, and we get that there is an inflection value at $x = 0$ and an inflection point at (0,0). Let's make a sign chart plugging in $f''(-1) = -6$ and $f''(1) = 6$ to get:

$$(-\infty) - - - (0) + + + (\infty)$$

Another way to understand where a curve is concave up or down is that the shape of the curve holds water where it is concave up and does not where it is concave down. Verify the information about concavity and inflection points on the following graph.

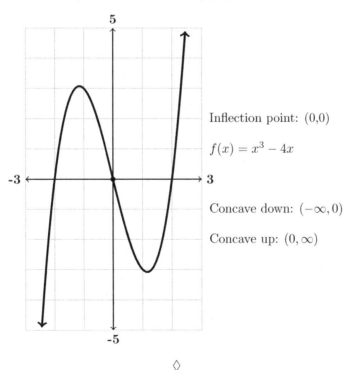

Inflection point: (0,0)

$f(x) = x^3 - 4x$

Concave down: $(-\infty, 0)$

Concave up: $(0, \infty)$

$\Diamond$

Concavity can also change at vertical asymptotes. So they are included on the second derivative sign chart for concavity. Let's do an example that shows this phenomenon.

Example 3.13. Find the inflection points, concave up, and concave down ranges for

$$f(x) = \frac{x^2 - 1}{x^2 - 4}.$$

Solution:

We've already computed, in Example 3.11, that

$$f'(x) = \frac{6x}{(x^2 - 4)^2}.$$

For inflection and concavity we need $f''(x)$.

$$
\begin{aligned}
f''(x) &= \frac{(x^2 - 4)^2 (6) - 6x \cdot 2 (x^4 - 4)^1 (2x)}{(x^2 - 4)^4} \\
&= \frac{(x^2 - 4)^{\not{2}1} (6) - 6x \cdot 2(x^4 - 4)(2x)}{(x^2 - 4)^{\not{4}3}} \\
&= \frac{6 (x^2 - 4) - 24x^2}{(x^2 - 4)^3} \\
&= \frac{6x^2 - 24 - 24x^2}{(x^2 - 4)^3} \\
&= -\frac{18x^2 + 24}{(x^2 - 4)^3}
\end{aligned}
$$

Which is zero when $18x^2 + 24 = 0$, which never happens. This means the function has no inflection points. We have already computed that this function has vertical asymptotes at $x = \pm 2$. The sign chart for the second derivative is thus based on $x = \pm 2$. Plugging in $x = \pm 3$ to $f''(x)$ we get a positive result; plugging in zero we get a negative result. The sign chart is thus:

$$f''(x) : (-\infty) + + + (-2) - - - (2) + + + (\infty)$$

Check this result against the graph.

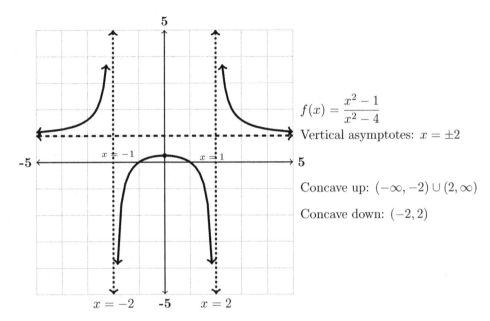

$$f(x) = \frac{x^2 - 1}{x^2 - 4}$$

Vertical asymptotes: $x = \pm 2$

Concave up: $(-\infty, -2) \cup (2, \infty)$

Concave down: $(-2, 2)$

It's worth noting that the "holds water" analogy for concave up breaks down on this example.

◊

Problems

Problem 3.15. *For each of the following functions, find the critical point(s).*

(a) $f(x) = \dfrac{x^2 - 4}{x}$

(b) $g(x) = \dfrac{1}{x^2 - 4x + 5}$

(c) $h(x) = \sin(x)$

(d) $r(x) = x^2 + \dfrac{1}{x}$

(e) $s(x) = \dfrac{x^3 - 1}{x}$

(f) $q(x) = \sec(x)$

Problem 3.16. *For each of the following functions, find the critical points and the vertical asymptotes, if any. Based on these, make a sign chart for the first derivative and report the ranges on which the function is increasing and decreasing.*

(a) $f(x) = x^2 + 4x + 5$

(b) $g(x) = \dfrac{x^2 - 1}{x^2}$

(c) $h(x) = \dfrac{x}{x^2 - x - 6}$

(d) $r(x) = \ln(x^2 + 3)$

(e) $s(x) = \sin(x)$

(f) $q(x) = \dfrac{1}{x} - \dfrac{1}{x^2}$

Problem 3.17. *Find a cubic polynomial that is never decreasing.*

Problem 3.18. *Find a cubic polynomal function that has critical values at $x = \pm 5$.*

Problem 3.19. *Find a cubic polynomal function that has critical values at $x = -2$ and $x = 4$.*

Problem 3.20. *Show that a polynomial of degree n has at most $n - 1$ critical points.*

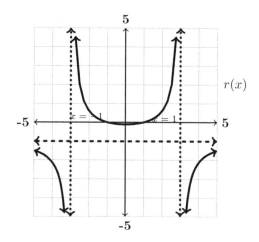

Problem 3.21. *For the graph above, make the best assessment you can of the critical points, vertical asymptotes, increasing ranges, and decreasing ranges.*

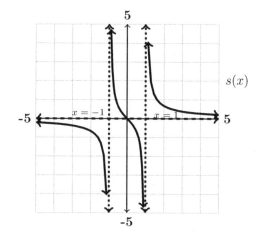

Problem 3.22. *For the graph above, make the best assessment you can of the critical points, vertical asymptotes, increasing ranges, and decreasing ranges.*

Problem 3.23. *For each of the following functions, find the inflection point(s).*

(a) $f(x) = \dfrac{1}{x^2 + 1}$

(b) $g(x) = x^3 - 3x^2 + 7x + 152$

(c) $h(x) = \cos(x)$

(d) $r(x) = \ln(x^2 + 1)$

(e) $s(x) = \dfrac{x^2 - 1}{x^2 + 1}$

(f) $q(x) = \tan^{-1}(x)$

Problem 3.24. *For each of the following functions, find the inflection point(s) and vertical asymptotes, make a sign chart for concavity, and give the concave up and down ranges.*

(a) $f(x) = x^4 - 3x^2 + 2$

(b) $g(x) = x^3 - 6x^2 + 5x + 6$

(c) $h(x) = x^4 - x^2 + 6$

(d) $r(x) = \ln(x^2 + 4)$

(e) $s(x) = \dfrac{x^2 - 4}{x^2 + 4}$

(f) $q(x) = \tan^{-1}(x^2)$

Problem 3.25. *Find a quadratic polynomial that is concave down everywhere.*

Problem 3.26. *Find a polynomal function of degree more than two that has only one concavity value – excluding inflection points.*

Problem 3.27. *Find a polynomal function that has inflection values at $x = -4$ and $x = 2$.*

Problem 3.28. *Show that a polynomial of degree n has at most $n - 2$ inflection points.*

Problem 3.29. *Suppose that $f(x)$ is a polynomial function with ranges that are both concave up and concave down. What is the smallest degree $f(x)$ can have?*

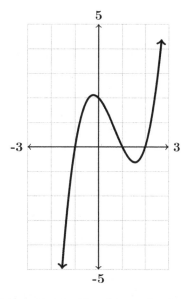

Problem 3.30. *For the graph above, do the best you can to find the inflection points and concave up and concave down intervals.*

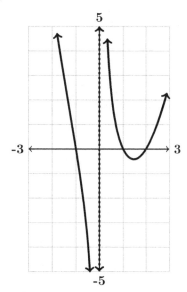

Problem 3.31. *For the graph above, do the best you can to find the inflection points and concave up and concave down intervals.*

3.3. The full report for curve sketching

In this chapter so far we have found nine types of information that help us sketch the curve generated by a function. These are the roots, both sorts of asymptotes, the critical points together with the increasing and decreasing ranges, and the inflection points together with the concave up and concave down ranges. In this section all we are going to do is pull them together into a single type of report. In addition to a good, annotated sketch of a curve, the following information is used to make a full report on a function.

Knowledge Box 3.9.

The table for a full report

Roots:

Vertical asymptotes:

Horizontal asymptotes:

Critical points:

Increasing on:

Decreasing on:

Inflection points:

Concave up on:

Concave down on:

With all the pieces in place, let's do a full example on a fairly simple curve. This first example will skip the issue of asymptotes by not having any.

Example 3.14. Make a full report and sketch of the curve

$$y = \frac{1}{3}x^3 - 3x.$$

Solution:

Factor to find roots:

$$\frac{1}{3}x^3 - 3x = \frac{1}{3}x(x^2 - 9) = \frac{1}{3}x(x-3)(x+3) = 0$$

So we have roots at $x = 0, \pm 3$. Since the function is a polynomial, we know it has no division by zero and so no vertical asymptotes; similarly as $x \to \pm\infty$ the function diverges, so no horizontal asymptotes. Now we are ready for derivative information.

A quick derivative and we see $f'(x) = x^2 - 3$. So we have critical values at $x = \pm\sqrt{3}$ and critical points at $(\pm\sqrt{3}, \mp 2\sqrt{3})$. Plugging the values $x = 0, \pm 2$ into $f'(x)$ gives us the increasing/decreasing sign chart:

$$(-\infty) + + + (-\sqrt{3}) - - - (\sqrt{3}) + + + (\infty)$$

This tells us the function is increasing on $(-\infty, -\sqrt{3}) \cup (\sqrt{3}, \infty)$ and decreasing on $(-\sqrt{3}, \sqrt{3})$.

Another quick derivative and we see $f''(x) = 2x$. So there is an inflection value at $x = 0$ and an inflection point at $(0,0)$. Plugging in ± 1 to the second derivative yields the sign chart:

$$(-\infty) - - - (0) + + + (\infty)$$

So, the function is concave down on $(-\infty, 0)$ and concave up on $(0, \infty)$.

Roots: $x = 0, \pm 3$

Vertical asymptotes: none

Horizontal asymptotes: none

Critical points: $(\pm\sqrt{3}, \mp 2\sqrt{3})$

Increasing on: $(-\infty, -\sqrt{3}) \cup (\sqrt{3}, \infty)$

Decreasing on: $(-\sqrt{3}, \sqrt{3})$

Inflection points: $(0, 0)$

Concave up on: $(-\infty, 0)$

Concave down on: $(0, \infty)$

The following picture displays all the information. Notice how roots, critical points, and inflection points are all displayed.

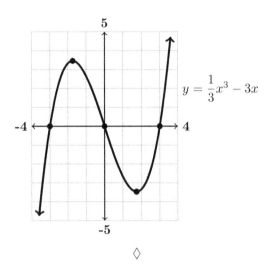

$$y = \frac{1}{3}x^3 - 3x$$

$\Diamond$

Let's move on to an example with both vertical and horizontal asymptotes.

Example 3.15. Make a full report and sketch for

$$y = \frac{x^2 - 4}{x^2}.$$

Solution:

To find the roots, we solve the numerator equal to zero, $x^2 - 4 = 0$. This gives us roots $x = \pm 2$. The denominator is zero at $x = 0$, giving us a vertical asymptote at $x = 0$. The rule for ratios of polynomials tells us:

$$\lim_{x \to \infty} \frac{x^2 - 4}{x^2} = 1$$

So we have a horizontal asymptote at $y = 1$.

Computing the first derivative we get:

$$f'(x) = \frac{x^2(2x) - (x^2 - 4)2x}{(x^2)^2} = \frac{8x}{x^4} = \frac{8}{x^3}$$

Since the numerator is a constant, there are no critical values and hence no critical points. We build a sign chart on the vertical asymptote, plugging in $f'(\pm 1) = \pm 8$, and get:

$$(-\infty) - - - (0) + + + (\infty)$$

This gives us decreasing on $(-\infty, 0)$ and increasing on $(0, \infty)$. Since $f'(x) = 8x^{-3}$, we can use the power rule to get that $f''(x) = -24x^{-4} = -\frac{24}{x^4}$. Again the top is a constant so there are no inflection values or points. Building a sign chart on the vertical asymptote at $x = \pm 1$ we get:

$$(-\infty) - - - (0) - - - (\infty)$$

In fact $f''(x)$ is negative everywhere it exists. This yields concave up: nowhere; concave down: $(-\infty, 0) \cup (0, \infty)$. And, we are done. Filling in the chart gives us:

Roots: $x = \pm 2$

Vertical asymptotes: x=0

Horizontal asymptotes: y=1

Critical points: none

Increasing on: $(0, \infty)$

Decreasing on: $(-\infty, 0)$

Inflection points: none

Concave up on: never

Concave down on: $(-\infty, 0) \cup (0, \infty)$

The corresponding sketch looks like this:

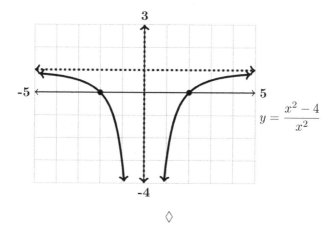

$$y = \frac{x^2 - 4}{x^2}$$

$\Diamond$

Example 3.16. Make a full report and sketch for

$$y = \tan^{-1}(3x).$$

Solution:

We already know that $\tan^{-1}(x)$ is zero only at $x = 0$. So, solving $3x = 0$ we get a root at $x = 0$. We also know that $-\pi/2 < \tan^{-1}(x) < \pi/2$. So, there are no points where a divide by zero and hence a vertical asymptote can form. We have that $\lim_{x \to \pm\infty} \tan(x) = \pm\infty$ The quantity $3x$ heads for ∞ faster than x, but it will not change the asymptotes of $y = \pm\pi/2$.

Using the chain rule we see that $f'(x) = \dfrac{3}{1 + 9x^2}$. The numerator is constant, so there are no critical points. The first derivative is the ratio of two positive quantities, and so the function

increases everywhere. Using the reciprocal rule we get that $f''(x) = 3 \cdot \dfrac{-18x}{\left(1+9x^2\right)^2} = \dfrac{-54x}{\left(1+9x^2\right)^2}$.
The numerator is zero at $x = 0$ and the denominator is not, yielding a single inflection value of $x = 0$ and an inflection point at $(0,0)$. Building a sign chart by plugging in $f''(\pm 1) = \mp 0.27$, we get $(-\infty) + + + (0) - - - (\infty)$. So, the function is concave up on $(-\infty, 0)$ and concave down on $(0, \infty)$. Putting all this in the table yields:

Roots: $x = 0$

Vertical asymptotes: none

Horizontal asymptotes: $y = \pm\pi/2$

Critical points: none

Increasing on: $(-\infty, \infty)$

Decreasing on: never

Inflection points: $(0,0)$

Concave up on: $(-\infty, 0)$

Concave down on: $(0, \infty)$

The resulting graph is:

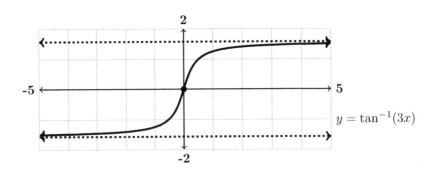

$$y = \tan^{-1}(3x)$$

$\Diamond$

Example 3.17. Make a full report and sketch for

$$y = \frac{x^2 - 9}{x}.$$

Solution:

The roots are easy, $x = \pm 3$; the vertical asymptote is clearly $x = 0$; and, since the top is higher degree, there are no horizontal asymptotes. Notice that:

$$y = \frac{x^2 - 9}{x} = \frac{x^2}{x} - \frac{9}{x} = x - \frac{9}{x}$$

From this, it is easy to see that $f'(x) = 1 + \dfrac{9}{x^2}$, which means there are no critical points, and $f'(x) > 0$ where it exists. This makes the first derivative sign chart

$$(-\infty) + + + (0) + + + (\infty)$$

The second derivative is $f''(x) = \dfrac{-18}{x^3}$, so there are no inflection points. Plugging in $f''(\pm 1) = \mp 18$, we get a second derivative sign chart like this:

$$(-\infty) + + + (0) - - - (\infty)$$

These sign charts make the various ranges obvious and we get:

Roots: $x = \pm 3$

Vertical asymptotes: x=0

Horizontal asymptotes: none

Critical points: none

Increasing on: $(-\infty, 0) \cup (0, \infty)$

Decreasing on: never

Inflection points: none

Concave up on: $(-\infty, 0)$

Concave down on: $(0, \infty)$

The corresponding sketch looks like this:

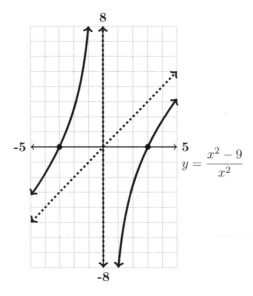

Alert students will notice that a non-standard asymptote appears in this picture. Recall that $y = x - \dfrac{9}{x}$. When $|x|$ is large, $y = x - $ something very small, which means that as $|x|$ grows large, the function gets very close to the line $y = x$. This is called a **diagonal asymptote**.

$\Diamond$

Knowledge Box 3.10.

Diagonal asymptotes

If $f(x) = ax + b + g(x)$ and
$$\lim_{x \to \pm\infty} g(x) = 0,$$
then the graph of $f(x)$ approaches the line $y = ax + b$ as $|x|$ gets large.

Problems

Problem 3.32. *Make a full report with a sketch for each of the following functions.*

(a) $f(x) = x^3 - 5x$

(b) $g(x) = x^3 - 6x^2 + 11x - 6$

(c) $h(x) = \dfrac{x^3 - 2x^2 - x + 2}{x}$

(d) $r(x) = \dfrac{x^2 - 9}{x^2 - 1}$

(e) $s(x) = x + \dfrac{1}{x-1} + \dfrac{1}{x+1}$

(f) $q(x) = \ln(x^2 + 5)$

(g) $a(x) = \tan^{-1}(x^2)$

(h) $b(x) = \dfrac{x^2 - 1}{3 - x^2}$

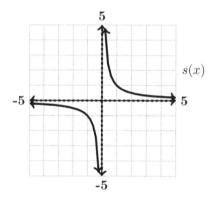

Problem 3.33. *Based on the sketch above, do your best to fill out a full report.*

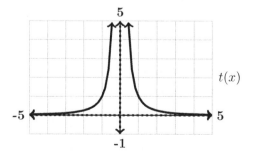

Problem 3.34. *Based on the sketch above, do your best to fill out a full report.*

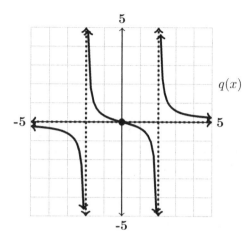

Problem 3.35. *Based on the sketch above, do your best to fill out a full report.*

Problem 3.36. *Make a full report with a sketch for:*

$$f(x) = \frac{3x}{x^2 + 1}$$

Problem 3.37. *Find a function that is concave down everywhere that it exists and has a range of $(-\infty, \infty)$.*

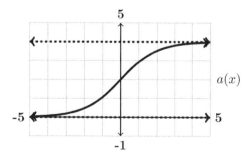

Problem 3.38. *Based on the sketch above, do your best to fill out a full report.*

Problem 3.39. *Make a full report with a sketch for each of the following functions. Include diagonal asymptotes if any.*

(a) $f(x) = \dfrac{x^2 + 2x + 1}{x - 1}$

(b) $g(x) = \dfrac{2x^2 + 3x + 1}{x}$

(c) $h(x) = 5x - 3 + \dfrac{1}{x^2}$

(d) $r(x) = \dfrac{x^3}{x^2 + 2}$

(e) $s(x) = \dfrac{x^2}{x^2 + 1}$

(f) $q(x) = \dfrac{x^2}{x^2 - 4}$

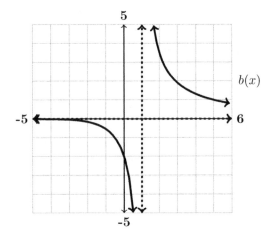

Problem 3.40. *Based on the sketch above, do your best to fill out a full report.*

Problem 3.41. *Make a full report with a sketch for*

$$f(x) = \dfrac{e^x}{1 + e^x}$$

Problem 3.42. *Find a function that has two vertical asymptotes and approaches the diagonal asymptote $y = 2x + 1$.*

Problem 3.43. *Find a function with two horizontal and two vertical asymptotes and demonstrate that your solution is correct.*

Problem 3.44. *Suppose that a function is the ratio of two polynomials in which the numerator has degree n and the denominator has degree m. What is the range of the possible numbers of vertical asymptotes of this function?*

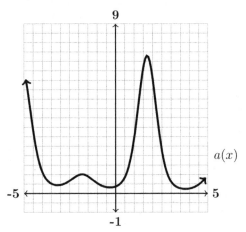

Problem 3.45. *Based on the sketch above, do your best to fill out a full report.*

Problem 3.46. *Find a function that has two horizontal asymptotes and explain why your solution works.*

Problem 3.47. *Find a function that has two diagonal asymptotes and explain why your solution is correct. Hint – start with a function that has two horizontal asymptotes.*

Problem 3.48. *What is the relationship between the graph of*

$$f(x) = \dfrac{x^4}{x^2 + 1}$$

and the graph of $y = x^2 - 1$?

Problem 3.49. *Suppose that*

$$f(x) = \dfrac{(x - a)(x - b)(x - c)}{x^4 + d}$$

where all four of $a, b, c,$ and d are positive constants. What items that go into a report can you deduce from this information?

3.4. Optimization: maximum-minimum problems

In this section we will start on **optimization**, the process of finding the largest or smallest value a function can take on as well as values that are larger (or smaller) than all nearby values. As usual, our first step is to define our terms.

Definition 3.7. If a function $f(x)$ is defined on an interval or collection of intervals $\mathcal{I}$, then the **global maximum** of $f(x)$ is the largest value $f(x)$ takes on anywhere on $\mathcal{I}$.

Definition 3.8. If a function $f(x)$ is defined on an interval or collection of intervals $\mathcal{I}$, then the **global minimum** of $f(x)$ is the smallest value $f(x)$ takes on anywhere on $\mathcal{I}$.

Definition 3.9. If a function $f(x)$ is defined near and including $x = c$ and, in some interval $I = (c - \delta, c + \delta)$, we have that, for all $a \in I$, $f(c) \geq f(a)$, then we say $f(x)$ has a **local maximum** at $x = c$.

Definition 3.10. If a function $f(x)$ is defined near and including $x = c$ and, in some interval $I = (c - \delta, c + \delta)$, we have that, for all $a \in I$, $f(c) \leq f(a)$, then we say $f(x)$ has a **local minimum** at $x = c$.

We use the term **optima** for values that are maxima or minima. Figure 3.4 demonstrates the different sorts of optima.

The key to finding optima is the behavior of the derivative. Notice that, at an optima, the function changes between increasing and decreasing. This means that optima, at least ones that don't occur at the beginning and end of an interval, are a kind of point we've seen before: they are critical points.

Knowledge Box 3.11.

Critical points in optimization

Except at the boundaries of an optimization problem, the optima of a continuous, differentiable function occur at critical points.

If we consider the tangent line at a critical point, we get a geometric description of the behavior of the derivative. A line with a slope of 0 is a horizontal tangent line. This tells us that **optima can occur at horizontal tangent lines.**

A function that diverges to infinity as $|x|$ gets large may fail to have a global maximum or minimum. Similarly, vertical asymptotes inside the area where you are optimizing can cause trouble. Having

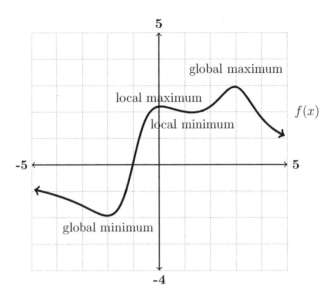

Figure 3.4. This picture shows examples of local and global optima.

said that, the other place besides critical points where optima can occur is at the boundaries of the region where you are optimizing. In general, when optimizing a continuous, differentiable function, you check the critical points and the points at the boundaries of the area you are optimizing.

The next example asks you to optimize a quadratic equation – something that should be easy – but it asks you to do it on a bounded domain. A quadratic that opens upward has a unique minimum. As well as the critical points, we must also check the x values at its endpoints.

Example 3.18. Find the global maximum and minimum of $g(x) = x^2 - 4$ on the interval [-2,3].

Solution:

The critical point of this function is easy: $g'(x) = 2x$. So $x = 0$ is the critical value, and there is a critical point at (0,-4). The boundaries of the optimization area are $x = -2$ and $x = 3$. Since $g(-2) = 0$ and $g(3) = 5$, we have three candidate points: (-2,0), (0,-4), and (3,5). The smallest y-value is -4 making (0,-4) the point where the global minimum occurs, while 5 is the largest y-value making (3,5) the point where the global maximum occurs. Let's look at a picture.

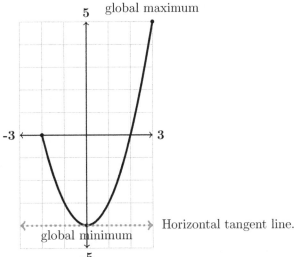

global maximum

5

-3   3

$\leftarrow$ $\cdots\cdots\cdots\cdots\cdots\cdots\cdots\cdots$ $\rightarrow$ Horizontal tangent line.

global minimum

-5

The function $g(x) = x^2 - 4$ on the interval [-2,3]

$\Diamond$

The following summarizes what we have so far.

Knowledge Box 3.12.

The Extreme Value Theorem

The global maximum and minimum of a continuous, differentiable function must occur at critical points (horizontal tangents) or at the boundaries of the domain where optimization is taking place.

An important point is that critical points don't have to be maxima or minima. Let's look at an example of this. Consider the function

$$h(x) = \frac{x^3}{9},$$

shown in Figure 3.5.

Since $f'(x) = \frac{1}{3}x^2$, it's easy to see there is a critical point at $(0,0)$ – but it's *not* a maximum or a minimum. This then opens the question, **how do we tell if a critical point is a maximum, a minimum, or neither?** We have already developed one useful tool: the sign chart for the first derivative. Let's look at the sign chart for $h(x) = \frac{x^3}{9}$:

$$(-\infty) + + + (0) + + + (\infty)$$

The function increases to zero and then continues to increase. There are four ways a sign chart can shake out which together are the **first derivative test** for the nature of an optimum.

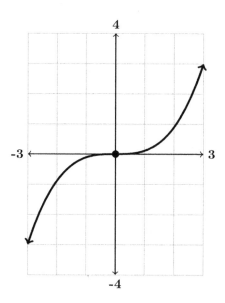

Figure 3.5. The function $h(x) = \dfrac{x^3}{9}$.

Knowledge Box 3.13.

The first derivative test for optima

$f'(x)$ chart near p	Conclusion:
$+++(p)+++$	no optimum at p
$+++(p)---$	maximum at p
$---(p)+++$	minimum at p
$---(p)---$	no optimum at p

Another method of spotting a potential maximum or minimum is to look at the second derivative. A function that is curved downward at a critical point has a maximum at that critical point; a function that is curved upward at a critical point has a minimum at that critical point. This technique is called the **second derivative test**.

Knowledge Box 3.14.

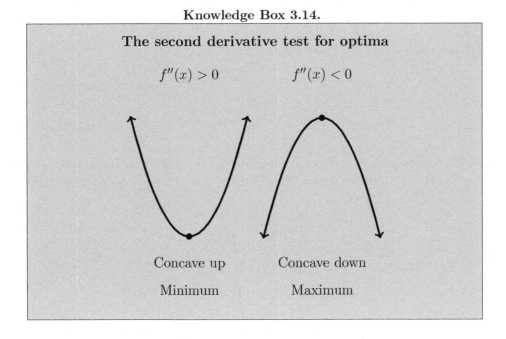

This gives us the tools we need to optimize things. Now it is time to do some examples.

Example 3.19. Find the global maximum and minimum of

$$f(x) = \frac{1}{2}x^3 - 2x$$

on the interval $[-1, 3]$.

Solution:

First we find the critical points. Solving $f'(x) = \frac{3}{2}x^2 - 2 = 0$ we get critical values of $\pm\frac{2}{\sqrt{3}}$, but only the positive value is in the interval [-1,3]. This means we need to check this value and the ends of the interval:

$$f(-1) = 3/2 = 1.5$$
$$f\left(\frac{2}{\sqrt{3}}\right) = -\frac{8}{3\sqrt{3}} \cong -1.54$$
$$f(3) = 15/2 = 7.5$$

This means that the global maximum is 7.5 at $x = 3$, and the global minimum is $-\dfrac{8}{3\sqrt{3}} \cong -1.54$ at $x = \dfrac{2}{\sqrt{3}}$. Let's look at the sign chart and the graph. The chart:

$$(-1) - - - (\tfrac{2}{\sqrt{3}}) + + + (3)$$

shows that the critical point is, in fact, a minimum. The maximum occurs at a boundary point. Notice that if we change the interval on which we are optimizing we can, in fact, change the results.

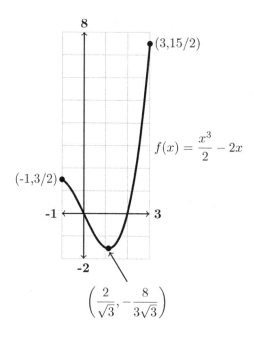

$$\left(\dfrac{2}{\sqrt{3}}, -\dfrac{8}{3\sqrt{3}}\right)$$

We can also look at the outcome of the second derivative test.

$$f''(x) = 3x$$

So if we plug in $x = \dfrac{2}{\sqrt{3}}$ we get a positive value, about 3.46. This means that the graph is concave up and so, again, we have determined that there is a minimum at the critical point.

◇

Example 3.20. Find the global optima of

$$g(x) = \frac{x}{1+x^2}.$$

Solution:

Since no interval is given, we use the interval $(-\infty, \infty)$. Our rule about the ratio of polynomials tells us that the limits of this function at $\pm\infty$ are zero. So the optima occur at critical points, if they occur. Using the quotient rule,

$$g'(x) = \frac{(1+x^2)(1) - x(2x)}{(1+x^2)^2} = \frac{1-x^2}{(1+x^2)^2}$$

Remembering that a fraction is zero when its numerator is zero, we get critical values where $x^2 - 1 = 0$ or $x = \pm 1$. Examining a sign chart, plugging in $0, \pm 2$ to $f'(x)$ we get

$$(-\infty) - - - (-1) + + + (1) - - - (\infty)$$

So the global minimum is at (-1,-1/2) and the global maximum is at (1,1/2). The graph is:

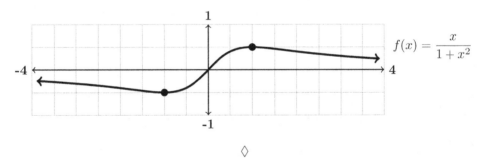

$$f(x) = \frac{x}{1+x^2}$$

We now turn to optimization story problems. These problems describe a situation and ask the reader to optimize some quantity. Common tasks are designing containers and laying out fences, but there are actually a huge number of possible story problems.

Example 3.21. Suppose we have 120 meters of fence that we want to use to enclose two pens that share a wall as shown. If one of the pens is $a \times b$ meters, what values of a and b maximize the area of the pens? What is the area of the optimal pen?

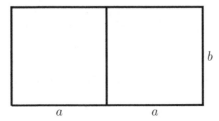

Solution:

There are four segments of length a and three segments of length b, so we have $4a + 3b = 120$. We want to maximize the area of the pens, $\text{Area}(a, b) = ab$. Since $b = (120 - 4a)/3$ we can maximize

$$\text{Area} = \frac{a(120 - 4a)}{3} = \frac{120}{3}a - \frac{4}{3}a^2$$

Take a derivative and we see that $\text{Area}' = \frac{120}{3} - \frac{8}{3}a$. Solving $\text{Area}' = 0$, we obtain a critical value of $a = \frac{120}{8} = 15\text{m}$ and so $b = \frac{120}{6} = 20\text{m}$. The area function is a quadratic with a negative squared term that opens downward – so the single critical point is the maximum. The answer is thus $a = 15\text{m}$, $b = 20\text{m}$, and $\text{Area} = 300\text{m}^2$.

Knowledge Box 3.15.

Steps for an optimization story problem

- Draw a picture of the problem

- Label the picture with reasonable variables

- Write out the quantity you are optimizing in terms of those variables

- Write out the equation for additional information

- Use the additional information to remove a variable from the quantity you are optimizing

- Figure out what the interval for that variable is, based on the problem

- Optimize the resulting single variable formula

It may help to match these steps against the previous example.

Example 3.22. Suppose that we have an open-topped, square-bottomed box with a volume of 1m^3. What side length s for the base and height h minimize the amount of material needed to make the box?

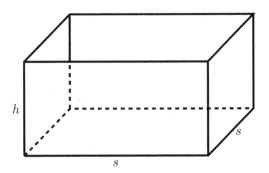

Solution:

We are minimizing the surface area A of the box to minimize the material. The surface area is the square bottom and four rectangular sides, making $A = s^2 + 4sh$. We also know the volume of the box is 1m^3, so $1 = s^2 h$ or $h = \dfrac{1}{s^2}$. Substituting this value for h into the area formula we get

$$A = s^2 + 4s\frac{1}{s^2} = s^2 + \frac{4}{s}$$

We know $s > 0$, so the interval is $(0, \infty)$. Taking the derivative we get

$$A' = 2s - \frac{4}{s^2} = \frac{2s^3 - 4}{s^2}$$

Solving this for zero, we get that $2s^3 - 4 = 0$ or that $s = \sqrt[3]{2}$ is the sole critical value. The second derivative is easy to compute:

$$A'' = 2 + \frac{8}{s^2}$$

This function is positive for any positive s. So we get $A''(\sqrt[3]{2}) > 0$, meaning that the critical value is a minimum. This means we have $s = 2^{1/3} \cong 1.26\text{m}$ and, plugging into the formula for h, $h = 2^{-2/3} \cong 0.630\text{m}$. These are the side length and height that minimize materials.

$\Diamond$

Example 3.23. Suppose we are fencing in a rectangular pen with 200m of fence, but we are putting the pen against the side of an existing building. What are the dimensions that enclose the maximum area?

Solution:

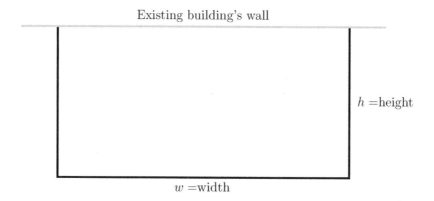

We are optimizing area $A = wh$, and we know that $w + 2h = 200$. So $w = 200 - 2h$. Substituting this into our area formula we get:

$$A = (200 - 2h)h = 200h - 2h^2$$

This function is a quadratic that opens downward. So we know that it will have a single critical value that corresponds to a maximum. Taking the derivative $A' = 200 - 4h$ gives us a critical value of $h = 50$, meaning $w + 100 = 200$ or $w = 100$, and we have height=50m and width=100m giving us the best possible area for the pen.

$$\Diamond$$

An important point is that some functions don't have optima. The next example illustrates this.

Example 3.24. Find the global optima, if any, of $y = e^x$.

Solution:

We know that as $x \to \infty$ this function diverges to ∞, so there is no hope that there is a global maximum. As $x \to -\infty$ the function goes to zero. What, however, is the range of the function? We already know that it is $(0, \infty)$. This means that $y = e^x$ can take on every positive value, but it can never be zero. This means that the function does not have a global minimum. For any value that e^a takes on, there is some b so that $e^b < e^a$.

$$\Diamond$$

This is a fairly weird way to fail to have an optimum. It's related to the strict inequality

$$0 < e^x < \infty$$

that defines the range of $y = e^x$. This sort of thing can also happen when the interval over which a function is optimized has open ends, (a, b) instead of closed ones $[a, b]$.

Example 3.25. What radius and height minimize the material needed to make a can with a volume of 200cc?

Solution:

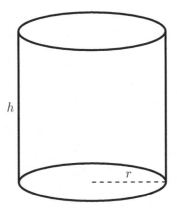

Again, assume we are minimizing surface area. The area of the can is the two circular ends with area πr^2 and the sides which, if flattened, form a $2\pi r \times h$ rectangle. So

$$A = 2\pi r^2 + 2\pi rh$$

The volume of a can is the area of the bottom times the height, which tells us

$$V = \pi r^2 h = 200$$

From this we deduce that $h = \dfrac{200}{\pi r^2}$. Plugging this in to get a single-variable problem, we get

$$A = 2\pi r^2 + 2\pi r \cdot \frac{200}{\pi r^2} = 2\pi r^2 + \frac{400}{r}$$

Setting the derivative equal to zero gives us:

$$4\pi r - \frac{400}{r^2} = 0$$

$$\frac{4\pi r^3 - 400}{r^2} = 0$$

$$4\pi r^3 - 400 = 0$$

$$r^3 = \frac{100}{\pi}$$

$$r \cong 3.17\text{cm}$$

We know $r > 0$. Looking at the sign chart for the derivative, plugging in $r = 2$ and $r = 4$, we get:

$$(0) - - - (3.17) + + + (\infty)$$

so the critical value $x = 3.17$ is a minimum. Plugging the r value into the formula for h, we get that $h \cong 6.34$. Our answer is $r = 3.17$cm and $h = 6.34$cm.

$$\diamond$$

Let's see how the second derivative test shakes out in the previous example.

$$A'' = 4\pi + \frac{800}{r^3}$$

which is positive for any $r > 0$. So the curve is concave up in the possible region, and our critical value is a minimum; the second derivative test agrees with the sign chart for the first derivative test.

When you are working optimization story problems, it is critical to make sure the values you get make sense. Negative lengths, for example, probably mean you made a mistake. These problems also have the property that some of what you did may be needed again in a later step. Using a neat layout, possibly informed by the steps given for optimization, will help.

Problems

Problem 3.50. *For each of the following functions, find the global maximum and minimum of the function, if they exist, on the stated interval.*

(a) $f(x) = 2x - 1$ on $(-2, 2)$

(b) $g(x) = 3x + 1$ on $[-3, 1]$

(c) $h(x) = x^2 + 4x + 3$ on $[-1, 4]$

(d) $r(x) = \ln(x)$ on $[1, e^3]$

(e) $s(x) = e^{-x^2}$, on $[-2, 3]$

(f) $q(x) = x^3 - 16x + 1$, on $[-4, 4]$

Problem 3.51. *How many different horizontal tangent lines do the following functions have? Be sure to justify your answer.*

(a) $f(x) = \cos(x)$

(b) $g(x) = xe^{-x}$

(c) $h(x) = x(x-5)(x+5)$

(d) $r(x) = \cos(x) + x$

(e) $s(x) = \ln(x^4 + 4x^3 + 5)$

(f) $q(x) = \tan^{-1}(x)$

Problem 3.52. *Construct a function that has exactly three horizontal tangent lines – all different from one another.*

Problem 3.53. *What is the largest number of horizontal tangent lines that a polynomial of degree n can have?*

Problem 3.54. *If $y = ax^2 + bx + c$ with $a \neq 0$, give a set of steps for finding the global optimum (there is exactly one), and determining the type of optima it is.*

Problem 3.55. *For each of the following functions, find the global maximum and minimum of the function, if they exist, on the stated interval.*

(a) $f(x) = \dfrac{4x}{x^2 + 2}$ on $(-\infty, \infty)$

(b) $g(x) = e^{x^3 - 5x + 12}$ on $[-5 : 5]$

(c) $h(x) = xe^{-2x}$ on $(0, \infty)$

(d) $r(x) = x^2 e^{-x}$ on $(0, \infty)$,

(e) $s(x) = \dfrac{\sqrt{x^2 + 1}}{x}$, on $[-4, -1] \cup [1, 4]$

(f) $q(x) = 25 - x^4$, on $[-1, 2]$.

Problem 3.56. *Suppose for $m(x)$ that when $a \leq b$ we have that $m(a) \leq m(b)$, and assume that $m(x)$ is continuous, differentiable, and not constant.*

(a) *What do we know about $m'(x)$? Explain.*

(b) *Prove that the critical values of $m(f(x))$ and $f(x)$ are the same.*

(c) *Is this a problem relevant to other problems in this section? Why?*

Problem 3.57. *Find the domain of*

$$h(x) = \sqrt{6 - 3x - x^2}$$

and find its global maximum and minimum on the domain.

Problem 3.58. *If an open-topped can holds 400cc, what radius and height minimize the amount of material needed to make the can?*

Problem 3.59. *What are the ratio of the sides of a rectangle of perimeter P that maximizes the area over all rectangles with that perimeter?*

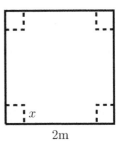

Problem 3.60. *Suppose that we cut square corners out of a 2m × 2m square of pasteboard and tape up the sides to make an open-topped box. What side length x of the square maximizes the volume of the box?*

Problem 3.61. *What point on the line*

$$y = x - 6$$

is closest to the origin?

Problem 3.62. *What point on the line*

$$y = 3x + 5$$

is closest to the origin?

Problem 3.63. *What point on the line*

$$y = 4x + 1$$

is closest to the origin?

Problem 3.64. *If we use 240m of fence to lay out three pens like those shown above, what length and width* for one pen *maximize the area enclosed?*

Problem 3.65. *Suppose that the top of a can is made of a material that costs twice as much as the bottom or sides. Find the radius and height of a can containing 200cc that minimizes the cost.*

Problem 3.66. *One side of a rectangular pen must be made of opaque material that costs three times as much as the material used to make the other three sides. What dimensions minimize the cost if the pen must have an area of $60m^2$?*

Problem 3.67. *A rocket takes off, straight up, so that for $t \geq 0$ the height of the rocket is $\dfrac{4t^2}{m}$. A camera 40m from the rocket tracks its takeoff.*

(a) *Make a sketch of the situation.*

(b) *Find an expression for the rate in rad/sec for the camera's rate of spin.*

(c) *When is the camera spinning fastest?*

(d) *What is the camera's rate of spin when it is spinning fastest?*

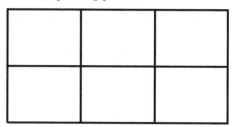

Problem 3.68. *Suppose we are optimizing the area of cells of a rectangular grid like the one shown above fixing the total length of the sides of the grids. Prove that the solution always places the same amount of material into vertical and horizontal cell sides.*

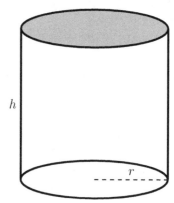

Problem 3.69. *What radius and height minimize the material needed to make an open-topped can with a volume of 600cc?*

Problem 3.70. *Maximize*

$$q(x) = x^2 e^{-2x}$$

for $x \geq 0$.

Problem 3.71. *What is the global maximum of the function*

$$y = e^{4 - x^2}$$

Can you do this problem without calculus? Explain.

Chapter 4

Integration, Area, and Initial Value Problems

The two main concepts in calculus are derivatives and integrals. So far we have studied derivatives. This chapter introduces the concept of **integrals**. Derivatives and integrals were developed independently, and later it was discovered that they are closely related. We will start with this relationship and define integrals as a sort of backwards derivative. Then, we will introduce some of the primary uses of integration – finding areas under curves and solving boundary value problems. Finally, we will show how integrals were originally defined. This chapter does not cover how to integrate most functions. That we've saved for later in Chapter 7.

4.1. Anti-derivatives

The anti-derivative of a function is another function whose derivative is the function you started with. More precisely:

Definition 4.1. If $f(x)$ and $F(x)$ are functions so that

$$F'(x) = f(x),$$

we say that $F(x)$ is an **anti-derivative** of $f(x)$.

The terminology **anti-derivative** is fairly modern. The original terminology is to call an anti-derivative of $f(x)$ an **integral** of $f(x)$. Integrals also have a special notation.

Knowledge Box 4.1.

Integral notation

If $F(x)$ is an anti-derivative of $f(x)$, we write

$$F(x) = \int f(x) \cdot dx$$

and call $F(x)$ an integral of $f(x)$.

Example 4.1. Find an anti-derivative of $f(x) = 2x$.

Solution:

We know that the derivative of $F(x) = x^2$ is $f(x) = 2x$. So, one possible answer is:

$$F(x) = x^2$$

$\Diamond$

One of the issues that arises when we work with anti-derivatives is that they are not unique. The derivative of

$$G(x) = x^2 + 5$$

is *also* 2x. This means that anti-derivatives (integrals) are known only up to some constant value. We will develop techniques for dealing with this in Section 4.3, but for now we will simply use **unknown constants** or **constants of integration**.

Example 4.2. Use integral notation, with an unknown constant C, to express the fact that x^2 is an anti-derivative of $2x$.

Solution:

$$\int 2x \cdot dx = x^2 + C$$

$\Diamond$

We also need to explain dx. The job of dx is to tell us which symbol is the variable.

Definition 4.2. The symbol dx is spoken "the differential of x" and is used to designate the active variable in an integral.

We have seen dx and other differentials before. The symbol $\dfrac{dy}{dx}$ was an alternate way of saying y', for example. There is another use of differential symbols – as an operator.

Knowledge Box 4.2.

The differential operator

Another way to say "take the derivative" is with the symbol $\dfrac{d}{dx}$.
This symbol denotes **the derivative with respect to** x. So:

$$f'(x) = \frac{d}{dx}f(x)$$

We will need this symbol, the differential operator, in the next section.

In order to compute $\displaystyle\int 2x \cdot dx$ we used the fact we knew that $\dfrac{d}{dx}x^2 = 2x$. In fact, each derivative rule is also an anti-derivative rule, just used backwards. Let's start by giving the reverse of the power rule for derivatives.

Knowledge Box 4.3.

The power rule for integration

$$\int x^n \cdot dx = \frac{1}{n+1}x^{n+1} + C$$

The rules for constant multiples and sums for derivatives also apply for integrals.

Knowledge Box 4.4.

Integrals of constant multiples and sums

$$\int a \cdot f(x) \cdot dx = a \cdot \int f(x) \cdot dx$$

$$\int (f(x) + g(x)) \cdot dx = \int f(x) \cdot dx + \int g(x) \cdot dx$$

This gives us enough machinery to be able to compute the integrals of polynomial functions.

Example 4.3. Find

$$\int \left(x^3 + 4x^2 + 5x + 3 \right) \cdot dx$$

Solution:

$$\int \left(x^3 + 4x^2 + 5x + 3 \right) \cdot dx = \int x^3 \cdot dx + 4 \int x^2 \cdot dx + 5 \int x^1 \cdot dx + 3 \int dx$$

$$= \frac{1}{4}x^4 + 4 \cdot \frac{1}{3}x^3 + 5 \cdot \frac{1}{2}x^2 + 3 \cdot x + C$$

$$= \frac{1}{4}x^4 + \frac{4}{3}x^3 + \frac{5}{2}x^2 + 3x + C$$

Notice that an anti-derivative of a constant a is ax and that, since the sum of several unknown constants is some other unknown constant, we can get away with one "+C".

To check an integral rule, you take the derivative of the result and see if you get back where you started.

Example 4.4. Check the preceding example by taking the derivative.

Solution:

$$\left(\frac{1}{4}x^4 + \frac{4}{3}x^3 + \frac{5}{2}x^2 + 3x + C \right)' = \frac{1}{4}4x^3 + \frac{4}{3}3x^2 + \frac{5}{2}2x^1 + 3(1) + 0 = x^3 + 4x^2 + 5x + 3$$

and we see the integral in the last example was correct.

All the derivative rules we learned in Chapter 2 have corresponding integral rules. Let's go through them.

Knowledge Box 4.5.

Integrals of logs and exponentials

$$\int \frac{1}{x} \cdot dx = \ln(x) + C$$

$$\int e^x \cdot dx = e^x + C$$

One problem with the laundry-list of integrals in this section is that, until we get the techniques in Chapter 7, the possible integrals are a bit contrived. Nevertheless, let's do an example.

Example 4.5. Compute:

$$\int \left(3e^x + \frac{4}{x} \right) \cdot dx$$

Solution:

$$\int \left(3e^x + \frac{4}{x} \right) \cdot dx = 3 \int e^x \cdot dx + 4 \int \frac{dx}{x}$$
$$= 3e^x + 4\ln(x) + C$$

◊

Next are the integrals arising from the trigonometric derivatives.

Knowledge Box 4.6.

Integrals of trigonometric derivatives

$$\int \sin(x) \cdot dx = -\cos(x) + C$$

$$\int \cos(x) \cdot dx = \sin(x) + C$$

$$\int \sec^2(x) \cdot dx = \tan(x) + C$$

$$\int \csc^2(x) \cdot dx = -\cot(x) + C$$

$$\int \sec(x)\tan(x) \cdot dx = \sec(x) + C$$

$$\int \csc(x)\cot(x) \cdot dx = -\csc(x) + C$$

In the next example, the integral we are asked to perform doesn't look like one of our known forms, but it can be rearranged into one of the known forms with a few trig identities.

Example 4.6. Compute:

$$\int \frac{\sin(x)}{\cos^2(x)} \cdot dx$$

Solution:

$$\int \frac{\sin(x)}{\cos^2(x)} \cdot dx = \int \frac{1}{\cos x} \cdot \frac{\sin(x)}{\cos(x)} \cdot dx$$

$$= \int \sec(x) \cdot \tan(x) \cdot dx$$

$$= \sec(x) + C$$

◊

Trigonometric identities can be used to make really easy integrals look really hard.

Example 4.7. Compute:

$$\int \left(\sin^2(x) + 2\sin(x) + \cos^2(x) \right) \cdot dx$$

Solution:

$$\int \left(\sin^2(x) + 2\sin(x) + \cos^2(x) \right) \cdot dx = \int \left(\sin^2(x) + \cos^2(x) \right) \cdot dx + 2 \int \sin(x) \cdot dx$$

$$= \int 1 \cdot dx + 2 \int \sin(x) \cdot dx$$

$$= x + 2(-\cos(x)) + C$$

$$= x - 2\cos(x) + C$$

◊

We also have some integrals arising from the inverse trigonometric functions.

Knowledge Box 4.7.

More integrals of trigonometric derivatives

$$\int \frac{1}{\sqrt{1 - x^2}} \cdot dx = \sin^{-1}(x) + C$$

$$\int \frac{1}{1 + x^2} \cdot dx = \tan^{-1}(x) + C$$

$$\int \frac{1}{x\sqrt{x^2 - 1}} \cdot dx = \sec^{-1}(|x|) + C$$

This example is another one in which we can set up a known form with a little bit of algebra.

Example 4.8. Compute:

$$\int \left(\frac{1}{3x^2 + 3} \right) \cdot dx$$

Solution:

$$\int \frac{1}{3x^2 + 3} \cdot dx = \int \frac{1}{3} \left(\frac{1}{x^2 + 1} \right) \cdot dx$$
$$= \frac{1}{3} \int \frac{dx}{1 + x^2}$$
$$= \frac{1}{3} \tan^{-1}(x) + C$$

$\Diamond$

It is quite common for integrals to require some algebraic setup before it becomes apparent how to do them. Let's do another.

Example 4.9. Compute:

$$\int \left(\frac{x^2}{1 + x^2} \right) \cdot dx$$

Solution:

$$\int \frac{x^2}{1 + x^2} \cdot dx = \int \frac{1 + x^2 - 1}{1 + x^2} \cdot dx$$

$$= \int \frac{1 + x^2}{1 + x^2} \cdot dx - \int \frac{dx}{1 + x^2}$$

$$= \int dx - \int \frac{dx}{1 + x^2}$$

$$= x - \tan^{-1}(x) + C$$

$\Diamond$

Example 4.10. Compute:

$$\int \frac{x\sqrt{1-x^2}+1}{\sqrt{1-x^2}} \, dx$$

Solution:

$$\int \frac{x\sqrt{1-x^2}+1}{\sqrt{1-x^2}} \, dx = \int \left(\frac{x\sqrt{1-x^2}}{\sqrt{1-x^2}} + \frac{1}{\sqrt{1-x^2}} \right) dx$$

$$= \int \left(x + \frac{1}{\sqrt{1-x^2}} \right) dx$$

$$= \frac{1}{2}x^2 + \sin^{-1}(x) + C$$

$\Diamond$

All of the integrals we've done in this section lead to formulas, not numbers. That will change in the next section. But before we're done, let's get the terminology for these "formula only" integrals.

Definition 4.3. The integral formulas given in this section of the form:

$$\int f(x) \cdot dx = F(x) + C$$

are called **indefinite integrals** in honor of the unknown constant C.

One way to think of indefinite integrals is they store patterns that we will use in the integrals that compute specific quantities.

The integrals presented in this section are all simple anti-derivatives, except for the ones that have been lightly disguised by the use of algebra. The next three sections explore the applications and origins of integration without expanding the tool set for actually performing integration. Chapter 7 develops u-substitution, integration by parts, partial fractions, and many other clever methods of doing integrals. It also develops some methods based on trig identities that turn calculus into a puzzle solving activity.

First introducing integration and later presenting the starter tool-kit is something that is done to hand you integration as a useful tool for physics as soon as possible. The later exploration of integration techniques is mind-expanding, but the fundamental concept that permits you to manipulate the formulas that describe natural law is more important in preparing you to study physics.

Problems

Problem 4.1. *Find the indefinite integral of each of the following polynomial functions.*

(a) $f(x) = x^2 + 1$

(b) $g(x) = (x+1)^2$

(c) $h(x) = x^4 - 3x^3 + 5x^2 - 7x + 6$

(d) $r(x) = \dfrac{x^3 - 1}{x + 1}$

(e) $s(x) = 7x^7 + 6x^6 + x - 2$

(f) $q(x) = (x+1)(x+2)(x+3)$

Problem 4.2. *Perform each of the following integrals.*

(a) $\displaystyle \int \frac{x^4}{x^2 + 1} dx$

(b) $\displaystyle \int \frac{x^2 + 3x + 5}{x} dx$

(c) $\displaystyle \int \frac{dx}{\cos^2(x)}$

(d) $\displaystyle \int \frac{dx}{\sin^2(x) + 4 + 5x^2 + \cos^2(x)}$

(e) $\displaystyle \int \left(\frac{1}{\cos^2(x)} + \frac{\sin(x)\sec(x)}{\cos(x)} \right) dx$

(f) $\displaystyle \int \big(\sin^2(x) + \cos^2(x) + \sin(x) + \cos(x) + 1 \big) \cdot dx$

(g) $\displaystyle \int 2\sin(x/2)\cos(x/2) \cdot dx$

(h) $\displaystyle \int \sin\left(x + \frac{\pi}{3} \right) \cdot dx$

Problem 4.3. *For each of the following statements, verify that the integral is correct, by taking a derivative.*

(a) $\ln(x^2 + 1) + C = \displaystyle \int \frac{2x}{x^2 + 1} \cdot dx$

(b) $\dfrac{1}{2}\sin(2x) + C = \displaystyle \int \cos(2x) \cdot dx$

(c) $\ln(\sec(x)) + C = \displaystyle \int \tan(x) \cdot dx$

(d) $\dfrac{1}{10}\left(x^2 + 1\right)^5 + C$
$= \displaystyle \int x \cdot (x^2 + 1)^4 \cdot dx$

(e) $\ln(\sec(x) + \tan(x)) + C$
$= \displaystyle \int \sec(x) \cdot dx$

(f) $(x - 1)\, e^x + C = \displaystyle \int x e^x \cdot dx$

Problem 4.4. *Verify that*
$$\int \frac{dx}{ax + b} = \frac{1}{a}\ln(ax + b) + C$$

Problem 4.5. *Verify that*
$$\int e^{ax} = \frac{1}{a}e^{ax} + C$$

Problem 4.6. *Demonstrate that if $p(x)$ is a polynomial, then we can compute*
$$\int \frac{p(x)}{x^k} \cdot dx$$
with the techniques in this section.

Problem 4.7. *Compute*
$$\int \frac{x^2 + 2}{x^2 + 1} \cdot dx$$

Problem 4.8. *Compute*
$$\int \frac{\sqrt{1 - x^2}}{(1 - x)(1 + x)} \cdot dx$$

4.2. The Fundamental Theorem

Before we can formally state the relationship between the integral and the derivative, we need to define the **definite integral**.

Definition 4.4. Suppose that $F(x) = \int f(x) \cdot dx$. In other words, $F(x)$ is an anti-derivative of $f(x)$. Then the definite integral from $x = a$ to $x = b$ of $f(x)$ is defined to be:

$$\int_a^b f(x) \cdot dx = F(b) - F(a)$$

One nice thing about the definite integral is that it removes the unknown constant. If we write $F(x) + C$ for the anti-derivative, then

$$(F(b) + C) - (F(a) + C) = F(b) - F(a) + C - C = F(b) - F(a)$$

With the definite integral defined we can now state the first form of the fundamental theorem.

Knowledge Box 4.8.

The First Fundamental Theorem of Calculus

$$\frac{d}{dx} \int_a^x f(t) \cdot dt = f(x)$$
for any constant a.

This form of the fundamental theorem tells us that the derivative of the integral of a function is the same function, although the variable of integration (t in the above statement) may be different from the variable appearing in the final expression.

Example 4.11.

$$\frac{d}{dx} \int_0^x (t^2 + 1) \cdot dt = x^2 + 1$$

$$\Diamond$$

The integral of the derivative is also the same function – almost. The ubiquitous unknown constant causes us to answer: except for the "+ C." Later we will see that this unknown constant is where we place the starting point (position, velocity, etc.) into the formula when solving an applied problem.

The second form of the fundamental theorem has more applications. It relates integrals to the area under the graph of a function.

Knowledge Box 4.9.

The Second Fundamental Theorem of Calculus

Suppose that $f(x) \geq 0$ on the interval [a,b]. Then, if A is the area under the graph of $f(x)$ between a and b,

$$A = \int_a^b f(x) \cdot dx$$

Example 4.12. Find the area A under the curve of $y = x^2$ between $x = 0$ and $x = 3$.

Solution:

Start with a picture:

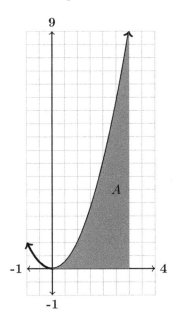

Guided by the picture, compute the definite integral.

$$\int_0^3 x^2 \cdot dx = \frac{1}{3}x^3 \Big|_0^3$$
$$= \frac{1}{3}3^3 - \frac{1}{3}0^3$$
$$= \frac{1}{3}27 - \frac{1}{3}0$$
$$= 9 - 0$$
$$= 9 \text{ units}^2$$

$\Diamond$

Notice the vertical bar notation, used to hold the limits until we plug them into the anti-derivative.

At this point let's check the intuition on this one. Why would the anti-derivative of a function be the area under it? The first form of the fundamental theorem tells us that a function is the derivative of its integral – but that means that a function is the **rate of change** of its integral.

The larger a function is, the faster the area under it is changing. The smaller a function is, the slower the area under it is changing. So – a function is the rate of change of the area under the function. If you're unconvinced, wait for Section 4.4. We will use another approach to show that the integral gives the area under the curve.

What meaning does the restriction $f(x) \geq 0$ in the second fundamental theorem have? The short answer is: the area below the x axis, for which $f(x) \leq 0$, comes out negative. This actually makes sense if we remember that the derivative is a rate of change. Positive derivatives represent increases, negative ones represent decreases. Since there is no such thing as negative area, we have to be careful when computing the total area between a graph and the x-axis.

Example 4.13. Compare the definite integral and the area between the curve and the x-axis for $f(x) = x^2 - 1$ from $x = -1$ to $x = 2$.

Solution:

This picture shows the areas above and below the x-axis.

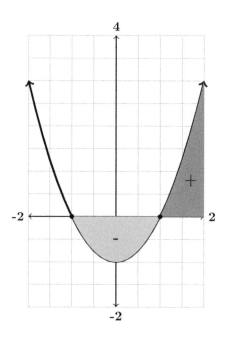

Notice the points where $f(x)$ crosses the axis are $x = \pm 1$.

First the integral:

$$\int_{-1}^{2} (x^2 - 1) \cdot dx = \frac{1}{3}x^3 - x \Big|_{-1}^{2}$$

$$= \frac{1}{3}2^3 - 2 - \left(\frac{1}{3}(-1)^3 - (-1)\right)$$

$$= \frac{8}{3} - 2 + \frac{1}{3} - 1$$

$$= \frac{9}{3} - 3$$

$$= 0$$

So, even though they are different shapes, the areas above and below the curve are equal. Now we need to compute the areas separately and take the positive area minus the "negative" one:

$$\int_{1}^{2} (x^2 - 1) \cdot dx - \int_{-1}^{1} (x^2 - 1) \cdot dx = \frac{1}{3}x^3 - x \Big|_{1}^{2} + \frac{1}{3}x^3 - x \Big|_{-1}^{1}$$

$$= \left(\frac{8}{3} - 2 - \frac{1}{3} + 1\right) - \left(\frac{1}{3} - 1 + \frac{1}{3} - 1\right)$$

$$= \frac{7}{3} - 1 - \frac{2}{3} + 2$$

$$= \frac{8}{3} \text{ units}^2$$

$\Diamond$

Notice that the integral is a *number* and so has no units, while the area between the curve and the x-axis has Cartesian units squared as its units. It is very important to keep clear in your mind the context in which you are using an integral. The meaning of the result is different for different procedures.

Also notice that we could have found the total area as

$$A = 2 \int_{1}^{2} (x^2 - 1) \cdot dx$$

once we knew the areas above and below the curve were equal. Since it's not obvious until *after* you do the integral, this isn't all that useful in this case. We will look at a useful version of this phenomenon later.

Example 4.14. Find the area under $y = \dfrac{1}{x}$ from $x = 1$ to $x = 4$.

Solution:

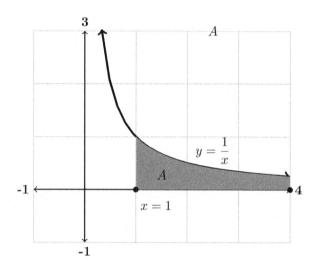

$$\int_1^4 \frac{1}{x} \cdot dx = \ln(x) \,\Big|_1^4$$

$$= \ln(4) - \ln(1)$$

$$= \ln(4) - 0$$

$$= \ln(4)$$

$$\Diamond$$

This example pays off on explaining a mystery – why we use e $\cong 2.71828\ldots$ as the base of the "natural" logarithm. It is because the area under $y = \dfrac{1}{x}$ from $x = 1$ to $x = a$ is $\ln(a)$. This gives us a method of computing logs, and it shows a place where logarithms arise naturally from the rest of mathematics. A much better way to choose a base than "we have ten fingers."

Example 4.15. Find the area bounded by $y = 4 - x^2$ and the x-axis.

Solution:

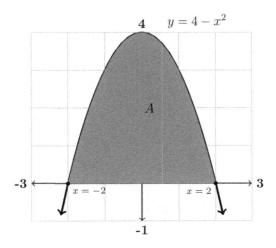

This problem does not give explicit limits – instead it tells us what the objects bounding the area are. Since the x-axis is where $y = 0$, we need to solve $0 = 4 - x^2$, which gives us $x = \pm 2$ as the points at the left and right end of the area. This then gives us a nice everything-above-the-axis integral.

$$A = \int_{-2}^{2} \left(4 - x^2\right) \cdot dx$$

$$= \left. 4x - \frac{1}{3}x^3 \right|_{-2}^{2}$$

$$= 4(2) - \frac{1}{3}8 - 4(-2) + \frac{1}{3}(-8)$$

$$= 8 + 8 - \frac{8}{3} - \frac{8}{3}$$

$$= 8\left(2 - \frac{2}{3}\right)$$

$$= 8 \cdot \frac{4}{3}$$

$$= \frac{32}{3} \text{ units}^2$$

◇

Finding the area bounded by two different curves requires solving for those x where

$$curve\ 1 = curve\ 2$$

in order to find the limits of integration. This will come up a lot in the future. This problem can be rephrased as $curve\ 1 - curve\ 2 = 0$ making it a root finding problem.

Example 4.16. Find the area bounded by $y = x^2$ and $y = \sqrt{x}$.

Solution:

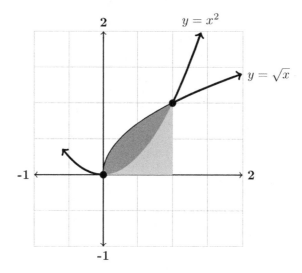

This one is a little tricky. The picture shows us that if we solve $x^2 = \sqrt{x}$ to get the points where the area begins and ends, then those points are $(0,0)$ and $(1,1)$. The area we want is shown in dark gray. It is the area under $y = \sqrt{x}$ that is not under $y = x^2$. If we computed $\int_0^1 \sqrt{x}\, dx$, we would get the dark gray area *and* the light gray area. The light gray area is the area under $y = x^2$. This means the dark gray area is:

$$\int_0^1 \sqrt{x}\, dx - \int_0^1 x^2 \cdot dx$$

Compute:

$$\int_0^1 \sqrt{x}\, dx - \int_0^1 x^2 \cdot dx = \int_0^1 \left(x^{1/2} - x^2 \right) \cdot dx$$

$$= \frac{2}{3} x^{3/2} - \frac{1}{3} x^3 \Big|_0^1$$

$$= \frac{2}{3} - \frac{1}{3} - 0 + 0$$

$$= \frac{1}{3}\ \text{units}^2$$

$\Diamond$

One thing that might be a little tricky in this example is the slightly odd version of the power rule:

$$\int x^{1/2} \cdot dx = \frac{2}{3}x^{3/2} + C$$

All that is going on is that $\frac{1}{2} + 1 = \frac{3}{2}$ and $\frac{1}{3/2} = \frac{2}{3}$.

To find the area bounded by the curves we first found their intersections and then subtracted the area under the lower curve from the area under the upper curve. Another point of view on this is that we integrated the upper curve minus the lower curve. Let's formulate this as a rule.

Knowledge Box 4.10.

Area between two curves

If $f(x) > g(x)$ on an interval [a,b], then the area between the graphs of the two functions on that interval is:

$$\int_a^b f(x) \cdot dx - \int_a^b g(x) \cdot dx = \int_a^b (f(x) - g(x)) \cdot dx$$

This rule is handy, but it does leave you with the problem of finding the appropriate intervals.

Example 4.17. Find the area between $y = x^2$ and $y = x + 2$.

Solution:

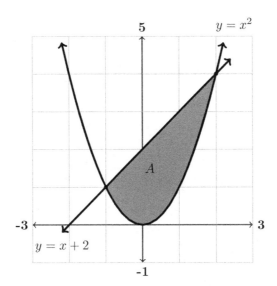

Solving $x^2 = x + 2$, or equivalently $x^2 - x - 2 = 0$, we see that $(x - 2)(x + 1) = 0$. So $x = -1, 2$, making the intersection points (-1,1) and (2,4). This also gives us the limits of integration and so

$$A = \int_{-1}^{2} \left(x + 2 - x^2 \right) \cdot dt$$

$$= \int_{-1}^{2} \left(2 + x - x^2 \right) \cdot dt$$

$$= 2x + \frac{1}{2}x^2 - \frac{1}{3}x^3 \Big|_{-1}^{2}$$

$$= 4 + \frac{4}{2} - \frac{8}{3} + 2 - \frac{1}{2} - \frac{1}{3}$$

$$= \frac{24 + 12 - 16 + 12 - 3 - 2}{6}$$

$$= 27/6$$

$$= 9/2 \text{ units}^2$$

$$\lozenge$$

Let's do an example that moves beyond polynomial functions.

Example 4.18. Find the area under the sine function from $x = 0$ to $x = \pi$.

Solution:

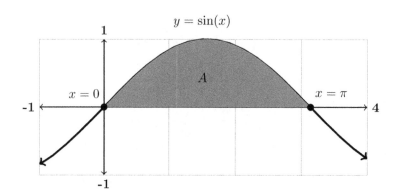

The sine function is non-negative on the entire interval so we may compute:

$$A = \int_0^\pi \sin(x) \cdot dx$$

$$= -\cos(x) \Big|_0^\pi$$

$$= -\cos(\pi) + \cos(0)$$

$$= -(-1) + 1$$

$$= 2 \text{ units}^2$$

◇

4.2.1. Even and odd functions. We defined even and odd functions in Definitions 1.5 and 1.6. It turns out that we can save some effort when integrating these functions, sometimes, because of special geometric properties of these functions.

Knowledge Box 4.11.

Integrating an even function on a symmetric interval

If $f(x)$ is an even function, then

$$\int_{-a}^a f(x) \cdot dx = 2 \int_0^a f(x) \cdot dx$$

Example 4.19. Find the integral of $f(x) = x^2$ on $[-2, 2]$.

Solution:

Remember that $x^2 = (-x)^2$. So $f(x)$ is an even function.

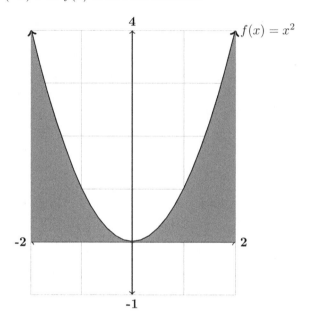

The area on either side of the y-axis is the same so:

$$\int_{-2}^{2} x^2 \cdot dx = 2 \int_{0}^{2} x^2 \cdot dx = 2 \cdot \frac{1}{3} x^3 \Big|_{0}^{2} = \frac{16}{3} \text{ units}^2$$

Not having to plug in the negative number avoids chances to make arithmetic mistakes.

Knowledge Box 4.12.

Integrating an odd function on a symmetric interval

If $f(x)$ is an odd function, then

$$\int_{-a}^{a} f(x) \cdot dx = 0$$

Example 4.20. Find the integral of $g(x) = \sin(x)$ on $[-2, 2]$.

Solution:

Remember that $\sin(x) = -\sin(-x)$. So $g(x)$ is an odd function.

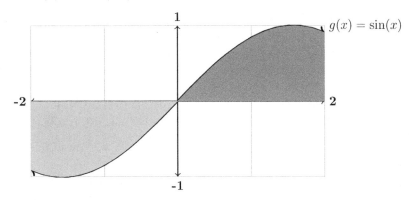

The area on either side of the y-axis is the same, but half is above the x axis, and half is below the x-axis. So:

$$\int_{-2}^{2} \sin(x) \cdot dx = 0$$

Wow, is that easier than plugging in numbers.

The odd and even function results are fairly special purpose. One of the big applications of them is writing test questions that you can do much faster if you notice something is an odd function on a symmetric interval, for example. But symmetric intervals *do* also show up in some application problems. If you're good at shifting functions sideways, you can also extend the application of these rules.

The set of even functions and the set of odd functions are closed under addition and also under multiplication by a constant. This fact is occasionally useful. If you're feeling ambitious, try and prove it's true.

Problems

Problem 4.9. *Simplify each of the follow-ing. Assume Snerp(x) has a domain of $(-\infty, \infty)$.*

(a) $\dfrac{d}{dx} \displaystyle\int_1^x \dfrac{t}{t^2 + 1} \cdot dt$

(b) $\dfrac{d}{dx} \displaystyle\int_\pi^x \cos(6t + 1) \cdot dt$

(c) $\dfrac{d}{dx} \displaystyle\int_0^x te^{t^2} \cdot dt$

(d) $\dfrac{d}{dx} \displaystyle\int_{-6}^x \dfrac{e^s + 1}{e^s - 1} \cdot ds$

(e) $\dfrac{d}{dx} \displaystyle\int_2^x \dfrac{e^y}{\ln(y) + 4} \cdot dy$

(f) $\dfrac{d}{dx} \displaystyle\int_0^x \mathrm{Snerp}(t + 2) \cdot dt$

Problem 4.10. *Compute the following def-inite integrals.*

(a) $\displaystyle\int_{-1}^1 \left(x^2 + x + 1\right) \cdot dx$

(b) $\displaystyle\int_{-\pi/2}^{\pi/2} \cos(x) \cdot dx$

(c) $\displaystyle\int_1^3 \left(\dfrac{x^2 + 1}{x}\right) \cdot dx$

(d) $\displaystyle\int_0^4 x^n \cdot dx$

(e) $\displaystyle\int_0^1 (x + e^x - 2) \cdot dx$, and

(f) $\displaystyle\int_0^{\sqrt{3}} \dfrac{dx}{x^2 + 1}$

Problem 4.11. *Compute*

$$\int_{-5}^5 \left(x^5 + 3x^3 + 7x^2 - x + 1\right) \cdot dx$$

Problem 4.12. *Compute the area between the following curves and the x axis on the stated interval. Be careful – some of these have area above and below the x-axis. Pic-tures may help.*

(a) $f(x) = x^2$ on $[-1, 1]$.

(b) $g(x) = x^3$ on $[-1, 1]$.

(c) $h(x) = \cos(x)$ on $[-\pi/2, 3\pi/2]$.

(d) $r(x) = \tan^{-1}(x)$
     on $[-1/\sqrt{3}, 1/\sqrt{3}]$.

(e) $s(x) = e^{-x}$ on $[0, 5]$.

(f) $q(x) = 4 - x^2$ on $[-3, 3]$.

Problem 4.13. *Compute the area bounded by the specified curves. You will need to know which curve is higher. Again: pic-tures may help.*

(a) $y = x^2$ and $y = x^3$

(b) $y = x^2$ and $y = 6 - x$

(c) $y = \sin(x)$ and $y = \cos(x)$ on $[0, 3\pi/2]$

(d) $y = x^2$ and $y = 4x - 3$

(e) $y = x^2$ and $y = 9 - x^2$, and

(f) $y = x^3$ and $y = 4x$.

Problem 4.14. *Find the area bounded by*

$$y = x^n \text{ and } y = x^m$$

for all positive whole numbers $m < n$. There will be different categories of answer based on whether n and m are even or odd. That's four categories: $++$, $+-$, $-+$, and $--$.

Problem 4.15. *Compute the following definite integrals.*

(a) $\int_{-3}^{3} \left(x^5 + x^3 + x + 1 \right) \cdot dx$

(b) $\int_{0}^{2\pi} \left(\cos(x) + \sin(x) \right) \cdot dx$

(c) $\int_{-1}^{1} \left(x^6 + x^5 + x^4 + x^3 + x^2 + \right.$ $\left. x + 1 \right) \cdot dx$

(d) $\int_{-3\sqrt{3}}^{3\sqrt{3}} \tan^{-1}(x) \cdot dx$

(e) $\int_{0}^{2\pi} \left(\sin(x) + \sin(2x) \right.$ $\left. + \sin(3x) \right) \cdot dx$

(f) $\int_{-a}^{a} x^{2n} \cdot dx$

(g) $\int_{-a}^{a} x^{2n+1} \cdot dx$, and

(h) $\int_{-a}^{a} \left(\sin(x) + tan^{-1}(x) + \right.$ $\left. x^5 \right) \cdot dx.$

Problem 4.16. *Classify each of the following functions as being odd, even, or neither.*

(a) $y = \sin(x)$

(b) $y = \cos(x)$

(c) $y = \tan^{-1}(x)$

(d) $y = \ln(x^2 + 1)$

(e) $y = x \cdot \ln(x^2 + 1)$

(f) $y = e^{-x^2/2}$

(g) $y = x^2 + x + 1$

(h) $y = \sin(x^2)$, and

(i) $p(x^2)$, where $p(x)$ is any polynomial.

Problem 4.17. *Suppose that $f(x)$ is an even function. Prove that $y = x \cdot f(x)$ is an odd function.*

Problem 4.18. *Suppose that $f(x)$ is an odd function. Prove that $y = x \cdot f(x)$ is an even function.*

Problem 4.19. *Suppose that $f(x)$ is a function and $g(x)$ is an even function. Is $f(g(x))$ an even function? Demonstrate your answer is correct.*

Problem 4.20. *If*

$$\int_{0}^{b} x^2 \cdot dx = 14,$$

what is b?

Problem 4.21. *If*

$$\int_{0}^{b} x^4 \cdot dx = 1,$$

what is b?

Problem 4.22. *If*

$$\int_{0}^{\theta} \sin(x) \cdot dx = 2,$$

what is the smallest possible value for θ?

Problem 4.23. *Find a constant c so that the area bounded by $y = c$ and $y = x^2$ is exactly 4 units^2.*

Problem 4.24. *Compute the slope m so that the area bounded by the curves $y = x^3$ and $y = mx$ is exactly 4 units.*

Problem 4.25. *Find the largest possible value of $\int_{a}^{b} \cos(x) \cdot dx$.*

Problem 4.26. *Compute*

$$\int_{-5}^{5} x^3 \cdot \ln \left(x^2 + 1 \right) \, dx$$

Problem 4.27. *Suppose*

$$F(x) = \int_{0}^{x} e^{t^2} \cdot dt$$

What is $F'(x)$?

4.3. Initial Value Problems

In this section we come to grips with the constant of integration and figure out what its value is. This requires that we have a bit of additional information. Our motivating example is to build up the position function $s(t)$ from the acceleration function $a(t)$ in steps, with the velocity function $v(t)$ as an intermediate object. The mathematical model of motion in one dimension under constant acceleration is:

$$s(t) = \frac{1}{2}at^2 + v_0t + s_0$$

In English – the distance an object is from a reference point is equal to half the acceleration times the time squared plus the initial velocity times the time plus the initial distance from that reference point. If we break this into integrals we get:

$$v(t) = \int_{t_0}^{t} a \cdot dx$$

$$= a \cdot (t - t_0) + C$$

$$v(t_0) = a \cdot 0 + C$$

$$v_0 = C$$

So the constant of integration when we transform acceleration into velocity is the initial velocity. Similarly:

$$s(t) = \int_{t_0}^{t} v \cdot dx$$

$$= v \cdot (t - t_0) + C$$

$$s(t_0) = v \cdot 0 + C$$

$$s_0 = C$$

The constant of integration for velocity is initial distance. This shows how the constant of integration can be solved for if we know the initial value of the quantity we are calculating. In fact, all we need is the value of the thing we are calculating *anywhere* in the interval we are integrating on. The initial value just has neater algebra.

Example 4.21. Suppose we fire a cannonball directly upward with a velocity of 120m/sec with a gravitational acceleration of 10m/sec^2. If the cannon is at a height of 160m above sea level, find an expression for the distance above sea level of the cannonball after the cannon is fired at $t = 0$.

Solution:

Plug into the equation of motion given above.

$$s(t) = -\frac{1}{2}10t^2 + 120t + 160 = 160 + 120t - 5t^2$$

Since we have a model for this situation, the calculus is all done. Let's look at a situation where we need calculus.

Example 4.22. Suppose a missile has an acceleration that builds gradually so that $a(t) = 100 + 0.1t$. If it is launched from a fixed position with a charge that gives it an initial velocity of 20m/sec, find an expression for the distance the missile has traveled t seconds after launch and find its position and velocity after $t = 20$ seconds when its fuel runs out.

Solution:

First we find the velocity function. Assume that we start at $t = 0$.

$$v(t) = \int a(t) \cdot dt$$
$$= \int (100 + 0.1t) \cdot dt$$
$$= 100t + 0.05t^2 + C$$

Now solve for C:
$$20 = v(0)$$
$$20 = 100(0) + 0.05(0) + C$$
$$20 = C$$

And we obtain:
$$v(t) = 0.05t^2 + 100t + 20$$

This gives us the velocity after $t = 20$ seconds: $V(20) = 2040$m/sec. Now we find the distance function.

$$s(t) = \int v(t) \cdot dt$$

$$= \int \left(0.05t^2 + 100t + 20 \right) \cdot dt$$

$$= \frac{1}{60}t^3 + 50t^2 + 20t + C$$

As before...

$$s(0) = 0 + 0 + 0 + C = 0$$

$$C = 0$$

So...

$$s(t) = \frac{1}{60}t^3 + 50t^2 + 20t$$

This is the expression for the the distance the missile has traveled t seconds after launch, and its position after $t = 20$ seconds is $s(20) \cong 20,533$m.

Example 4.23. Suppose that

$$f(x) = \int \left(3x^2 + 1 \right) dx$$

and we know that $f(2) = 3$. Find an expression for $f(x)$ with no unknown constants.

Solution:

$$f(x) = \int \left(3x^2 + 1 \right) dx$$

$$= x^3 + x + C$$

Now use the added information.

$$f(2) = 2^3 + 2 + C$$

$$3 = 10 + C$$

$$-7 = C$$

Combine

$$f(x) = x^3 + x - 7$$

◇

Knowledge Box 4.13.

Solving initial value problems

When a function resulting from integration has the form

$$f(x) + C,$$

an additional piece of information is needed to determine a value for C.

Example 4.24. Suppose that $g(x) = \int e^x \cdot dx$. For $g(1) = 12.2$, find an expression for $g(x)$ that does not involve any unknown constants.

Solution:

The integral is trivial – e^x is its own integral – and so

$$g(x) = e^x + C$$

Plug in the additional information and solve.

$$g(1) = 12.2$$
$$e^1 + C = 12.2$$
$$C = 12.2 - e$$
$$C \cong 9.48$$

$\diamond$

If we need to do more than one integral, we will need one piece of added information per unknown constant that arises.

Example 4.25. Suppose that the second derivative of $h(x)$ is $h''(x) = 1.2x - 1$. For $h(0) = 4$ and $h'(2) = 2$, find an expression for $h(x)$ that is free of unknown constants.

First we find the first derivative of $h(x)$:

$$h'(x) = \int (1.2\,x - 1)dx$$
$$= 0.6x^2 - x + C_1$$
$$2 = h'(2)$$
$$= 0.6(2)^2 - 2 + C_1$$
$$= 0.4 + C_1$$
$$1.6 = C_1$$
$$h'(x) = 0.6x^2 - x + 1.6$$

Now we move on to the function itself:

$$h(x) = \int (0.6x^2 - x + 1.6)dx$$
$$= 0.2x^3 - 0.5x^2 + 1.6x + C_2$$
$$4 = h(0)$$
$$4 = 0 + 0 + 0 + C$$
$$4 = C$$
$$h(x) = 0.2x^3 - 0.5x^2 + 1.6x + 4$$

And we have our answer. Notice that, since two unknown constants appeared, we gave them different names: C_1 and C_2. Also notice that it is much easier to deal with added information or initial conditions that happen at time zero.

We will revisit initial value problems in Chapter 7 where we learn more integration techniques and in Chapter 9 where we learn about differential equations, which are all about initial conditions.

Problems

Problem 4.28. *Assume we are describing the upward motion of a projectile. Given the constant acceleration, initial velocity, and initial position, determine the greatest height of the projectile and the time it has that height.*

(a) $a = -5\text{m/s}^2$, $v_0 = 20\text{m/sec}$, and $s_0 = 5\text{m}$

(b) $a = -5 \text{ m/s}^2$, $v_0 = 10 \text{ m/sec}$, and $s_0 = -4 \text{ m}$

(c) $a = -5\text{m/s}^2$, $v_0 = 12\text{m/sec}$, and $s_0 = 3\text{m}$

(d) $a = -8 \text{ m/s}^2$, $v_0 = 30 \text{ m/sec}$, and $s_0 = -4.1 \text{ m}$

(e) $a = -2 \text{ m/s}^2$, $v_0 = 11 \text{ m/sec}$, and $s_0 = 1.4 \text{ m}$

(f) $a = -14 \text{ m/s}^2$, $v_0 = 14.2 \text{ m/sec}$, and $s_0 = 7.33 \text{ m}$

Problem 4.29. *Find $s(t)$ given the following rates of acceleration and initial velocity and position.*

(a) $a(t) = 3.2 + 1.1 \cdot t \text{ m/s}^2$, $v_0 = 3 \text{ m/s}$, $s_0 = 0 \text{ m}$

(b) $a(t) = \sqrt{t} \text{ m/s}^2$, $v_0 = -1 \text{ m/s}$, $s_0 = 5 \text{ m}$

(c) $a(t) = t^{1.5} \text{ m/s}^2$, $v_0 = 4 \text{ m/s}$, $s_0 = -2 \text{ m}$

(d) $a(t) = t^2 - t \text{ m/s}^2$, $v_0 = 2 \text{ m/s}$, $s_0 = 0.4 \text{m}$

(e) $a(t) = \cos(t) \text{ m/s}^2$, $v_0 = 1 \text{ m/s}$, $s_0 = 0 \text{ m}$

(f) $a(t) = \dfrac{1}{t} \text{ m/s}^2$, $v_0 = -1 \text{ m/s}$, $s_0 = 0 \text{ m}$

Problem 4.30. *Solve the following initial value problems.*

(a) $f(x) = \displaystyle\int x^2 - x \cdot dx$ when $f(0) = 3$

(b) $g(x) = \displaystyle\int -\sin(x) \cdot dx$ when $g(0) = 0.7071$

(c) $h(x) = \displaystyle\int (0.04x^3 - 2x + 4) \cdot dx$ when $h(1) = 2.4$

(d) $r(x) = \displaystyle\int \left(0.08x + \dfrac{2.3}{x}\right) \cdot dx$ when $r(2) = 0.4$

(e) $s(x) = \displaystyle\int \dfrac{dx}{x^2 + 1}$ when $s(1) = 1.2$

(f) $q(x) = \displaystyle\int \dfrac{x^2 + 1}{x} \cdot dx$ when $q(1) = 8.2$

Problem 4.31. *Suppose that*
$$f''(x) = 2x + 1$$
If $f(1) = 3$ and $f'(1) = 0.5$, find an expression for $f(x)$ with no unknown constants.

Problem 4.32. *Suppose that*
$$g''(x) = 0.03x^2 - 4x + 1$$
If $g(2) = 7$ and $g'(1) = 3$, find an expression for $g(x)$ with no unknown constants.

Problem 4.33. *Suppose that*
$$h''(x) = 1.4x + 5.6$$
If $h(1) = 2$ and $h(2) = 4$, find an expression for $h(x)$ with no unknown constants. Notice that you will have to integrate an unknown constant before you can solve.

Problem 4.34. *Suppose that*
$$r''(x) = 2e^x$$
If $r(0) = 2$ and $r'(0) = 8$, find an expression for $r(x)$ with no unknown constants.

Problem 4.35. *Suppose that*
$$s''(x) = e^x$$
If $s(0) = 1.2$ and $s(1) = 5$, find an expression for $s(x)$ with no unknown constants. Notice that you will have to integrate an unknown constant before you can solve.

Problem 4.36. *Suppose that*
$$q''(x) = \sin(x) + \cos(x)$$
If $q(0) = 1.0$ and $q(\pi/2) = 0.5$, find an expression for $q(x)$ with no unknown constants. Notice that you will have to integrate an unknown constant before you can solve.

4.4. Induction and Sums of Rectangles

In this section we will study the theory of integration that was developed before the fundamental theorem that doesn't need the concept of the derivative. Let's start with an example that could be solved with calculus but need not be.

Example 4.26. Find, without using calculus, the area under $f(x) = 2x + 1$ from $x = 0$ to $x = 3$.

Solution:

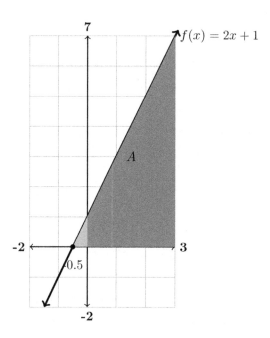

The area of the large triangle is $\frac{1}{2}bh = \frac{1}{2}3.5 \times 7 = 12.25$. The area of the light gray triangle is also $\frac{1}{2}bh = \frac{1}{2}0.5 \times 1 = 0.25$. The area we want, the dark gray, is thus $12.25 - 0.25 = 12$ units^2.

$\Diamond$

The reason we don't need calculus for this example is that the shape of the area is really simple. As long as we remember that the area of a triangle is one-half base times height, we're fine. As soon as the integral is under a function that is curved, we will need calculus. Where does this calculus come from? At this point we need a digression to create the mathematical machinery that permits us to synthesize the calculus.

The symbol

$$\sum$$

means "add up the things following the symbol." So

$$\sum_{i=1}^{5} i$$

is a symbolic way of saying "add up the numbers i from $i = 1$ to $i = 5$. The name of the symbol is **upper case sigma** – not to be confused with σ which is called **lower case sigma**. Applying $\sum$ we get

$$\sum_{i=1}^{5} i = 1 + 2 + 3 + 4 + 5 = 15$$

Which isn't too bad. The problem is when we get something like

$$\sum_{i=1}^{200} i = 1 + 2 + \cdots + 199 + 200 = 20,100$$

The incomparable German mathematician Carl Friedrich Gauss found the following shortcut:

$$\sum_{i=1}^{n} i = \frac{1}{2}n(n+1)$$

How do you prove a formula like that is correct?

The usual technique is called **mathematical induction**.

Knowledge Box 4.14.

Mathematical induction

Suppose we wish to prove a proposition $P(n)$ is correct for all $n \geq c$. Then the following steps suffice:

- Verify that $P(c)$ is true.

- Assume that, for some n, $P(n)$ is true.

- Show that, if $P(n)$ is true, then so is $P(n+1)$

The assumption in the second step is easy, because $P(n)$ is true when $n = c$, due to your work on the first step. The key step is the third one. Once you've got it, $P(c)$ implies $P(c+1)$ is true, which in turn implies $P(c+2)$ is true, and so on, until you hit any particular $n \geq c$.

Example 4.27. Use mathematical induction to prove Gauss' formula.

Solution:

The proposition is

$$P(n) \ : \ \sum_{i=1}^{n} i = \frac{1}{2}n(n+1)$$

Let's start with $c = 1$.

$$P(c) = P(1) : \sum_{i=1}^{1} i = 1 = \frac{1}{2}c(c+1) = \frac{1}{2}(1)(1+1) = 1$$

Since 1=1, this is true, and we have the first step of the induction. We now assume that $P(n)$ is true (for some n) and look at $P(n+1)$.

$$\sum_{i=1}^{n+1} i = \sum_{i=1}^{n} i + (n+1)$$

$$= \frac{1}{2}n(n+1) + (n+1) \qquad\qquad \text{Apply } P(n)$$

$$= \frac{n(n+1) + 2(n+1)}{2}$$

$$= \frac{n^2 + 3n + 2}{2}$$

$$= \frac{1}{2}(n+1)(n+2)$$

$$= \frac{1}{2}(n+1)((n+1)+1) \qquad\qquad \text{Which is } P(n+1)$$

So if $P(n)$ is true, we can show using algebra that $P(n+1)$ is true. This tells us, by mathematical induction, that Gauss' formula is true for all $n \geq 1$.

Some other examples of proof with mathematical induction appear in the homework problems. Now we know enough to calculate an integral or a curved function *without* using the fundamental theorem of calculus. Our goal is to approximate the area under the curve with shapes we can compute.

Examine the two pictures of $f(x) = x^2$ in Figure 4.1. The gray area is $\int_{0}^{2} x^2 \cdot dx$; the four rectangles are a (bad) approximation of the area under the curve. How can we make the approximation better? Use more rectangles! See Figure 4.2. The more rectangles, the better the area is approximated.

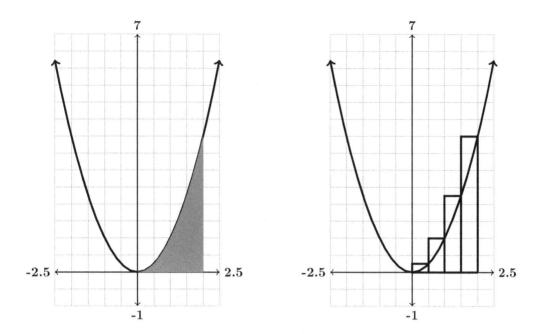

Figure 4.1. Finding the area under $f(x) = x^2$ on $[0, 2]$.

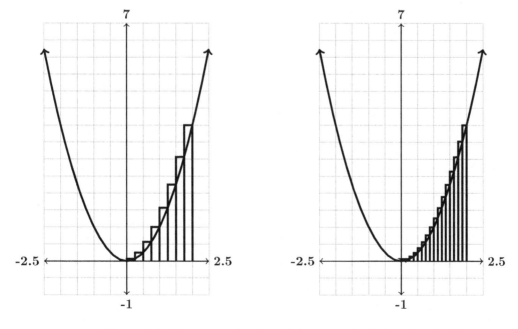

Figure 4.2. Better approximations of the area under $f(x) = x^2$ on $[0, 2]$.

It turns out that summing and integrating share two algebraic properties.

Knowledge Box 4.15.

Algebraic properties of $\sum$

$$\sum_{i=a}^{b} f(i) + g(i) = \sum_{i=a}^{b} f(i) + \sum_{i=a}^{b} g(i)$$

$$\sum_{i=a}^{b} c \cdot f(i) = c \cdot \sum_{i=a}^{b} f(i)$$

for any constant c

There are an infinite number of summation formulas – but a few of them are especially useful. All of these can be proved by induction (and you are asked to do so in the homework).

Knowledge Box 4.16.

Useful sum formulas

- $$\sum_{i=1}^{n} 1 = n$$

- $$\sum_{i=1}^{n} i = \frac{n(n+1)}{2}$$

- $$\sum_{i=1}^{n} i^2 = \frac{n(n+1)(2n+1)}{6}$$

- $$\sum_{i=1}^{n} i^3 = \frac{n^2(n+1)^2}{4}$$

If we use n rectangles to approximate $\displaystyle\int_{1}^{2} x^2 \cdot dx$, then each rectangle has a width of $w = \dfrac{2}{n}$. The right side of each rectangle (which determines its height) is $x_i = i \cdot w$ for $i = 1, 2, \ldots n$. The height of the ith rectangle is $(x_i)^2 = (iw)^2 = \left(\dfrac{2i}{n}\right)^2$. Summing the areas we get that:

$$A \cong \sum_{i=1}^{n} W \times H = \sum_{i=1}^{n} \frac{2}{n}\left(\frac{2i}{n}\right)^2$$

Let's simplify this with the algebraic rules for sums.

$$\sum_{i=1}^{n} \frac{2}{n} \left(\frac{2i}{n}\right)^2 = \sum_{i=1}^{n} \frac{2}{n} \frac{4i^2}{n^2}$$

$$= \sum_{i=1}^{n} \frac{8i^2}{n^3}$$

$$= \frac{8}{n^3} \sum_{i=1}^{n} i^2$$

$$= \frac{8}{n^3} \frac{n(n+1)(2n+1)}{6} \qquad \text{Use } \sum i^2 \text{ formula}$$

$$= \frac{16n^3 + 24n^2 + 8n}{6n^3}$$

$$= \frac{8n^3 + 12n^2 + 4n}{3n^3}$$

Now we have a formula for the approximate area with n rectangles – the approximation gets better as n grows. This means that

$$\int_{0}^{2} x^2 \cdot dx = \lim_{n \to \infty} \frac{8n^3 + 12n^2 + 4n}{3n^3} = \frac{8}{3} \text{ units}^2$$

which is the same result we get if we do the integral in the usual way.

This is a **very cumbersome** method of computing integrals – not used in practice – but it shows that there is a theory for integrals, just as there is for derivatives. The fundamental theorem is a godsend. Imagine if you did not know that integrals and anti-derivatives were the same thing. Every integral would be a limit of sums of rectangles (or some other shape).

Many integrals cannot be done symbolically – with formulas and algebra. The discipline of **numerical analysis** studies how to use things like rectangle-sum approximations to get useful values for integrals that cannot be calculated with pure calculus.

A final note: we will use summation notation *a lot* in Chapter 13. Remember the facts in this section.

Problems

Problem 4.37. *Compute the following sums. You may use the formulas you are asked to prove in Problem 4.38.*

(a) $\displaystyle\sum_{i=1}^{40} i^2$

(b) $\displaystyle\sum_{i=20}^{60} i$

(c) $\displaystyle\sum_{i=1}^{30} (2i + 5)$

(d) $\displaystyle\sum_{i=5}^{23} 2^i$

(e) $\displaystyle\sum_{i=14}^{28} i^3$

(f) $\displaystyle\sum_{i=18}^{37} (2i - 1)$

Problem 4.38. *Use mathematical induction to demonstrate that the following formulas are correct.*

(a) $\displaystyle\sum_{i=1}^{n} (2i - 1) = n^2$

(b) $\displaystyle\sum_{i=1}^{n} 1 = n$

(c) $\displaystyle\sum_{i=1}^{n} i^2 = \frac{n(n+1)(2n+1)}{6}$

(d) $\displaystyle\sum_{i=1}^{n} i^3 = \frac{n^2(n+1)^2}{4}$

(e) $\displaystyle\sum_{i=0}^{n} 2^i = 2^{n+1} - 1$

(f) $\displaystyle\sum_{i=0}^{n} x^i = \frac{x^{n+1} - 1}{x - 1}$

Problem 4.39. *Explain why the area under a line $y = mx + b$ can always be found without calculus.*

Problem 4.40. *A formula for approximating*

$$\int_0^2 x^2 \cdot dx$$

with n rectangles was computed in this section. For $n = 4, 6, 8, 12, 20,$ and 50 rectangles, compute the error of the approximation.

Problem 4.41. *Find the formula for the sum of rectangles for $y = x^3$ to approximate the integral of $y = x^3$ from $x = 0$ to $x = c$. Having found the formula, find the integral by taking a limit.*

Problem 4.42. *The sum of rectangles used in this section was based on the right side of the intervals. How would an approximation that used the left side of the interval be different? Could it still be used with a limit, to compute integrals? Explain.*

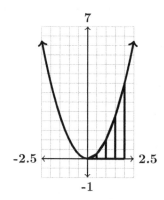

Problem 4.43. *The approximation sketched above uses trapezoids instead of rectangles to approximate the area under the curve. The area of each trapezoid is the average of the height of the points on either side of the interval. Write out the approximation formula for n trapezoids instead of n rectangles. Make it as simple as you can.*

Problem 4.44. *Is the trapezoid method more accurate than the rectangle method? Explain or justify your answer.*

Chapter 5

Parametric, Polar, and Vector Functions

Consider the plot in Figure 5.1. By the rules we've come up with so far this is not even close to being a function. It also looks like the graph of nothing we have done so far. It is a spiral. In this chapter we will look at methods for using calculus – which wants things to act like functions at least locally – on curves like this. The underlying idea is to treat the position of a point as having coordinates that are individually functions of a **parameter**. These are called **parametric functions**.

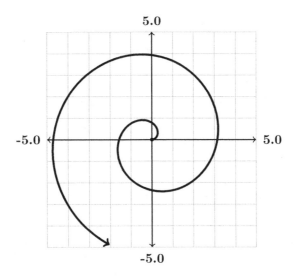

Figure 5.1. Example of a parametric function – a spiral.

5.1. Parametric Functions

The spiral shown in Figure 5.1 is not a function of the form $y = f(x)$. It violates the vertical line test and, if we had shown the entire spiral instead of only its beginning, would have intersected every vertical line an infinite number of times. In order to make the spiral – and many other curves – into functions, we will make the x and y coordinates of the curve functions of a parameter t in their own right. This is how parametric curves are created.

Knowledge Box 5.1.

Parametric curves

If we specify a set of points by
$$(x(t),\ y(t)),$$
where $x = x(t)$ and $y = y(t)$, then the resulting structure is called a **parametric curve**. The variable t is called the parameter.

Example 5.1. Graph the curve $(x(t), y(t))$ if $x(t) = 3\cos(t)$ and $y(t) = 2\sin(t)$.

Solution:

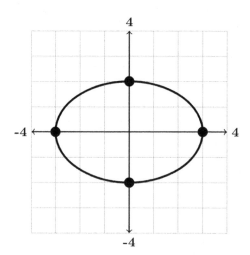

In order to graph this curve, assuming we don't know ahead of time that it is an ellipse with major axis of length 6 and minor axis of length 4, we plot points. We know that $-1 \le \sin(x), \cos(x) \le 1$. So, it's not hard to figure out where the bounds are. The points on the curve where the extreme values are have been plotted with dots.

◊

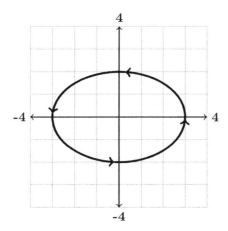

Figure 5.2. Orientation of ellipse from Example 5.1.

Definition 5.1. The **major axis** of an ellipse is its largest diameter. The **minor axis** of an ellipse is its smallest diameter.

The major axis of the ellipse in Example 5.1 is 6; the minor axis is 4. In general

$$(a \cdot \cos(t), b \cdot \sin(t))$$

is an ellipse with major and minor axes in the direction of the coordinate axes, centered at the origin. The major and minor axes have size $2a$ and $2b$ with the major/minor order determined by which is larger.

One thing that is very different about a parametric curve is that the points are **ordered** by the parameter. Points generated by a larger value of the parameter are thought of as coming *after* points generated by a smaller value of the parameter.

Definition 5.2. The **positive orientation** of a parametric curve is the direction, along the curve, in which the parameter increases.

The orientation of the ellipse in Example 5.1, shown in Figure 5.2, is counterclockwise.

It turns out that any function of the usual sort can be put into the form of a parametric curve by the simple technique of starting with $y = f(x)$ and defining $x(t) = t$ and $y(t) = f(t)$. This will make the parametric function exactly trace the graph of $y = f(x)$.

One disadvantage of parametric functions is that they give you many, many ways to specify the same function. The lines $(t, 2t + 1)$ and $(3t + 1, 6t + 3)$ are two different parametric forms of the line $y = 2x + 1$. If $x(t)$ and $y(t)$ are both linear functions, then $(x(t), y(t))$ is the parametric form of *some* line.

Example 5.2. Put the line
$$(2t + 5, 4t - 1)$$
into $y = mx + b$ form.

Solution:
$$x = 2t + 5$$
$$2x = 4t + 10$$
$$2x - 11 = 4t - 1$$
$$2x - 11 = y$$

So we see $y = 2x - 11$ is the standard form of this parametric line.

$\Diamond$

The real advantage of parametric curves is that they let you specify things that are hard to specify in other ways and that they let you turn things that were not functions into functions of the parameter.

Example 5.3. Consider the parametric curve $(\sin(t), t\cos(t))$. Graph the curve for $0 \le t \le 2\pi$.

Solution:

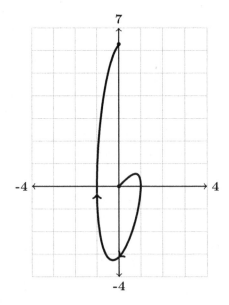

This one is done by plotting points.

$\Diamond$

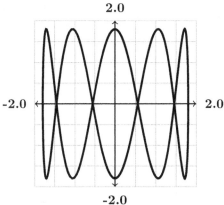

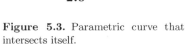

Figure 5.3. Parametric curve that intersects itself.

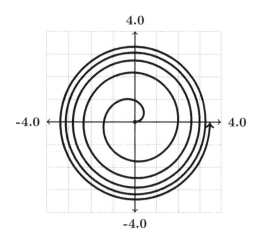

Figure 5.4. Spiral parametric curve.

Parametric curves can intersect themselves, possibly many times. If we graph $x(t) = 1.8\cos(\frac{4}{3}\pi t)$, $y(t) = 1.8\sin(\frac{20}{3}\pi t)$, we get the curve in Figure 5.3.

We have not yet explained how to make a spiral. If $f(t)$ is a function that is increasing on $[0, \infty)$, then the parametric curve

$$(f(t)\cos(t), f(t)\sin(t))$$

is a spiral with $f(t)$ controlling how fast it spirals outward. If we take $f(t) = \ln(t+1)$, for $t \geq 0$, then we get the curve in Figure 5.4.

5.1.1. The Derivative of a Parametric Curve. Way back in Chapter 2 we learned that we could algebraically cancel differentials. The ordinary derivative of a curve is $\dfrac{dy}{dx}$, which lets us calculate the derivative by doing this:

$$\frac{dy}{dx} = \frac{dy/dt}{dx/dt} = \frac{y'(t)}{x'(t)}$$

Let's highlight that in a Knowledge Box.

Knowledge Box 5.2.

Derivatives of parametric curves

The slope m of the tangent line to a parametric curve $(x(t), y(t))$ at parameter value t is:

$$m = \frac{y'(t)}{x'(t)}$$

Example 5.4. Find the derivative $\dfrac{dy}{dx}$ of the parametric curve with $x(t) = 5\cos(t)$ and $y(t) = 3\sin(t)$.

Solution:

First compute $x'(t) = -5\sin(t)$ and $y'(t) = 3\cos(t)$.

$$\frac{dy}{dx} = \frac{y'(t)}{x'(t)}$$

$$= \frac{-5\sin(t)}{3\cos(t)}$$

$$= -\frac{5}{3}\tan(t)$$

$$\Diamond$$

Example 5.5. Find the tangent line to $(t^2, 2t - 3)$ at (4,1).

Solution:

Our first job is to check if the point is on the curve. If $2t - 3 = 1$, then $t = 2$, and the point for $t = 2$ is (4,1). So, no problem there.

$$\frac{dy}{dx} = \frac{y'(t)}{x'(t)} = \frac{2}{2t} = \frac{1}{t}$$

so we have $m = \dfrac{1}{t} = \dfrac{1}{2}$. Computing the line using the point-slope formula, we get:

$$y - 1 = \frac{1}{2}(x - 4)$$

$$y = \frac{1}{2}x - 2 + 1$$

$$y = \frac{1}{2}x - 1$$

$$\Diamond$$

Example 5.6. Find the tangent line to $(4\cos(t), 2\sin(t))$ at $(2\sqrt{3}, 1)$.

Solution:

Our first job is to check if the point is on the curve. If $2\sin(t) = 1$, then $\sin(t) = 1/2$, and so $t = \dfrac{\pi}{6}$. Then $4\cos\left(\dfrac{\pi}{6}\right) = 4\dfrac{\sqrt{3}}{2} = 2\sqrt{3}$. So the point is on the curve, no problem. Compute the slope of the tangent line:

$$\frac{dy}{dx} = \frac{y'(t)}{x'(t)} = \frac{2\cos(t)}{-4\sin(t)} = -\frac{1}{2}\cot(t)$$

At $t = \dfrac{\pi}{6}$ this is $m = -\dfrac{1}{2}\sqrt{3} = -\dfrac{\sqrt{3}}{2}$. Apply the point-slope formula and simplify:

$$y - 1 = -\frac{\sqrt{3}}{2}(x - 2\sqrt{3})$$

$$y = -\frac{\sqrt{3}}{2}x + 3 + 1$$

$$y = -\frac{\sqrt{3}}{2}x + 4$$

and we have the tangent line.

$$\diamond$$

Example 5.7. Find the parameter values for which the tangent line to $(2\sin(t), \cos(t))$ has slope $m = -1/2$.

Solution:

$$\frac{dy}{dx} = \frac{y'(t)}{x'(t)} = \frac{-\sin(t)}{2\cos(t)} = -\frac{1}{2}\tan(t)$$

Solve:

$$-\frac{1}{2}\tan(t) = -\frac{1}{2}$$
$$\tan(t) = 1$$
$$t = \frac{\pi}{4} + n\pi \qquad\qquad n = 0, \pm 1, \pm 2, \ldots$$

Remember that the angles with tangent equal to 1 are the ones with *equal* sine and cosine.

$$\diamond$$

This section has been a small sampler platter of the many sorts of parametric curves that are possible. We've also restricted ourselves to only two coordinates. In Section 5.3 we will revisit this with the formalism of vectors rather than parameters.

Problems

Problem 5.1. *Graph each of the following parametric curves.*

(a) $(\cos(t), \sin(t) + 2)$, $0 \leq t \leq 2\pi$

(b) $(1 - t, 2t - 4)$, $-5 \leq t \leq 5$

(c) $(\sin(t), t \cos(t))$,
    $-2\pi \leq y \leq 2\pi$

(d) $(t \cos(t), t \sin(t))$, $0 \leq t \leq 4\pi$

(e) $(2 \cos(3t), 3 \sin(2t))$,
    $0 \leq t \leq 2\pi$

(f) (t^2, t), $-3 \leq t \leq 3$.

Problem 5.2. *Find a parametric equation for an ellipse with major axis of length 12 in the direction of the y-axis and minor axis of length 2 in the direction of the x-axis.*

Problem 5.3. *Find a parametric equation for an ellipse with major axis of length 5 in the direction of the x-axis and minor axis of length 3 in the direction of the y-axis.*

Problem 5.4. *Find a parametric equation for an ellipse with major axis of length 14 in the direction of the y-axis and minor axis of length 7 in the direction of the x-axis.*

Problem 5.5. *Find the standard $y = mx + b$ form for each of the following parametric lines.*

(a) $(t, 2t + 1)$

(b) $(2t, 1 - t)$

(c) $(3t + 1, 5t - 1)$

(d) $(t + 7, 7 - t)$

(e) $(3t - 1, 3 - 2t)$

(f) $(2 - 3t, 3 - 2t)$

Problem 5.6. *Graph and carefully describe the parametric curve*

$$(2 \sin(t), 3 \sin(t) + 1)$$

Problem 5.7. *Find y' for each of the following parametric curves.*

(a) $(\sin(2t), \cos(3t))$

(b) $(1 - t^2, 3t + 1)$

(c) $(t \cdot \sin(t), \cos(t))$

(d) $(t \sin(t), t \cos(t))$

(e) $(\cos(5t), 2 \sin(3t))$

(f) (t^3, t^2)

Problem 5.8. *Find the tangent line to (t^3, t^2) at (-8,4).*

Problem 5.9. *Find the tangent line to $(\sin(t), t)$ at $(1, \pi/2)$.*

Problem 5.10. *Find the tangent line to*

$$\left(\frac{1}{t^2 + 1}, \frac{1}{t} \right)$$

at the point $(\frac{1}{2}, 1)$.

Problem 5.11. *For which values of t does*

$$(\cos(t), \sin(2t))$$

have a tangent line with a slope of 2?

Problem 5.12. *Let $f(t) = \dfrac{e^t}{e^t + 1}$. Describe carefully the parametric curve*

$$(f(t) \cos(t), f(t) \sin(t))$$

for $-\infty < t < \infty$.

Problem 5.13. *The parametric curve $(af(t), bf(t))$ on $-\infty < t < \infty$ is a line or a line segment. Give explicit directions on how to figure out which line segment it is.*

Problem 5.14. *Plot carefully the parametric curve:*

$$(\sin(t) + \cos(2t)/2, \cos(t) + \sin(2t)/2)$$

Problem 5.15. *A function cannot assign multiple values to the same point. In light of this why are points where a parametric curve intersects itself not a problem?*

5.2. Polar Coordinates

Polar functions can be expressed as parametric functions, but what they really are is an alternate coordinate system for doing business, based on direction and distance instead of x-coordinate and y-coordinate. Examine Figure 5.5. The point $(\sqrt{3}, 1)$ can also be specified by going a distance (radius) of two from the origin in the direction of the angle $\theta = \dfrac{\pi}{6}$.

The circle of radius 2 is shown to supply clarity – polar graphs use circles and radial lines the way that a standard graph uses grids. Figure 5.6 shows an example of polar graph paper.

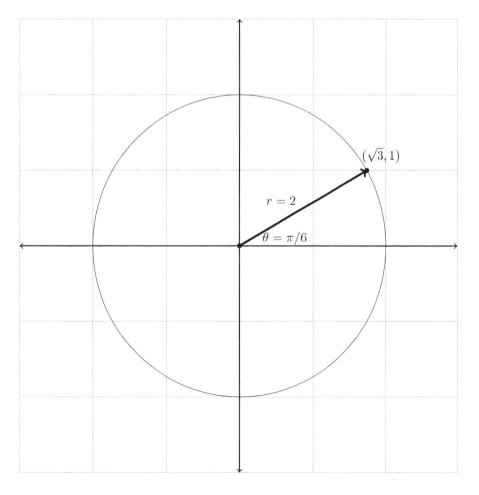

Figure 5.5. Point in both polar and standard coordinates.

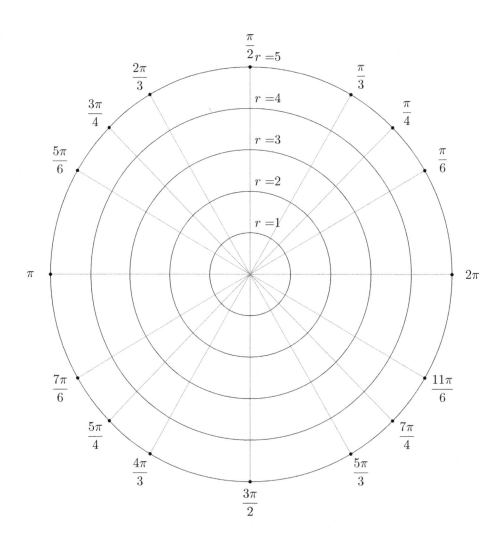

Figure 5.6. Polar graph paper.

Knowledge Box 5.3.

Polar to rectangular conversion formulas

In order to transform between (x, y) and (r, θ) coordinates we use the following formulas.

- $x = r \cdot \cos(\theta)$
- $y = r \cdot \sin(\theta)$
- $r = \sqrt{x^2 + y^2}$
- $\theta = \tan^{-1}(y/x)$

Where the formulas don't work, when x is zero and so θ is an odd multiple of $\pi/2$, use common sense – the directions are vertical.

For historical reasons there are two names for the standard coordinate system when we are comparing it to the polar coordinate system – **rectangular** coordinates and **Cartesian** coordinates. We will use the terms interchangeably.

Example 5.8. Find the polar version of the point $(2, 1)$.

Solution:

$$r = \sqrt{2^2 + 1^2} = \sqrt{5}, \text{ and}$$
$$\theta = \tan^{-1}\left(\frac{1}{2}\right) \cong 0.4636 \text{ rad}$$

So the polar point is $(r, \theta) = (\sqrt{5}, 0.4636)$.

$\Diamond$

Example 5.9. Find the rectangular coordinates for the polar point $\left(4, \frac{3\pi}{4}\right)$.

Solution:

In this case $r = 4$ and $\theta = \frac{3\pi}{4}$ so:

$$x = r \cdot \cos(\theta) = 4 \cdot \cos\left(\frac{3\pi}{4}\right) = 4 \cdot -\frac{\sqrt{2}}{2} = -2\sqrt{2}$$

$$y = r \cdot \sin(\theta) = 4 \cdot \sin\left(\frac{3\pi}{4}\right) = 4 \cdot \frac{\sqrt{2}}{2} = 2\sqrt{2}$$

and so the corresponding point in Cartesian coordinates is $(x, y) = (-2\sqrt{2}, 2\sqrt{2})$.

$\Diamond$

One of the major uses for converting between the two coordinate systems is to permit you to plot polar points on normal graph paper when you are graphing a polar function.

One of the really nice things about polar coordinates is that they let us deal very easily with circles centered at the origin. Circles centered at the origin are *constant functions* in polar coordinates.

Example 5.10. Graph the polar function $r = 2$.

Solution:

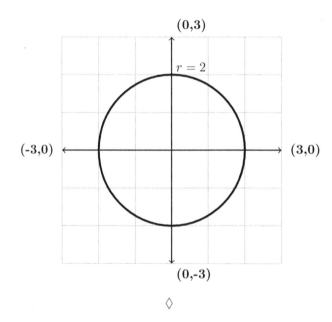

We usually write polar functions in the form

$$r = f(\theta)$$

making the angle the independent variable and the radius the dependent variable. This makes it very easy to give polar functions as parametric functions.

Knowledge Box 5.4.

Parametric form of polar curves

If $r = f(\theta)$ on $\theta_1 \le \theta \le \theta_2$ is a polar curve, then a parametric form for the same curve is:

$$(f(t) \cdot \cos(t), f(t) \cdot \sin(t))$$

for $t \in [\theta_1, \theta_2]$.

So far in this section we have established the connections between polar coordinates and the rest of the systems developed in the text. It is time to display polar curves that have unique characteristics that are most easily seen in the polar system.

Example 5.11. Graph the polar function $r = \cos(3\theta)$ on $[0, \pi)$.

Solution:

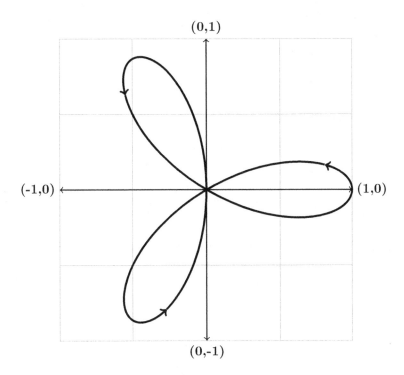

The first point on the curve is at (1,0) which happens when $\theta = 0$; the arrows show the drawing direction.

◊

Definition 5.3. Petal curves are curves with equations of the form:

$$r = \cos(n\theta) \text{ or } r = \sin(n\theta),$$

where n is an integer.

If no restriction is placed on n, then the curve is traced out an infinite number of times. For odd n, a domain of $\theta \in [0, \pi)$ traces the curve once; when n is even, $\theta \in [0, 2\pi)$ is needed to trace the entire curve once.

Example 5.12. Compare the curves $r = \sin(5\theta)$ and $r = \cos(5\theta)$ on the range $\theta \in [0, \pi)$.

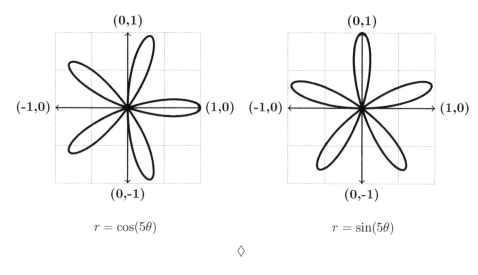

$$r = \cos(5\theta) \qquad\qquad r = \sin(5\theta)$$

◇

The odd fact, that the minimal domain (to hit all the points) is twice as large when n is even, is to some degree explained by the fact that, while petal curves with odd parameter n yield n petals, when n is even you get $2n$ petals.

Example 5.13. Plot the polar function $r = \cos(4\theta)$.

Solution:

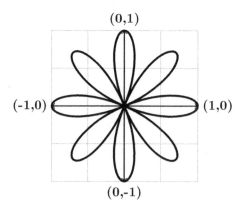

See? We have $n = 4$ but 8 petals.

◇

It is possible to use values of the petal-determining parameter for polar curves that are not integers, but then figuring out the minimal domain to plot the curve becomes problematic.

Example 5.14. Plot the polar function $r = \cos(1.5\,\theta)$ for $\theta \in [0, 4\pi)$.

Solution:

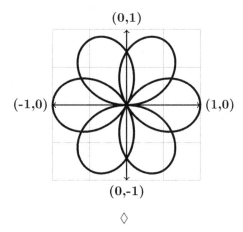

$\Diamond$

A fractional parameter yields more and fatter petals on the curve. If you have polar curve plotting software, experiment with these parameters and see what you can find.

5.2.1. Polar Calculus. There are many interesting shapes that can be made with polar coordinates, but at this point we are going to look into the calculus of polar curves. First, computing $\dfrac{dy}{dx}$ at a point (r, θ).

We already have a parametric form for polar curves:

$$x(\theta) = f(\theta) \cdot \cos(\theta), \quad y(\theta) = f(\theta) \cdot \sin(\theta)$$

This is the parametric curve:

$$(x = f(\theta)\cos(\theta), y = f(\theta)\sin(\theta))$$

We can then apply the technique for finding the derivative of a parametric curve and get:

$$\frac{dy}{dx} = \frac{y'(\theta)}{x'(\theta)} = \frac{f'(\theta) \cdot \sin(\theta) + f(\theta) \cdot \cos(\theta)}{f'(\theta) \cdot \cos(\theta) - f(\theta) \cdot \sin(\theta)}$$

Let's put this formula in a Knowledge Box.

Knowledge Box 5.5.

Computing $\dfrac{dy}{dx}$ for a polar curve

$$\frac{dy}{dx} = \frac{f'(\theta) \cdot \sin(\theta) + f(\theta) \cdot \cos(\theta)}{f'(\theta) \cdot \cos(\theta) - f(\theta) \cdot \sin(\theta)}$$

Example 5.15. Find the tangent line to $r = \cos(3\theta)$ at $\theta = \dfrac{\pi}{12}$.

Solution:

To use the formula in Knowledge Box 5.5 we need $f(\theta) = r = \cos(3\theta)$ and $f'(\theta) = -3\sin(3\theta)$. Computing the derivative we get:

$$\frac{dy}{dx} = \frac{-3\sin(3\theta)\sin(\theta) + \cos(3\theta)\cos(\theta)}{-3\sin(3\theta)\cos(\theta) - \cos(3\theta)\sin(\theta)}$$

$$= \frac{-3\sin\left(3\frac{\pi}{12}\right)\sin\left(\frac{\pi}{12}\right) + \cos\left(3\frac{\pi}{12}\right)\cos\left(\frac{\pi}{12}\right)}{-3\sin\left(3\frac{\pi}{12}\right)\cos\left(\frac{\pi}{12}\right) - \cos\left(3\frac{\pi}{12}\right)\sin\left(\frac{\pi}{12}\right)} \cong -0.06$$

When $\theta = \dfrac{\pi}{12}$ we get the point in polar coordinates $\left(\cos\left(\dfrac{\pi}{4}\right), \dfrac{\pi}{12}\right) = \left(\dfrac{\sqrt{2}}{2}, \dfrac{\pi}{12}\right)$. Now we need the Cartesian version of the point. Applying the polar-to-rectangular formulas we obtain: $\left(\dfrac{\sqrt{2}}{2}\cos\left(\dfrac{\pi}{12}\right), \dfrac{\sqrt{2}}{2}\sin\left(\dfrac{\pi}{12}\right)\right) \cong (0.683, 0.183)$. Find the line:

$$y - 0.183 = -0.06(x - 0.683)$$
$$y = -0.06x + 0.224$$

Let's actually take a look at the graph with the curve and the tangent line. The tangent line hits the curve at three points, but the point of tangency is also plotted and is clearly a tangent line; the other two intersections cut through the curve at sharp angles.

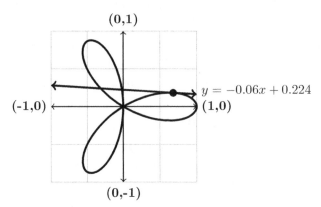

The curve $r = \cos(3\theta)$ and the tangent line at $\theta = \dfrac{\pi}{12}$.

◊

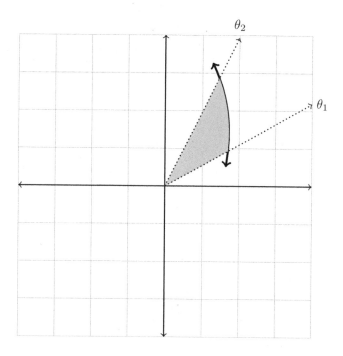

Figure 5.7. Area "under the curve" in polar coordinates.

The formula for the Cartesian derivative of a curve in polar coordinates is pretty cumbersome, but the Cartesian form of $r = \cos(3\theta)$ is really just not manageable at all. So, if we need tangent lines to these curves, this formula is our only hope.

What about integrals? If we have a sector of a circle subtending an angle of θ, then the area is just the fraction of the full circle that θ covers times the area formula of πr^2 so we get

$$\text{Area} = \frac{\theta}{2\pi} \cdot \pi r^2 = \frac{\theta}{2} r^2.$$

This means that the farther from the origin an area is, the larger it gets. This leads to a really different integral formula for area. First of all, it is area enclosed by the polar curve. Second, we need to be very careful about $r < 0$ because it's so much less obvious than $y < 0$.

The shaded area between the polar curve shown in Figure 5.7 and the origin, in the angular range $\theta_1 \leq \theta \leq \theta_2$ is

$$\text{Area} = \frac{1}{2} \int_{\theta_1}^{\theta_2} r(\theta)^2 \cdot d\theta$$

When we did integrals in Cartesian space, we approximated areas with rectangles. In polar space, we use pie-shaped slices, and so the area depends on the square of the functional value instead of just its value. The constant of one-half carries over from the area formulas for sectors of a circle.

Knowledge Box 5.6.

Finding the area between a polar curve and the origin

$$\text{Area} = \frac{1}{2} \int_{\theta_1}^{\theta_2} r(\theta)^2 \cdot d\theta$$

In Chapter 4, we had to be careful about area above and below the x-axis. When working with polar curves, the analogous problems are areas with $r \geq 0$ and $r \leq 0$. If we know r stays positive, then we still need to know the minimal angular domain that sweeps out the entire shape once. Let's do some examples.

Example 5.16. Find the area enclosed by $r = \theta$ from $\theta = 0$ to $\theta = \frac{4\pi}{3}$.

Solution:

The radius is positive on the domain of integration so we can just integrate.

$$A = \frac{1}{2} \int_0^{4\pi/3} r^2 \cdot d\theta$$

$$= \frac{1}{2} \int_0^{4\pi/3} \theta^2 \cdot d\theta$$

$$= \frac{1}{6} \theta^3 \Big|_0^{4\pi/3}$$

$$= \frac{1}{6} \left(\left(\frac{4\pi}{3} \right)^3 - 0 \right)$$

$$= \frac{32\pi^3}{81} \text{ units}^2$$

$$\cong 12.25 \text{ units}^2$$

Let's look at a picture of the area:

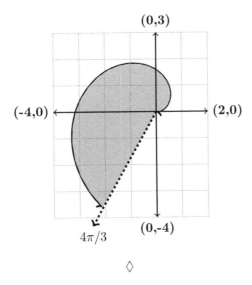

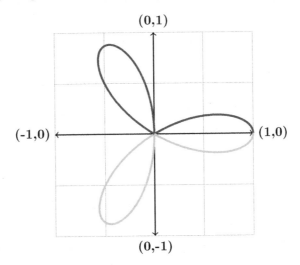

Let's look at a more challenging example. The function in Example 5.11 is $r = \cos(3\theta)$ on $[0, \pi)$. It has both positive and negative radii in the course of the graph.

Example 5.17. Find the area enclosed by $r = \cos(3\theta)$.

Solution:

Let's graph this again, using different shades for the positive and negative radii.

Notice the first dark gray arc of the graph is from $\theta = 0$ to $\theta = \pi/6$. This arc encompasses half of one of the three petals and so is $\dfrac{1}{6}$ of the area. Since on this angular interval the radius is positive, computing six times the integral of $r = \cos(3\theta)$ on $[0, \pi/6]$ will give us the full area of the curve.

$$\text{Area} = 6 \cdot \frac{1}{2} \int_0^{\pi/6} \cos^2(3\theta) \cdot d\theta$$

$$= 3 \int_0^{\pi/6} \frac{1 + \cos(6\theta)}{2} \cdot d\theta$$

$$= \frac{3}{2} \int_0^{\pi/6} d\theta + \int_0^{\pi/6} \cos(6\theta) \cdot d\theta$$

$$= \frac{3}{2} \int_0^{\pi/6} d\theta + \frac{3}{2} \int_0^{\pi} \cos(\theta) \cdot d\theta \qquad\qquad \text{Cosine was rescaled}$$

$$= \frac{3}{2} \theta \Big|_0^{\pi/6} + \frac{3}{2} \sin(\theta) \Big|_0^{\pi}$$

$$= \frac{3}{2} \left(\frac{\pi}{6} - 0 \right) + \frac{3}{2}(0 - 0)$$

$$= \frac{\pi}{4} \text{ units}^2$$

$\Diamond$

Problems

Problem 5.16. *For each of the following points in Cartesian coordinates, give the corresponding points in polar coordinates.*

(a) (1,1)

(b) (3,4)

(c) $(4, 4\sqrt{3})$

(d) (-3,7)

(e) (-5,-5)

(f) (0,8)

Problem 5.17. *For each of the following points in polar coordinates, give the corresponding points in rectangular coordinates. The points are in the form (r, θ) with the angular coordinate in radians.*

(a) $(4, \pi/3)$

(b) $(2, \pi/5)$

(c) $(-5, \pi/4)$

(d) $(7, 4\pi/3)$

(e) $(0.2, 3\pi/4)$

(f) $(2.6, 1.14)$

Problem 5.18. *Plot the following polar curves on the given domain.*

(a) $r = \cos(2\theta)$ on $[0, 2\pi)$

(b) $r = \sin(7\theta)$ on $[0, \pi)$

(c) $r = \sin(3\theta) + 2$ on $[0, 2\pi)$

(d) $r = 4(\cos(\theta) + 1)$ on $[0, 2\pi)$

(e) $r = \theta/4$ on $[0, 6\pi)$

(f) $r = 5\cos(\theta)$ on $[0, \pi)$

Problem 5.19. *Find a parametric form for $3 = 5\cos(\theta)$ on $[0, \pi)$.*

Problem 5.20. *Find the tangent line to $r = 6$ at $\theta = \dfrac{3\pi}{4}$.*

Problem 5.21. *Find the tangent line to $r = \theta$ at $\theta = \dfrac{\pi}{4}$.*

Problem 5.22. *Find the tangent line to $r = \cos(4\theta)$ at $\theta = \dfrac{\pi}{16}$.*

Problem 5.23. *Using the polar integral, verify that a circle of radius R has area equal to πR^2.*

Problem 5.24. *Graph $r = \sin(5\theta)$ on $[0, \pi)$ using two colors or line styles to show where the radius is positive or negative.*

Problem 5.25. *Find the area enclosed by $r = \sin(5\theta)$. Worry about the issue of negative radii.*

Problem 5.26. *Graph $r = \cos(2\theta)$ on $[0, 2\pi)$ using two colors or line styles to show where the radius is positive or negative.*

Problem 5.27. *Find the area enclosed by $r = \cos(2\theta)$. Worry about the issue of negative radii.*

Problem 5.28. *Using software of some sort, graph $r = \cos(1.1 \cdot \theta)$ on $[0, 20\pi)$.*

Problem 5.29. *Using software of some sort, graph $r = \sin(5.1 \cdot \theta)$ on $[0, 20\pi)$.*

Problem 5.30. *Describe completely all polar curves of the form $r = C \cdot \sin(\theta)$ in simple English.*

Problem 5.31. *For the polar curve $r = \theta$ find the angle α so that the area between the curve and the origin on the interval $[0, \alpha]$ is 2.*

Problem 5.32. *Carefully graph the polar curve $r = \cos^2(\theta)$.*

Problem 5.33. *Carefully graph the polar curve $r = \sin^2(\theta)$.*

Problem 5.34. *Write a paragraph accurately explaining where the term "Cartesian," as used in describing a coordinate system, comes from.*

5.3. Vector Functions

One of the most unsatisfying things about the treatment in Section 2.5 of position, velocity, and acceleration was that these things happened in a weird one-dimensional way. This situation can be amended by using **vectors**. A vector is a lot like a point – it has coordinates and a dimension – but a vector is thought of as specifying a direction and a magnitude rather than just a point in space.

Example 5.18. A vector:

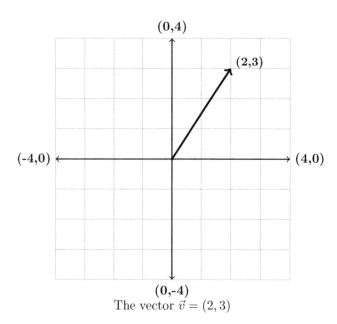

The vector $\vec{v} = (2,3)$

We use a small arrow over a variable name to denote that an object is a vector. So, the vector shown above is $\vec{v} = (2,3)$. In a way, vectors are like multi-dimensional numbers that let us work in an many dimensions as we need to.

Vectors have their own arithmetic. It's called vector arithmetic.

Knowledge Box 5.7.

Vector Arithmetic

If $\vec{v} = (v_1, v_2, \ldots, v_n)$ and $\vec{w} = (w_1, w_2, \ldots, w_n)$ are vectors in n-dimensions and c is a constant, then we define the following arithmetic on vectors.

- $c \cdot \vec{v} = (cv_1, cv_2, \ldots, cv_n)$ (scalar multiplication)
- $\vec{v} + \vec{w} = (v_1 + w_1, v_2 + w_2, \ldots, v_n + w_n)$ (vector addition)
- $\vec{v} - \vec{w} = (v_1 - w_1, v_2 - w_2, \ldots, v_n - w_n)$ (vector subtraction)
- $\vec{v} \cdot \vec{w} = v_1 w_1 + v_2 w_2 + \ldots + v_n w_n$ (dot product)

Example 5.19. If $\vec{v} = (1, 2, -1)$ and $\vec{w} = (0, 2, 4)$, compute $5 \cdot \vec{w} - \vec{v}$.

Solution:

$$5 \cdot (0, 2, 4) - (1, 2, -1) = (0, 10, 20) - (1, 2, -1) = (0 - 1, 10 - 2, 20 - (-1)) = (-1, 8, 21)$$

$$\Diamond$$

Example 5.20. If $\vec{v} = (-2, 1, 3)$ and $\vec{w} = (1, 2, 1)$, compute $\vec{v} \cdot \vec{w}$.

Solution:

$$\vec{v} \cdot \vec{w} = -2 \cdot 1 + 1 \cdot 2 + 3 \cdot 1 = -2 + 2 + 3 = 3$$

$$\Diamond$$

There are a number of useful algebraic rules for vector arithmetic.

Knowledge Box 5.8.

Vector Algebra

- $c \cdot (\vec{v} + \vec{w}) = c \cdot \vec{v} + c \cdot \vec{w}$
- $c \cdot (d \cdot \vec{v}) = (cd) \cdot \vec{v}$
- $\vec{v} + \vec{w} = \vec{w} + \vec{v}$
- $\vec{u} \cdot (\vec{v} + \vec{w}) = \vec{u} \cdot \vec{v} + \vec{u} \cdot \vec{w}$

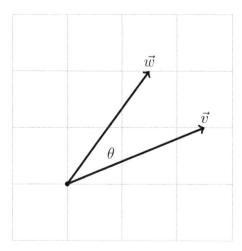

Figure 5.8. Two vectors.

Definition 5.4. The **length** or **magnitude** of a vector $\vec{v} = (v_1, v_2, \ldots, v_n)$, denoted $|\vec{v}|$, is given by

$$|\vec{v}| = \sqrt{v_1^2 + v_2^2 + \cdots + v_n^2}$$

Notice that this is similar to the formula for calculating distance. The length of a vector is the distance it spans, so this is a natural definition.

It turns out that there is a really useful property of one way we multiply two vectors, the dot product in Knowledge Box 5.7, that also uses the length of vectors. Suppose we have two vectors $\vec{v}$ and $\vec{w}$ as shown in Figure 5.8. Then, we can use the dot product together with the lengths of the vectors to calculate the angle between them.

Knowledge Box 5.9.

Formula for the angle between vectors

$$\cos(\theta) = \frac{\vec{v} \cdot \vec{w}}{|v||w|}$$

What makes this property so useful? If two vectors are at right angles, then the cosine of the angle between them is zero. This means we can detect right angles in a new and easy way and in any number of dimensions. In particular **two vectors are at right angles to one another if and only if their dot product is zero.**

Example 5.21. Which of the following pairs of vectors are at right angles to one another?

(a) $\vec{v} = (1, 1)$ and $\vec{w} = (1, -1)$

(b) $\vec{s} = (1, 2, 1)$ and $\vec{t} = (-1, 2, -3)$

(c) $\vec{a} = (3, 1, 2)$ and $\vec{b} = (2, -2, 3)$

Solution:

Compute dot products:

$$\vec{v} \cdot \vec{w} = (1, 1) \cdot (1, -1) = 1 - 1 = 0$$

So $\vec{v}$ and $\vec{w}$ are at right angles to one another.

$$\vec{s} \cdot \vec{t} = (1, 2, 1) \cdot (-1, 2, -3) = -1 + 4 - 3 = 0$$

So $\vec{s}$ and $\vec{r}$ are at right angles to one another.

$$\vec{a} \cdot \vec{b} = (3, 1, 2) \cdot (2, -2, 3) = 6 - 2 + 6 = 10$$

So $\vec{a}$ and $\vec{b}$ are not at right angles to one another.

$$\diamond$$

Definition 5.5. Objects at right angles to one another are also said to be **orthogonal** to one another.

So with our new word we can re-state this property of the dot product as follows. **Two vectors are orthogonal if and only if they have a dot product of zero.**

Definition 5.6. The **cross product** of $\vec{v} = (v_1, v_2, v_3)$ and $\vec{w} = (w_1, w_2, w_3)$ is defined to be:

$$\vec{v} \times \vec{w} = (\, v_2 w_3 - v_3 v_2, \quad v_3 w_1 - v_1 w_3, \quad v_1 w_2 - v_2 w_1 \,)$$

The cross product of two vectors is only defined in three dimensions. It is a very special purpose operator, but it is quite useful in physics. The following property is used to construct systems of directions.

Knowledge Box 5.10.

Orthogonality of the cross product

If $\vec{v}$ and $\vec{w}$ are non-zero vectors in three dimensions that are not scalar multiples of one another, then $\vec{v} \times \vec{w}$ is at right angles to (orthogonal to) both $\vec{v}$ and $\vec{w}$.

Example 5.22. Find a vector at right angles to both (1, 1, 2) and (3, 0, 4). Check your answer by taking the relevant dot products.

Solution:

Use the cross product:

$$(1, 1, 2) \times (3, 0, 4) = (1 \cdot 4 - 2 \cdot 0, 2 \cdot 3 - 1 \cdot 4, 1 \cdot 0 - 1 \cdot 3) = (4, 2, -3)$$

Check:

$$(1, 1, 2) \cdot (4, 2, -3) = 4 + 2 - 6 = 0 \checkmark$$
$$(3, 0, 4) \cdot (4, 2, -3) = 12 + 0 - 12 = 0 \checkmark$$

So the vector $(4, 2, -3)$ *is* orthogonal to $(1, 1, 2)$ and $(3, 0, 4)$.

$\Diamond$

Definition 5.7. A **unit vector** is a vector $\vec{v}$ with $|\vec{v}| = 1$.

This is an almost trivial definition, but, coupled with the next Knowledge Box rule, it captures an important feature of vectors.

Knowledge Box 5.11.

The unit vector in the direction of a given vector

If $\vec{v}$ is not zero, then

$$\frac{1}{|\vec{v}|} \cdot \vec{v}$$

is the unit vector in the direction of $\vec{v}$.

Unit vectors have a number of applications, but the one we will use the most often is that they capture a notion of **direction.** There are an infinite number of vectors in any given direction, but only two unit vectors, and they point in opposite directions from one another.

Example 5.23. Find the unit vector in the direction of $\vec{v} = (1, 2)$.

Solution:

Start with $|\vec{v}| = \sqrt{1^2 + 2^2} = \sqrt{5}$. Then the desired unit vector is

$$\frac{1}{\sqrt{5}} \cdot (1, 2) = \left(\frac{1}{\sqrt{5}}, \frac{2}{\sqrt{5}} \right)$$

$\Diamond$

We've now reviewed enough vector algebra to start working with vector functions.

Definition 5.8. A **vector function** is a vector $\vec{v}(t)$ whose coordinates are functions of a parameter t.

This is very close to the notion of a parametric curve. In fact, for some applications the two notions are interchangeable.

Example 5.24. Identify the points that appear on the graph of the vector function

$$\vec{v}(t) = (\cos(t), \sin(t))$$

Solution:

The unit circle centered at the origin.

Vector functions permit us to specify a large number of different types of curves through space. They share with parametric curves the property that they can encode systems that are not Cartesian functions at all.

Example 5.25. Plot the vector curve $\vec{w}(t) = (\sin(t) + t/5, \cos(t) + t/8)$.

Solution:

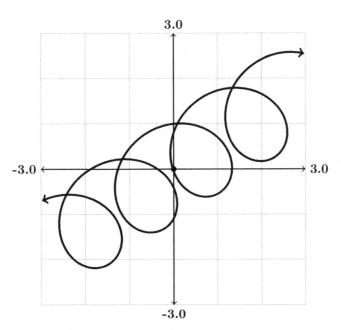

It is not too hard to understand this curve. The sine and cosine terms make a simple circle while the $t/5$ and $t/8$ terms create a line that moves the center of the circle. We can play a similar trick by combining two circles. We will make the circles both have distinct radii and have the particle travel around them at different speeds.

Example 5.26. Plot the vector curve $\vec{w}(t) = (2\sin(t/4) + \cos(t), 2\cos(t/4) + \sin(t))$.

Solution:

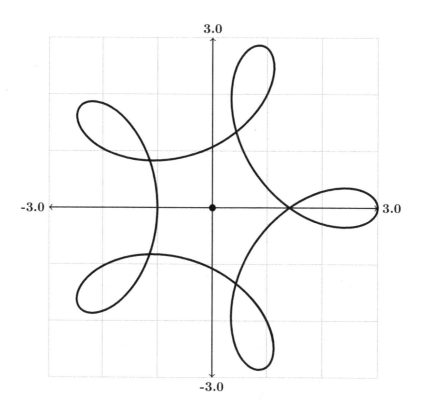

This curve is tracing a circle of radius 2 that moves at one fourth the rate of a circle of radius one that is circling the current point of the circle of radius two. It is potentially instructive to look at the previous example for several different speeds of motion of the particle on the smaller circle.

Example 5.27. Plot the vector curves:

- $\vec{w}(t) = (2\sin(t/4) + \cos(t), 2\cos(t/4) + \sin(t))$

- $\vec{v}(t) = (2\sin(t/4) + \cos(2t), 2\cos(t/4) + \sin(2t))$

- $\vec{s}(t) = (2\sin(t/4) + \cos(3t), 2\cos(t/4) + \sin(3t))$

- $\vec{t}(t) = (2\sin(t/4) + \cos(4t), 2\cos(t/4) + \sin(4t))$

Solution:

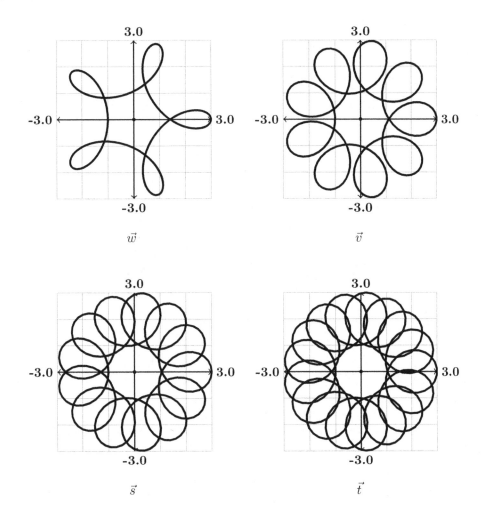

$\vec{w}$

$\vec{v}$

$\vec{s}$

$\vec{t}$

◊

5.3.1. Calculus with vector curves. A big advantage of vector functions is that they free us from the tyranny of abstract one-dimensional position, velocity, and acceleration.

Definition 5.9. If $\vec{v}(t) = (f_1(t), f_2(t), \ldots, f_n(t))$ then the **derivative** of $\vec{v}(t)$ is
$$\vec{v}'(t) = (f_1'(t), f_2'(t), \ldots, f_n'(t)).$$

Here is the payoff. If $\vec{s}(t)$ gives the position of a particle, then its velocity vector is $\vec{v}(t) = \vec{s}'(t)$, and its acceleration vector is $\vec{a}(t) = \vec{v}'(t)$. The derivative-based relationships between position, velocity, and acceleration hold for vector functions just as they do for ordinary functions. The difference is that we may now describe motion in complex two and three dimensional paths.

Example 5.28. Suppose the path of a particle is given by the vector function:
$$\vec{s}(t) = (\sin(t), \cos(2t))$$
Plot the particle's path and find its velocity and acceleration vectors.

Solution:

The plot looks like this:

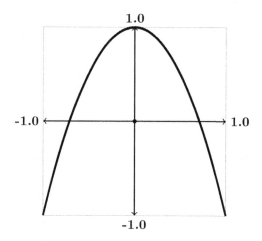

Take derivatives to find
$$\vec{v}(t) = (\cos(t), -2\sin(2t))$$
and
$$\vec{a}(t) = (-\sin(t), -4\cos(2t))$$

◇

What do the vector velocity and acceleration mean? They give (as their magnitude) the *amount* of velocity or acceleration, but they also tell us which direction in space the velocity or acceleration is going.

Example 5.29. If a particle has position $\vec{s}(t) = (2\cos(t), \sin(t))$, plot the position curve and show the velocity vectors – starting at the curve – at times $t = \pi/6, 2\pi/3, 7\pi/6$, and $5\pi/3$.

Solution:

Taking a derivative we get that the velocity vector is $\vec{v} = (-2\sin(t), \cos(t))$. So, let's plot the curve and the vectors at the specified times.

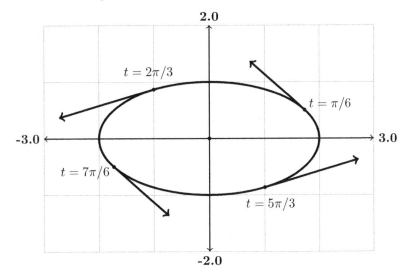

Notice that these vectors are tangent to the curve and also show which direction the curve is oriented. They show the instantaneous velocity of the particle following the curve.

$\diamond$

Example 5.30. Find the times when the position curve $\vec{s}(t) = (2\sin(t), \cos(t) + 1)$ has velocity of the greatest magnitude.

Solution:

First, take a derivative to obtain the velocity function:
$$\vec{v}(t) = (2\cos(t), -\sin(t))$$
The magnitude of the velocity is then
$$|(2\cos(t), -\sin(t))| = \sqrt{4\cos^2(t) + \sin^2(t)} = \sqrt{3\cos^2(t) + 1}$$

Note that the second step above used the Pythagorean identity, $\sin^2(t) + \cos^2(t) = 1$. Now we need to know when this is largest... Since $0 \le \cos^2(t) \le 1$ it will simply happen when $\cos^2(t) = 1$. This means that $\cos(t) = \pm 1$. So the times are $t = k\pi$ for any whole number k.

$\diamond$

Problems

Problem 5.35. *For each possible pair of the following four vectors, say which are orthogonal.*

- $\vec{v} = (1,1)$
- $\vec{r} = (1,2)$
- $\vec{s} = (2,-1)$
- $\vec{w} = (5,-5)$

Problem 5.36. *For each possible pair of the following four vectors, say which are orthogonal.*

- $\vec{v} = (1,1,1)$
- $\vec{r} = (-2,2,0)$
- $\vec{s} = (3,1,-4)$
- $\vec{w} = (1,-2,1)$

Problem 5.37. *Plot the following vector curves. You are free to use software.*

(a) $\vec{v}(t) = (3\sin(t/3) + \cos(t),\ 3\cos(t/2) + \sin(t))$

(b) $\vec{w}(t) = (3\sin(t/3) + \cos(2t),\ 3\cos(t/2) + \sin(2t))$

(c) $\vec{s}(t) = (2\sin(t/3) + 2\cos(2t),\ 2\cos(t/2) + 2\sin(2t))$

(d) $\vec{q}(t) = (\sin(t/2) + 3\cos(3t),\ \cos(t/2) + 3\sin(3t))$

(e) $\vec{a}(t) = (3\cos(2t), 3\sin(t))$

(f) $\vec{b}(t) = (t + \sin(t), \cos(t))$

Problem 5.38. *Example 5.28 shows the graph of the position function*

$$\vec{s}(t) = (\sin(t), \cos(2t))$$

and it looks like a part of a parabola. Demonstrate that it is a parabola by finding the Cartesian form, including the domain on which the curve is defined.

Problem 5.39. *If $\vec{s}(t) = (2t, \cos(t))$ is the position of a point, then plot the curve traced by the point on $[0, 2\pi]$ and show the velocity vectors at each multiple of $\pi/6$ in the interval.*

Problem 5.40. *For each of the following position vectors, find the velocity and acceleration vectors.*

(a) $\vec{s}(t) = (3t + 5, 2t + 6)$

(b) $\vec{s}(t) = (t^2 + t + 1, 5 - t)$

(c) $\vec{s}(t) = (\sin(t), \cos(2t))$

(d) $\vec{s}(t) = (\tan^{-1}(t), t^2)$

(e) $\vec{s}(t) = (1 + t + t^2, 1 - t + t^2)$

(f) $\vec{s}(t) = (t\cos(t), t\sin(t))$

Problem 5.41. *Find when a particle whose position is given by*

$$\vec{s}(t) = (\sin(t + \pi/3), \cos(t))$$

is traveling parallel to the y-axis.

Problem 5.42. *Find a Cartesian function with the same graph as:*

$$\vec{w}(t) = (3t + 1, 5t + 2)$$

Problem 5.43. *Find a Cartesian function with the same graph as*

$$\vec{u}(t) = (t^2, 1 - 3t)$$

or give a reason it is impossible.

Problem 5.44. *Show, by argumentation, that a vector function, all of whose components are linear functions of t, has the same graph as a Cartesian line.*

Problem 5.45. *The vector function $(\cos(t), \sin(t), t)$ is a twisting path. First explain what curve this vector function traces out of we ignore the third coordinate and then do your best to explain what the shape of the curve is.*

Problem 5.46. *If $f(t)$ is a function with domain $-\infty < t < \infty$, then describe the vector function $(f(t), f(t), f(t))$ as best you can.*

Chapter 6

The Arithmetic and Geometry of Polynomials

Polynomials are a rich family of functions. They include lines and quadratic equations – functions that we've studied in a good deal of detail because they are useful for so many things. Polynomials are defined everywhere on the real line; they are continuous, differentiable, and just generally nice to work with. As we will see in Chapter 13, polynomials can be used to approximate any continuous function. In this chapter we will learn to work with polynomials more closely as well as introducing a couple of techniques that are generally useful in the context of polynomials.

In Section 6.1 we give *Newton's method* which is a general technique for finding the roots of an equation. In Section 6.3 we use polynomials to demonstrate a rule, La'Hospital's rule, that is useful for evaluating limits in general. This section should be covered even if the other material on polynomials is not of interest.

6.1. Polynomial Arithmetic

Consider the problem of computing $(x^2 + x + 2)^2$. We might do this in the following fashion:

$$(x^2 + x + 2)(x^2 + x + 2) = x^2(x^2 + x + 2) + x(x^2 + x + 2) + 2(x^2 + x + 2)$$

$$= x^4 + x^3 + 2x^2 + x^3 + x^2 + 2x + 2x^2 + 2x + 4$$

$$= x^4 + 2x^3 + 5x^2 + 4x + 4$$

Which is a lot of algebra! If we made the polynomial higher degree it would get worse. Let's try this instead.

```
                        1   1   2

                        1   1   2

                        -   -   -

                        2   2   4

                    1   1   2   .

                1   1   2   .   .

                -   -   -   -   -

                1   2   5   4   4
```

Which also shows $(x^2 + x + 2)^2 = x^4 + 2x^3 + 5x^2 + 4x + 4$. But how? In the "long multiplication" table the positions correspond to powers of x. The rightmost place is the 1's place, the next is the x's place, then the x^2's place and so on. It is very similar to the algorithm for multiplying numbers with pencil-and-paper, but *there is no carry*. Why no carry? Because when you add up a whole lot of terms of the form ax^m you just get many x^m's; it never piles up to the point where you get some x^{m+1}'s. In other words, this is actually easier than the multiplication algorithm for numbers.

Knowledge Box 6.1.

Fast Polynomial Multiplication

- Place the coefficients of the polynomials in the first two rows.

- In the following rows, scale the coefficients of the first polynomial by each of the coefficients of the second, one per row.

- The first new row lines up with the ones above it; after that shift left one place for each row.

- Add up the columns for the rows you just generated.

- This creates a last row that is the coefficients of the product.

Example 6.1. Find
$$(x^3 + 4x^2 + 2x - 3) \times (x^2 + 3x + 2)$$

Solution:

```
              1   4   2  -3
                  1   3   2
              -   -   -   -
                  2   8   4  -6
              3  12   6  -9  .
          1   4   2  -3   .   .
          -   -   -   -   -   -
          1   7  16  11  -5  -6
```

So:
$$(x^3 + 4x^2 + 2x - 3) \times (x^2 + 3x + 2) = x^5 + 7x^4 + 16x^3 + 11x^2 - 5x - 6$$

$\Diamond$

This technique is enormously faster than imposing the distributive law and collecting terms, and because it's structured it is easier to avoid errors. There is one potential sand trap – when a term is missing you must fill in a zero for it. Let's do another example that demonstrates this.

Example 6.2. Find $(x^2 + x + 1) \times (x^2 + 4)$.

Solution:

```
              1   1   1
              1   0   4
              -   -   -
              4   4   4
          0   0   0   .
      1   1   1   .   .
      -   -   -   -   -
      1   1   5   4   4
```

So:
$$(x^2 + x + 1) \times (x^2 + 4) = x^4 + x^3 + 5x^2 + 4x + 4$$

$\Diamond$

A similar technique – that is more likely to be familiar to you – can also be used to divide polynomials. It is sometimes called **synthetic division**.

Example 6.3. Compute $(x^3 + 6x^2 + 11x + 6) \div (x + 2)$.

Solution:

$$
\begin{array}{rrrrr}
 & & 1 & 4 & 3 \\
 & & - & - & - & - \\
1 \ \ 2 \ \mid & 1 & 6 & 11 & 6 \\
 & 1 & 2 & & \\
 & & - & - & \\
 & & 4 & 11 & 6 \\
 & & 4 & 8 & \\
 & & & - & - \\
 & & & 3 & 6 \\
 & & & 3 & 6 \\
 & & & & - & - \\
 & & & & 0
\end{array}
$$

So:

$$(x^3 + 6x^2 + 11x + 6) \div (x + 2) = x^2 + 4x + 3$$

$\Diamond$

These techniques for multiplication and division of polynomials are essentially book-keeping devices. They don't give you new capabilities, but they make existing capabilities more reliable.

Remember that we often want to find the roots of a function. For instance, when optimizing or sketching curves we want to to be able to solve $f(x) = 0$, $f'(x) = 0$, and $f''(x) = 0$. Being able to rapidly multiply and divide polynomials makes these tasks easier. Back in Chapter 1 we learned that "If $f(x)$ is a polynomial and $f(c) = 0$ for some number c, then $(x - c)$ is a factor of $f(x)$." This is the famous **root-factor theorem**. One thing these new techniques do not do is tell us what value of c to try. Plugging in values of c and looking for zeros often works – if there is some whole number c that is a root.

Sir Issac Newton worked out a formula that, given a value close to a root, can move it closer.

Knowledge Box 6.2.

Newton's method for finding roots

If x_0 is close to a root of $f(x)$, then the sequence generated by using the formula $x_{i+1} = x_i - \dfrac{f(x_i)}{f'(x_i)}$ will approach the nearby root.

Example 6.4. Use Newton's method with an initial guess of $x_0 = 1$ to approximate a root of $f(x) = x^2 - 2$.

Solution:

To start with, we need to find the Newton's method formula. Since $f'(x) = 2x$ we get that

$$x_{i+1} = x_i - \frac{x_i^2 - 2}{2x_i}$$

Now compute:

$$x_1 = 1 - \frac{1 - 2}{2} = 1.5$$

$$x_2 = 1.5 - \frac{2.25 - 2}{3} \cong 1.4166666667$$

$$x_3 \cong 1.4142157$$

$$x_4 \cong 1.4142126$$

$$x_5 \cong 1.4142126$$

So by the fifth updating the approximation has stabilized at a value that agrees with $\sqrt{2}$ up to seven decimals. Since the roots of $f(x) = x^2 - 2$ are $\pm\sqrt{2}$ this is a nice demonstration that the technique works. Although we are demonstrating Newton's method on polynomials, it will work for any function that is continuous and differentiable near a root we are trying to find.

◇

Newton's method – or other more sophisticated root finding methods – are often built into a calculator. By showing the values for x_1, x_2, and so on, coding the formula into a spreadsheet is a relatively low work method of doing the calculations. Once you code the formula you can just let the spreadsheet perform the iterations.

Example 6.5. Find the Newton's method formula for approximating roots of:

$$f(x) = x^3 + 3x - 1$$

Solution:

Just plug into the form for the Newton's method formula:

$$x_{i+1} = x_i - \frac{x_i^3 + 3x_i - 1}{3x_i^2 + 3}$$

If we start with $x_0 = 1.0$ we get:

$x_0 = 1$
$x_1 = 0.5$
$x_2 = 0.36111111$
$x_3 = 0.32917086$
$x_4 = 0.32335786$
$x_5 = 0.32237954$
$x_6 = 0.32221744$
$x_7 = 0.32219065$
$x_8 = 0.32218623$
$x_9 = 0.32218549$
$x_{10} = 0.32218537$
$x_{11} = 0.32218535$
$x_{12} = 0.32218535$

At which point the number has stopped changing and so is the root to 8 decimals. Verify that $f(x)$ has only one root by graphing it.

$\Diamond$

The picture on the cover of this book demonstrates how Newton's method can be used to make pretty pictures. It plots points on the complex plane with the real part of a number plotted on the x-axis, and the imaginary part on the y-axis. It uses the polynomial:

$$x^6 + 2.43 \cdot x^4 - 5.8644 \cdot x^2 - 10.4976$$

This polynomial has six roots. Points are colored based on which root they converge to when used as a starting guess for Newton's method.

Problems

Problem 6.1. *Compute the following polynomial products.*

(a) $(x^3 + 2x^2 + x + 3) \times (x + 5)$

(b) $(x^2 + 4x - 1) \times (x^2 - 4x + 1)$

(c) $(x^3 + 6x^2 + 11x + 6) \times$
    $(x^3 - 6x^2 + 11x - 6)$

(d) $(x^3 + 3x + 7) \times (x^4 + x + 2)$

(e) $(x + 1)^3 \times (x + 2)^3$

(f) $(x^2 + x + 1)^3$

Problem 6.2. *Perform the following polynomial divisions.*

(a) $(x^4 + 4x^3 + 6x^2 + 4x + 1) \div$
    $(x + 1)$

(b) $(x^4 - x^3 + 2x^2 + x + 3) \div$
    $(x^2 + x + 1)$

(c) $x^4 + x^3 + 2x^2 + x + 1) \div (x^2 + 1)$

(d) $(x^5 + 8x^4 + 21x^3 + 35x^2 + 28x + 15) \div$
    $(x^2 + 2x + 3)$

(e) $(x^6 - x^5 + 2x^4 - x^3 + 2x^2 - x + 1)$
    $\div (x^2 - x + 1)$

(f) $(x^6 - 1) \div (x - 1)$

Problem 6.3. *Use Newton's method to approximate the roots of the polynomial*

$$f(x) = x^3 - 3x + 1$$

using initial guesses of $x_0 = -2, 0,$ and 2. These should generate three distinct roots.

Problem 6.4. *Use Newton's method to find a root of:*

$$g(x) = x^5 - 5x^2 - 6$$

Problem 6.5. *Earlier in the text it was asserted that a polynomial of odd degree must have at least one root. Explain why.*

Problem 6.6. *Find all real roots of the following polynomials.*

(a) $f(x) = x^4 - 2x^3 - 2x^2 - 2x - 3$

(b) $g(x) = x^3 - 6x^2 + 11x - 6$

(c) $h(x) = x^6 - 64$

(d) $q(x) = x^4 - 5x^2 + 4$

(e) $r(x) = x^4 - 8x^2 + 5x + 6$

(f) $s(x) = x^3 - 6x^2 + 12x - 8$

Problem 6.7. *Find a polynomial of the form*

$$x^4 + ax^3 + bx^2 + xc + d$$

with none of $a, b, c,$ or d zero that has no real roots at all.

Problem 6.8. *Suppose that $p(x)$ is a polynomial. How many roots does $p(x)^2 + 1$ have? Justify your answer with one or more sentences.*

Problem 6.9. *Consider the polynomial*

$$q(x) = x^3 - 25x$$

First find its three roots. Then figure out how many roots $q(x)^2 - 1$ has.

Problem 6.10. *Find a polynomial with roots at $x = 1, 2, 3, 4,$ and 5.*

Problem 6.11. *Find a polynomial with roots at $x = 1, 2, 3, 4,$ and 5 that does not take on any negative values.*

Problem 6.12. *Given the Newton's method formula for a polynomial:*

$$x_{i+1} = \frac{2x_i^3 - x_i^2 + 4}{3x_i^2 - 2x_i - 4}$$

what is the polynomial?

Problem 6.13. *If the Newton's method formula for a polynomial is*

$$x_{i+1} = \frac{3x_i^4 + x_i^2 - 7}{4x_i^4 + 10x_i}$$

what is the polynomial?

6.2. Qualitative properties of polynomials

Much of the work we do in a calculus class consists of calculating things. Some of the examples and homework problems are chosen to demonstrate the power of instead *deducing* things. In this section we will learn a number of properties of polynomials that help us deduce things about polynomials. Our first task is understanding how the roots of a polynomial affect the shape of its graph.

Knowledge Box 6.3.

Rolle's Theorem

Suppose a continuous, differentiable function $y = f(x)$ has roots at $x = a$ and at $x = b$. Then for some c, $a \leq c \leq b$, $f'(c) = 0$.

The best way to understand Rolle's theorem is with a pictorial example. The function in Figure 6.1 has a root at $x = -2$ and one at $x = 3$. At $c = -\frac{1}{3}$ it has a critical value and so a horizontal tangent line.

Colloquially, Rolle's theorem says there is (at least) one horizontal tangent between two roots. While we use it to describe polynomials, Rolle's theorem actually applies to all continuous differentiable functions.

The pictures in Figures 6.2 and 6.3 show examples of even and odd degree polynomials with the smallest possible number of roots. Knowledge Box 6.4 is a rephrasing of Knowledge Box 1.24.

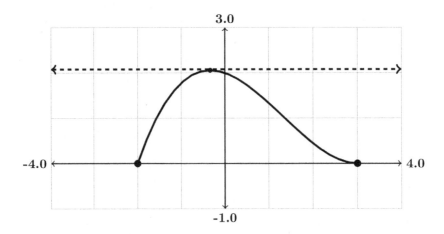

Figure 6.1. Illustration of Rolle's Theorem.

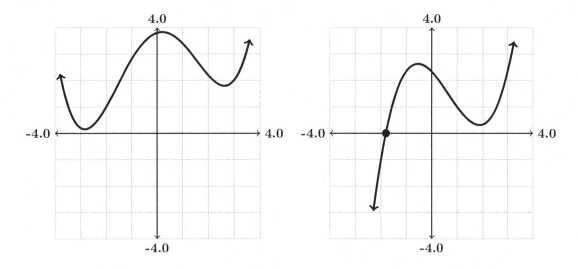

Figure 6.2. Fourth degree, no roots.   **Figure 6.3.** Third degree, one root.

Knowledge Box 6.4.

The number of roots of a polynomial

- An odd degree polynomial has at least one root.

- An even degree polynomial can have no roots.

- A polynomial of degree n has at most n roots.

- Any number of roots between the minimum and maximum number of roots are possible.

The fact that a polynomial of odd degree must have at least one root follows from the way its limits at infinity act. An odd degree polynomial must go to both of $\pm\infty$, which means that, on its way from one infinity to the other, it must pass through zero.

What does Rolle's theorem tell us about polynomials? Between any two roots there must be a horizontal tangent. Whenever there is a horizontal tangent, the function changes direction, so Rolle's theorem tells us that the polynomial must change directions between roots.

A polynomial of degree n has *at most* $n-1$ horizontal tangents. This follows from the fact that the derivative has degree $n-1$ and, so, at most $n-1$ roots. The hill-tops and valley-bottoms of a polynomial correspond to the roots of its derivative.

Example 6.6. What is the smallest number of horizontal tangents a polynomial $p(x)$ may have?

Solution:

Since horizontal tangents are roots of $p'(x)$, the odd/even degree rules tell us that an odd degree polynomial may have *no* horizontal tangents, while an even degree polynomial must have at least *one*.

In fact $p(x) = x^n$ exemplifies this minimum number of horizontal tangents. The even powers of x have a single critical value at $x = 0$ that yields a sign chart of

$$(+\infty) - - - (0) + + + (+\infty)$$

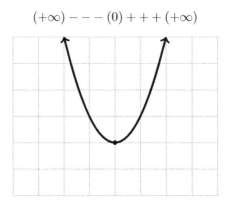

while the odd ones have a sign chart of

$$(-\infty) + + + (0) + + + (+\infty)$$

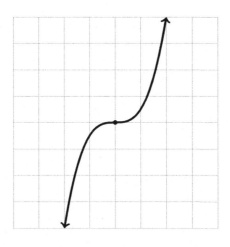

$\Diamond$

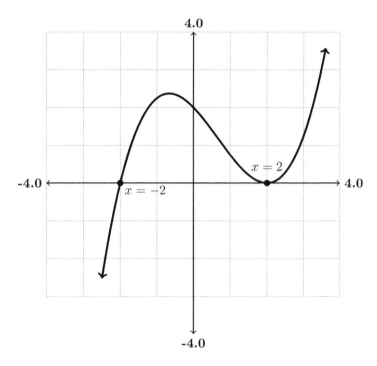

Figure 6.4. $p(x) = \dfrac{1}{4}(x+2)(x-2)^2$

6.2.1. Multiplicity of roots. What makes it possible for a polynomial of degree n to have less than n roots? Part of this is we can add constants to a polynomial so that it intersects the x-axis as little as possible. Another factor is that roots can be repeated. Look at the polynomial shown in Figure 6.4:

$$p(x) = \frac{1}{4}(x+2)(x-2)^2 = \frac{1}{4}x^3 - \frac{1}{2}x^2 - x + 2$$

The fact that the root at $x = 2$ is squared makes the graph bounce off of the x axis instead of passing through. This leads to a definition.

Definition 6.1. If $(x - a)$ divides a polynomial $p(x)$, then the highest power k so that $(x - a)^k$ divides $p(x)$ is the **multiplicity** of the root $x = a$ in $p(x)$.

The multiplicity of the root $x = 2$ in Figure 6.4 is two.

Knowledge Box 6.5.

The effect of root multiplicity

If $p(x)$ is a polynomial, and $x = a$ is a root of $p(x)$, then

- If the root at $x = a$ is of odd multiplicity, the graph of $p(x)$ passes through the x-axis at $x = a$.

- If the root at $x = a$ is of even multiplicity, the graph of $p(x)$ touches the x-axis and then bounces back on the same side it was on before at $x = a$.

Notice that, since the root at $x = 2$ in Figure 6.4 has multiplicity two, there is a local minimum and a critical value at $x = 2$. This effect is fairly general and can be summarized as in Knowledge Box 6.6.

Knowledge Box 6.6.

Roots with multiplicity and derivatives

If $p(x)$ is a polynomial, and $x = a$ is a root of $p(x)$ with multiplicity $k > 1$, then $x = a$ is also a root of $f'(x)$ with multiplicity $k - 1$.

Example 6.7. Verify that if

$$f(x) = (x - 2)^2(x + 2) = x^3 - 2x^2 - 4x + 8,$$

then $f'(2) = 0$.

Solution:

Compute:

$$f'(x) = 3x^2 - 4x - 4$$

Then $f'(2) = 12 - 8 - 4 = 0$, and 2 is a root of the derivative.

The examples in this section show us how we can construct polynomials with particular properties quite easily by putting in the roots we want with the correct multiplicities. This sort of information is made even more useful by the fact that polynomials can be used to approximate other functions on bounded intervals. This will be investigated in detail in Chapter 13.

Problems

Problem 6.14. *Show that the number of horizontal tangents between two roots – that do not have another root between them – must be odd.*

Problem 6.15. *Construct a polynomial that has more than one horizontal tangent between two roots that do not have another root between them.*

Problem 6.16. *Prove that a non-constant polynomial can have zero inflection values; do this by giving an example of one.*

Problem 6.17. *Show that a polynomial of degree n can have at most n − 2 inflection values.*

Problem 6.18. *Give an example of a non-constant polynomial with two roots that never takes on a negative values. Explain why your example is correct.*

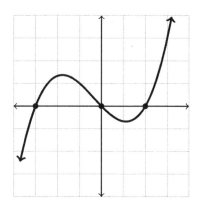

Problem 6.19. *Deduce as much as you can from the graph above. The grids are of length one unit. In particular what is the minimum degree and what is possible to deduce about the roots and their multiplicity?*

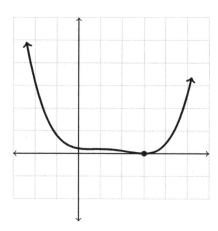

Problem 6.20. *Deduce as much as you can from the graph above. The grids are of length one unit. In particular what is the minimum degree and what is possible to deduce about the roots and their multiplicity?*

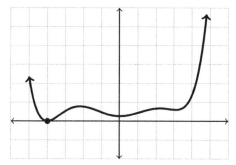

Problem 6.21. *Deduce as much as you can from the graph above. The grids are of length one unit. In particular what is the minimum degree and what is possible to deduce about the roots and their multiplicity?*

Problem 6.22. *What is the smallest possible degree for a polynomial that is an answer to Question 6.18.*

Problem 6.23. *Using the product rule, prove the statement: If $p(x)$ is a polynomial and $x = a$ is a root of $p(x)$ with multiplicity $k > 1$, then $x = a$ is also a root of $f'(x)$ (from Knowledge Box 6.6).*

6.3. L'Hôpital's rule; strange polynomials

In Chapter 3 we developed a rule of thumb about computing the limit at infinity of the ratio of two polynomials. If $p(x) = ax^n + \cdots$ is a polynomial of degree n and $q(x) = bx^m + \cdots$ is a polynomial of degree m the rule said:

$$\lim_{x \to \infty} \frac{p(x)}{q(x)} = \begin{cases} \pm\infty & n > m \\ a/b & n = m \\ 0 & n < m \end{cases}$$

This rule is a consequence of **L'Hôpital's rule**, a useful tool for computing limits that cannot be resolved other ways.

Knowledge Box 6.7.

L'Hôpital's rule at $x = c$

If $\lim_{x \to c} f(x) = 0$ and $\lim_{x \to c} g(x) = 0$ or
if $\lim_{x \to c} f(x) = \pm\infty$ and $\lim_{x \to c} g(x) = \pm\infty$, then

$$\lim_{x \to c} \frac{f(x)}{g(x)} = \lim_{x \to c} \frac{f'(x)}{g'(x)}$$

The hypotheses that the limits of the numerator and denominator are both going to zero or both going to some sort of infinity are critical. L'Hôpital's rule will give the wrong answer if the hypotheses are not satisfied. Let's do an example.

Example 6.8. Compute

$$\lim_{x \to 2} \frac{x^2 - 4}{x - 2}$$

with L'Hôpital's rule.

Solution:

We could resolve this limit by factoring and canceling, but, since the top and bottom of the fraction are both 0 at $x = 2$, L'Hôpital's rule applies.

$$\lim_{x \to 2} \frac{x^2 - 4}{x - 2} = \lim_{x \to 2} \frac{2x}{1} = \frac{4}{1} = 4$$

$\Diamond$

Knowledge Box 6.8.

L'Hôpital's rule as $x \to \infty$

If $\lim\limits_{x \to \infty} f(x) = 0$ and $\lim\limits_{x \to \infty} g(x) = 0$ or
if $\lim\limits_{x \to \infty} f(x) = \pm\infty$ and $\lim\limits_{x \to \infty} g(x) = \pm\infty$, then

$$\lim_{x \to \infty} \frac{f(x)}{g(x)} = \lim_{x \to \infty} \frac{f'(x)}{g'(x)}$$

This version of L'Hôpital's rule is the one that gives us the rule-of-thumb for the limit at infinity of ratios of polynomials. Let's do an example.

Example 6.9. Compute $\lim\limits_{x \to \infty} \dfrac{x^2 + 2x + 3}{3x^2 + 1}$.

Solution:

Both the numerator and denominator are going to infinity, so:

$$\lim_{x \to \infty} \frac{x^2 + 2x + 3}{3x^2 + 1} = \lim_{x \to \infty} \frac{2x + 2}{6x} \qquad \text{L'Hôpital's rule still applies}$$

$$= \lim_{x \to \infty} \frac{2}{6}$$

$$= \frac{1}{3}$$

And so the limit at infinity is $\dfrac{1}{3}$ – exactly what the rule of thumb gives us.

The reason the rule of thumb works for polynomials of equal degree is that L'Hôpital's rule applies until we have taken so many derivatives that the numerator and denominator are constants – at which point the accumulated values from bringing the power out front cancel – leaving the ratio of the highest degree coefficients as the limit.

At this point, we add a name to our mathematical vocabulary with the following definition.

Definition 6.2. A **rational function** is a function that is the ratio of polynomial functions.

We have been working with and graphing rational functions for some time now, so they are not new. We just have a name for them now. While L'Hôpital's rule was introduced to back-fill the rule of thumb for limits at infinity of rational functions (to use their new name), it in fact applies to all continuous, differentiable functions. Let's do a couple of examples.

Example 6.10. Compute $\lim\limits_{x\to 0} \dfrac{\sin(x)}{x}$.

Solution:

The numerator and denominator both go to zero, so L'Hôpital's rule applies.

$$\lim_{x\to 0} \frac{\sin(x)}{x} = \lim_{x\to 0} \frac{\cos(x)}{1} = 1$$

$\Diamond$

Example 6.11. Compute $\lim\limits_{x\to 0} \dfrac{\cos(x) - 1}{x}$.

Solution:

The numerator and denominator are both going to zero, so L'Hôpital's rule applies.

$$\lim_{x\to 0} \frac{\cos(x) - 1}{x} = \lim_{x\to 0} \frac{-\sin(x)}{1} = 0$$

$\Diamond$

As with every other technique we have learned, we may use algebraic rearrangement to permit us to extend the reach of L'Hôpital's rule.

Example 6.12. Compute $\lim\limits_{x\to 0} x \cdot \ln(x)$.

Solution:

This problem does not fit the form for L'Hôpital's rule, but we can coerce it into that form:

$$\lim_{x\to 0} x \cdot \ln(x) = \lim_{x\to 0} \frac{\ln(x)}{\dfrac{1}{x}} \qquad\qquad \text{Now in correct form } \frac{-\infty}{\infty} \text{ for L'Hôpital}$$

$$= \lim_{x\to 0} \frac{\dfrac{1}{x}}{\dfrac{-1}{x^2}} \qquad\qquad\qquad \text{Use L'Hôpital}$$

$$= \lim_{x\to 0} -\frac{x^2}{x} = \lim_{x\to 0} -\frac{x}{1} \qquad\qquad \text{Simplify}$$

$$= 0 \qquad\qquad\qquad\qquad \text{Done}$$

So x moves toward zero faster than $\ln(x)$ moves toward negative infinity as $x \to 0$.

$\Diamond$

It is possible for L'Hôpital to take you to a limit that clearly does not exist.

Example 6.13. Find $\displaystyle\lim_{x\to\infty} \frac{x^3+1}{x^2}$.

Solution:

The rule-of-thumb tells us this diverges to ∞, but let's work through this with L'Hôpital's rule and see where it goes.

$$\lim_{x\to\infty} \frac{x^3+1}{x^2} = \lim_{x\to\infty} \frac{3x^2}{2x}$$
$$= \lim_{x\to\infty} \frac{3}{2}x$$
$$\to \infty$$

L'Hôpital's rules also tells us this diverges to infinity.

$$\Diamond$$

Since this chapter is about polynomials, Knowledge Box 6.9 gives a number of useful polynomial identities.

Knowledge Box 6.9.

Useful polynomial identities

- $x^2 - a^2 = (x-a)(x+a)$
- $x^3 - a^3 = (x-a)(x^2 + ax + a^2)$
- $x^3 + a^3 = (x+a)(x^2 - ax + a^2)$
- $x^n - a^n = (x-a)(x^{n-1} + ax^{n-2} + \cdots a^{n-2}x + a^{n-1})$

That last identity is quite useful for adding up powers of a variable. Another form of the identity, with $a = 1$ is

$$\frac{x^n - 1}{x - 1} = x^{n-1} + x^{n-2} + \cdots + x + 1$$

How can this be applied?

Example 6.14. Compute

$$\sum_{k=0}^{20} 1.5^k$$

Solution:

$$\sum_{k=0}^{20} 1.5^k = \frac{1.5^{21} - 1}{1.5 - 1} = \frac{\dfrac{3^{21}}{2} - 1}{\dfrac{1}{2}} = \frac{3^{21} - 2^{21}}{2^{20}} = \frac{10458256051}{1048576} \cong 9973.8$$

$\Diamond$

Let's do another with slightly less insane numbers.

Example 6.15. Find

$$\sum_{k=0}^{11} 3^k$$

Solution:

$$\sum_{k=0}^{11} 3^k = \frac{3^{12} - 1}{3 - 1} = \frac{1}{2} \cdot 531440 = 265720$$

$\Diamond$

This will come up again when we study sequences and series in Chapter 13.

6.3.1. Strange polynomials. The goal of this section is to let us broaden the reach of what we know about polynomials to other functions. Suppose that we wish to solve the equation:

$$e^{2x} - 3e^x + 2 = 0$$

This is clearly not a polynomial equation, but we can find a polynomial with the following trick. Let $u = e^x$. Then, since $e^{2x} = (e^x)^2 = u^2$, we transform the original problem into:

$$u^2 - 3u + 2 = 0$$

This factors into $(u - 2)(u - 1) = 0$ so $u = 1, 2$. Reversing the transformation $e^x = 1, 2$. Take the log of both sides and the result becomes $x = \ln(1), \ln(2)$ or $x = 0, \ln(2)$.

This trick, called u-substitution, lets us turn one type of equation into another. This technique will become *very* useful in Chapter 7, but for now it permits us to solve a wider variety of equations. Some care is required.

Example 6.16. Find the solutions to the equation:

$$2\sin^2(x) - 5\sin(x) + 2$$

Solution:

Try $u = \sin(x)$ changing the problem to:

$$2u^2 - 5u + 2 = 0$$

As in the previous example, this factors, giving us $(2u - 1)(u - 2) = 0$, and we get that $u = 1/2$, 2. So far, so good. But when we reverse the transformation we get $\sin(x) = 1/2$, 2 and $\sin(x) = 2$ is impossible! We know that $-1 \le \sin(x) \le 1$. So even though there are two values for u, only one leads to a solution of the original equation.

$$\sin(x) = \frac{1}{2} \quad \text{or} \quad x = \frac{\pi}{6}$$

Giving the sole solution to the problem.

◊

Example 6.17. Solve

$$\ln(x)^2 - 3\ln(x) - 3 = 0$$

Solution:

Here the obvious substitution is $u = \ln(x)$, and we get $u^2 - 3u - 3 = 0$. Apply the quadratic equation and reverse the transformation:

$$u = \frac{3 \pm \sqrt{9 - 4(1)(-3)}}{2}$$

$$= \frac{3 \pm \sqrt{21}}{2} \qquad \qquad \text{Only one positive root}$$

$$\ln(x) = \frac{3 + \sqrt{21}}{2} \cong 3.79$$

$$x \cong 1.33$$

Note that we dismissed the negative root because only positive numbers have logs.

◊

This sometimes helps with polynomials as well.

Example 6.18. Solve $x^4 - 6x^2 + 8 = 0$.

Solution:

Take $u = x^2$. Then
$$u^2 - 6u + 8 = 0$$
Factor: $(u-2)(u-4) = 0$, and we get $u = 2, 4$. So $x^2 = 2, 4$ and thus $x = \pm\sqrt{2}, \pm 2$.

$\Diamond$

We can apply this idea to taking limits as well.

Example 6.19. Find
$$\lim_{x \to \infty} \frac{e^{2x} + 3e^x + 5}{2e^{2x} - 4}$$

Solution:

First, note that as $x \to \infty$, we also have $e^x \to \infty$. Knowing this, apply the transformation $u = e^x$ and we get:

$$\lim_{x \to \infty} \frac{e^{2x} + 3e^x + 5}{2e^{2x} - 4} = \lim_{u \to \infty} \frac{u^2 + 3u + 5}{2u^2 - 4} = \frac{1}{2}$$
$$\ldots\text{by using the rule of thumb.}$$

$\Diamond$

Problems

Problem 6.24. *For each of the following limits, use L'Hôpital's rule to resolve the limit.*

(a) $\lim\limits_{x \to 3} \dfrac{x-3}{x^2-9}$

(b) $\lim\limits_{x \to 2} \dfrac{x^3-x-6}{x^2-4}$

(c) $\lim\limits_{x \to 1} \dfrac{x^6-1}{x^2-1}$

(d) $\lim\limits_{x \to -2} \dfrac{x^3+8}{x^3+x^2-x+2}$

(e) $\lim\limits_{x \to 2} \dfrac{x^5-32}{x^2-4}$

(f) $\lim\limits_{x \to \pi} \dfrac{\cos(x)+1}{x-\pi}$

Problem 6.25. *Before, when we had to resolve a limit of the form $\frac{0}{0}$ by algebra using cancellation, we had to know the factorization of the numerator and denominator. Discuss: does L'Hôpital's rule make this process easier?*

Problem 6.26. *Find, in general, using L'Hôpital's rule*

$$\lim\limits_{x \to 1} \dfrac{x^n-1}{x^2-1}$$

Problem 6.27. *For each of the following limits, use L'Hôpital's rule to resolve the limit.*

(a) $\lim\limits_{x \to \infty} \dfrac{x^2+3x+5}{x^2-3x+4}$

(b) $\lim\limits_{x \to \infty} \dfrac{2e^x}{e^x+4}$

(c) $\lim\limits_{x \to \infty} \dfrac{x+e^x}{x+x^2+3e^x}$

(d) $\lim\limits_{x \to 1} (x-1) \cdot \ln(x^3-1)$

(e) $\lim\limits_{x \to 0} \dfrac{\sin(3x)}{2x}$

(f) $\lim\limits_{x \to 0} \dfrac{\cos(x)-1}{x^2}$

Problem 6.28. *Compute the following sums with a polynomial identity. Some of these may require using the identity twice.*

(a) $\sum\limits_{k=0}^{19} 0.9^k$

(b) $\sum\limits_{k=0}^{25} 1.2^k$

(c) $\sum\limits_{k=0}^{8} 7^k$

(d) $\sum\limits_{k=5}^{15} 1.5^k$

(e) $\sum\limits_{k=0}^{12} 0.1^k$

(f) $\sum\limits_{k=0}^{10} (-0.5)^k$

Problem 6.29. *Solve the following equations. Note that they are, in a sense, polynomials.*

(a) $2\sin^2(x) + \sin(x) = 0$

(b) $4\cos^2(x) - 3 = 0$

(c) $4\sin^3(x) - \sin(x) = 0$

(d) $e^{2x} - 7e^x + 12 = 0$

(e) $e^{3x} - 6e^{2x} + 11e^x - 6 = 0$

(f) $\ln^2(x) - 8\ln(x) + 15 = 0$

(g) $\cos^2(x) - \sin^2(x) - \cos^2(2x) = 0$

(h) $\tan^3(x) - \tan(x) = 0$

Problem 6.30. *Factor $x^6 - 729$ as far as you can.*

Problem 6.31. *Factor $x^8 - 256$ as far as you can.*

Problem 6.32. *If we know that $p(x)$ is a polynomial with n roots and $q(x)$ is a polynomial with m roots, with $n \le m$, then what do we know about the number of roots of $h(x) = p(x) \cdot q(x)$?*

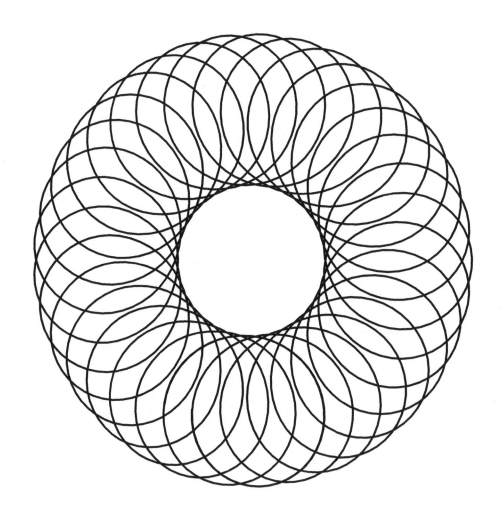

Chapter 7

Methods of Integration

The integration methods we have learned thus far are based on the fact that integrals are reversed derivatives. To some extent we can increase the reach of the "reverse derivative" technique by setting things up with algebra. This chapter on methods of integration will introduce the most useful and broadly applicable of thousands of integration methods that have been found over the last several centuries.

7.1. u-Substitution

Every integral method is a reversed derivative, and u-substitution is the reverse of the chain rule. We introduced u-substitution in Chapter 6 to permit us to solve more complex equations with polynomial techniques. For integrals the technique is much the same, except that we need to worry about the differential. Let's start with an example.

Example 7.1. Find $\displaystyle\int 2xe^{x^2} \cdot dx$.

Solution:

This is not an integral for which we already have a form. Set $u = x^2$ and then compute $du = 2x \cdot dx$. These pieces make up all of the integral and we can now solve the problem as follows.

$$\int 2xe^{x^2} \cdot dx = \int e^{x^2} \cdot (2x \cdot dx) \qquad\qquad \text{Set it up}$$

$$= \int e^u \cdot du \qquad\qquad \text{Substitute}$$

$$= e^u + C \qquad\qquad \text{Integrate}$$

$$= e^{x^2} + C \qquad\qquad \text{At the end, reverse the substitution.}$$

$\Diamond$

Recall that you can check an integral by taking a derivative, so

$$\left(e^{x^2} + C\right)' = e^{x^2} \cdot 2x = 2xe^{x^2}$$

and we see the substitution worked to let us do this integral. Notice that the chain rule re-created du for us.

Let's do some more examples.

Example 7.2. Find $\displaystyle\int \cos(2x) \cdot dx$.

Solution:
We know that the integral of cosine is sine. So the only problem is the $2x$. This is where to make our u-substitution. If we let $u = 2x$ then $du = 2dx$ so $dx = \dfrac{1}{2}du$. Now we substitute and integrate:

$$\int \cos(2x) \cdot dx = \int \cos(u)\frac{1}{2}du$$

$$= \frac{1}{2}\int \cos(u) \cdot du$$

$$= \frac{1}{2}\sin(u) + C$$

$$= \frac{1}{2}\sin(2x) + C$$

$$\Diamond$$

This last example has a very general form.

Example 7.3. Suppose that $F(x) + C = \displaystyle\int f(x) \cdot dx$. Then what is $\displaystyle\int f(ax + b) \cdot dx$?

Solution:

Let $u = ax + b$ and so $du = a \cdot dx$ and $dx = \dfrac{1}{a} \cdot du$. Then:

$$\int f(ax + b) \cdot dx = \int f(u) \cdot \frac{1}{a} \cdot du$$

$$= \frac{1}{a}\int f(u) \cdot du$$

$$= \frac{1}{a}F(u) + C$$

$$= \frac{1}{a}F(ax + b) + C$$

$$\Diamond$$

This is important enough to put in a Knowledge Box. What we are doing is correcting for composing a function with a linear function.

Knowledge Box 7.1.

u-substitution to correct a linear composition

If $\displaystyle\int f(x) \cdot dx = F(x) + C$, then

$$\int f(ax + b) \cdot dx = \frac{1}{a}F(ax + b) + C$$

Example 7.4. Find $\displaystyle\int e^{0.5x-1}dx$.

Solution:

$$2e^{0.5x-1} + C$$

by directly applying the linear correction with $a = 0.5$ and $b = -1$.

◇

This example shows how much calculation the linear correction technique saves.

Sometimes a u substitution is not obvious.

Example 7.5. Find

$$\int \frac{dx}{x\ln(x)}$$

Solution:

The key to this one is remembering that $\dfrac{d}{dx}\ln(x) = \dfrac{1}{x}$ making $u = \ln(x)$ a natural choice to try. Then $du = \dfrac{dx}{x}$ and we can integrate.

$$\int \frac{dx}{x\ln(x)} = \int \frac{1}{\ln(x)} \cdot \frac{dx}{x}$$
$$= \int \frac{1}{u} \cdot du$$
$$= \ln(u) + C$$
$$= \ln(\ln(x)) + C$$

◇

Sometimes the substitution is so unobvious that people clearly figured out the integral another way. The following example has this property.

Example 7.6. Find

$$\int \sec(x) \cdot dx.$$

Solution:

We begin with an algebraic transformation.

$$\int \sec(x) \cdot dx = \int \sec(x) \frac{\sec(x) + \tan(x)}{\sec(x) + \tan(x)} \cdot dx$$

$$= \int \frac{\sec^2(x) + \sec(x)\tan(x)}{\tan(x) + \sec(x)} \cdot dx$$

At this point let $u = \tan(x) + \sec(x)$ so that

$$du = \left(\sec^2(x) + \sec(x)\tan(x)\right) dx$$

which makes this all:

$$\int \sec(x) \cdot dx = \int \frac{du}{u}$$

$$= \ln(|u|) + C$$

$$= \ln(|\tan(x) + \sec(x)|) + C$$

The absolute value bars are needed because the value of these functions can be negative.

◊

We also now know enough to capture the integral of another of the basic functions.

Example 7.7. Compute $\int \tan(x) \cdot dx.$

Solution:

Since $\tan(x) = \dfrac{\sin(x)}{\cos(x)}$ we let $u = \cos(x)$ obtaining $du = -\sin(x)dx$ or $-du = \sin(x)dx$. Plugging in the substitution we get

$$\int \tan(x) \cdot dx = \int \frac{-du}{u} = -\int \frac{du}{u}$$

$$= -\ln(|u|) + C$$

$$= \ln\left(\frac{1}{|u|}\right) + C$$

$$= \ln\left(\frac{1}{|\cos(x)|}\right) + C$$

$$= \ln(|\sec(x)|) + C$$

Taking the minus sign inside the log is a negative first power or reciprocal, which is how we get to the secant.

◊

Knowledge Box 7.2.

Basic trig integrals from *u*-substitution

- $\displaystyle\int \tan(x) \cdot dx = \ln(|\sec(x)|) + C$

- $\displaystyle\int \sec(x) \cdot dx = \ln(|\sec(x) + \tan(x)|) + C$

It is possible to have fairly obvious *u*-substitution problems.

Example 7.8. Find

$$\int x(x^2 + 3)^6 \cdot dx.$$

Solution:

Let $u = x^2 + 3$, then $du = 2x \cdot dx$ and so $\dfrac{1}{2}du = x \cdot dx$. Integrate:

$$\int x(x^2 + 3)^6 \cdot dx = \int (x^2 + 3)^6 \cdot (x \cdot dx)$$

$$= \int u^6 \frac{1}{2} du$$

$$= \frac{1}{2} \int u^6 \cdot du$$

$$= \frac{1}{2} \times \frac{1}{7} u^7 + C$$

$$= \frac{1}{14}(x^2 + 3)^7 + C$$

$$\diamond$$

As always, it is possible to need algebra to set up a *u*-substitution, often in less apocalyptic ways than the integral of secant, earlier in this section.

Example 7.9. Compute

$$\int \frac{x^2 + 2x + 1}{x^2 + 1} \cdot dx$$

Solution:

$$\int \frac{x^2 + 2x + 1}{x^2 + 1} \cdot dx = \int \frac{x^2 + 1}{x^2 + 1} dx + \frac{2x}{x^2 + 1} dx$$
$$= \int dx + \int \frac{2x}{x^2 + 1} dx$$
$$= x + \int \frac{2x}{x^2 + 1} dx + C$$

At this point let $u = x^2 + 1$ so that $du = 2x \cdot dx$ and finish

$$\int \frac{x^2 + 2x + 1}{x^2 + 1} \cdot dx = x + \int \frac{du}{u} + C$$
$$= x + \ln(u) + C$$
$$= x + \ln(x^2 + 1) + C$$

Done! If you have a rational function whose denominator is not higher degree, dividing first is a good way to simplify things.

$$\Diamond$$

Another use of u-substitution is to integrate odd powers of sine and cosine.

Example 7.10. Find
$$\int \cos^3(x) \cdot dx.$$

Solution:

$$\int \cos^3(x) \cdot dx = \int \cos^2(x) \cdot \cos(x) \cdot dx$$

$$= \int (1 - \sin^2(x)) \cdot \cos(x) \cdot dx$$

$$= \int \cos(x) \cdot dx - \int \sin^2(x) \cos(x) \cdot dx$$

$$= \sin(x) - \int \sin^2(x) \cos(x) \cdot dx$$

Let $u = \sin(x)$ so that $du = \cos(x) \cdot dx$

$$= \sin(x) - \int u^2 \cdot du$$

$$= \sin(x) - \frac{1}{3}u^3 + C$$

$$= \sin(x) - \frac{1}{3}\sin^3(x) + C$$

$\Diamond$

We also have enough machinery to integrate even powers of sine and cosine now. We will learn how to integrate odd powers in Section 7.3.

Example 7.11. Find
$$\int \sin^2(x) \cdot dx.$$

Solution:

$$\int \sin^2(x) \cdot dx = \int \frac{1}{2}(1 - \cos(2x)) \cdot dx = \frac{1}{2}(x - \frac{1}{2}\sin(2x)) + C$$

We are using a trig identity for the second step and the answer is $\frac{1}{2}x - \frac{1}{4}\sin(2x) + C$.

$\Diamond$

7.1.1. Substitution in definite integrals. An issue we have not dealt with is substitution in definite integrals. There are two viable approaches to this topic.

- Substitute back to the original variable before plugging in the limits.

- Apply the substitution to the limits to get new limits.

So far we have always substituted back to the original variable to finish the integral.

Example 7.12. Compute

$$\int_0^3 x\sqrt{x^2 + 4} \cdot dx$$

Solution:

This is a simple u-substitution. Let $u = x^2 + 4$, so that $du = 2x \cdot dx$ and $\frac{1}{2}du = x \cdot dx$. Applying the substitution to the limits we get that 0 to 3 becomes $u(0) = 4$ and $u(3) = 13$. This means the integral is

$$\int_0^3 x\sqrt{x^2 + 4} \cdot dx = \int_4^{13} \sqrt{u} \cdot \frac{1}{2}du$$

$$= \frac{1}{2}\int_4^{13} u^{1/2} \cdot du$$

$$= \frac{1}{2} \cdot \frac{2}{3} u^{3/2}\Big|_4^{13}$$

$$= \frac{1}{3}\left(13^{3/2} - 4^{3/2}\right)$$

$$= \frac{1}{3}\left(\sqrt{2197} - 8\right)$$

$$\cong 12.96$$

$\Diamond$

Example 7.13. Compute

$$\int_0^1 \frac{e^x}{e^{2x} + 1} \cdot dx$$

Solution:

Let $u = e^x$ so that $du = e^x \cdot dx$. The limits become $u(0) = 1$ and $u(1) = e$. The integral becomes:

$$\int_0^1 \frac{e^x}{e^{2x} + 1} \cdot dx = \int_1^e \frac{du}{u^2 + 1}$$

$$= \tan^{-1}(u)\Big|_1^e$$

$$= \tan^{-1}(e) - \tan^{-1}(0)$$

$$= \tan^{-1}(e)$$

$$\cong 1.218$$

$\Diamond$

Problems

Problem 7.1. *Integrate the following by using u-substitution. Explicitly state u.*

(a) $\int \sin(3x) \cdot dx$

(b) $\int \sec^2(5x+1) \cdot dx$

(c) $\int_0^4 \frac{x}{x^2+1} \cdot dx$

(d) $\int_1^3 4x \cdot (2x^2+1)^5 \cdot dx$

(e) $\int_1^4 \frac{1}{x \cdot \ln^2(x)} \cdot dx$

(f) $\int (\sin(x) + \cos(x))^2 \cdot dx$

Problem 7.2. *Integrate the following.*

(a) $\int \cos(\pi x) \cdot dx$

(b) $\int e^{3x+2} \cdot dx$

(c) $\int \sin(14x-5) \cdot dx$

(d) $\int_0^1 (3x+4)^{12} \cdot dx$

(e) $\int \sec(2x) \tan(2x) \cdot dx$

(f) $\int_0^1 \frac{dx}{3x+5}$

Problem 7.3. *Show that if*
$$F(x) + C = \int f(x) \cdot dx, \text{ then}$$
$$\int x \cdot f(x^2+a) \cdot dx = \frac{1}{2}F(x^2+a) + C$$

Problem 7.4. *Compute*
$$\int \cot(x) \cdot dx.$$

Problem 7.5. *Compute*
$$\int \csc(x) \cdot dx.$$

Problem 7.6. *Compute*
$$\int_0^{\pi/4} \sin^4(x) \cos(x) \cdot dx.$$

Problem 7.7. *Compute*
$$\int_0^{\pi/2} \cos^4(x) \cdot dx.$$

Problem 7.8. *Compute*
$$\int \cos^5(x) \cdot dx.$$

Problem 7.9. *Compute*
$$\int \sin^3(x) \cos^3(x) \cdot dx.$$

Problem 7.10. *Compute*
$$\int \tan^3(x) \sec^2(x) \cdot dx.$$

Problem 7.11. *Compute*
$$\int \frac{\sin^3(x)}{\cos^5(x)} \cdot dx.$$

Problem 7.12. *Integrate the following by using u-substitution, using algebra as needed. Explicitly state u.*

(a) $\int xe^{x^2} \cdot dx$

(b) $\int_0^1 \frac{x^2+4x+2}{x^2+2} \cdot dx$

(c) $\int \frac{x^5}{x^2+1} \cdot dx$

(d) $\int \frac{\ln^2(x) + 4\ln(x) + 2}{x} \cdot dx$

(e) $\int \frac{\sin^2(x)}{\cos^2(x)+1} \cdot dx$

(f) $\int_0^2 \frac{2x+1}{(x^2+x+1)^5} \cdot dx$

(g) $\int \cos^7(2x) \sin(x) \cos(x) \cdot dx$

(h) $\int \frac{e^x}{e^x+1} \cdot dx$

(i) $\int \frac{\tan^{-1}(x)}{x^2+1} \cdot dx$

7.2. Integration by Parts

Probably the easiest of the complex derivative rules is the product rule. The reversed product rule is a technique called **integration by parts**. Let's derive it.

$$(UV)' = V \cdot dU + U \cdot dV \qquad \text{This is just the product rule}$$
$$U \cdot dV = (UV)' - V \cdot dU \qquad \text{Rearrange}$$
$$\int U \cdot dV = \int (UV)' - V \cdot dU \qquad \text{Integrate both sides}$$
$$\int U \cdot dV = UV - \int V \cdot dU \qquad \text{This is the formula}$$

Knowledge Box 7.3.

Integration by parts

$$\int U \cdot dV = UV - \int V \cdot dU$$

The technique for using this rule is not obvious from its statement. The "parts" are U and V together with their derivatives. Examples are needed.

Example 7.14. Compute $\int x \cdot \cos(x) \cdot dx$.

Solution:

Choose $U = x$ and $dV = \cos(x) \cdot dx$. Then $dU = dx$ and $V = \int \cos(x) \cdot dx = \sin(x)$. When doing integration by parts we add the $+C$ as the last step of the integration process. Now that we have the parts and their derivatives, we apply the integration by parts formula.

$$\int U \cdot dV = UV - \int V \cdot dU$$
$$\int x \cdot \cos(x) \cdot dx = x \cdot \sin(x) - \int \sin(x) \cdot dx$$
$$= x \cdot \sin(x) - (-\cos(x)) + C$$
$$= x \cdot \sin(x) + \cos(x) + C$$

Let's check this result by taking the derivative:

$$(x \cdot \sin(x) + \cos(x) + C)' = \sin(x) + x \cdot \cos(x) - \sin(x) = x \cdot \cos(x)$$

So the method worked.

$\diamond$

There is a substantial strategic component to choosing the parts U and dV when doing integration by parts. You take the derivative of U, and you must integrate dV, and when you're done it would be lovely if the result could be integrated without too much difficulty. In general you choose U so that differentiation will make a problem go away. Let's do another simple example.

Example 7.15. Find
$$\int x\,e^x \cdot dx.$$

Solution:
We can integrate e^x, and x goes away if we take its derivative. So, choose $U = x$, $dV = e^x \cdot dx$. This means $dU = dx$ and $V = e^x$. Apply the formula:
$$\int U \cdot dV = UV - \int V \cdot dU$$
$$\int xe^x \cdot dx = xe^x - \int e^x \cdot dx$$
$$= xe^x - e^x + C$$

$$\Diamond$$

Sometimes the choice of parts is not obvious. Let's pick up another basic integral using integration by parts.

Example 7.16. Compute
$$\int \tan^{-1}(x) \cdot dx$$

Solution:
At first it might look like there are no parts, but there are. Choose $U = \tan^{-1}(x)$ and $dV = dx$. Then $dU = \dfrac{dx}{x^2 + 1}$ and $V = x$. Apply the formula:
$$\int U \cdot dV = UV - \int V \cdot dU$$
$$\int \tan^{-1}(x) \cdot dx = x \cdot \tan^{-1}(x) - \int \frac{x}{x^2 + 1} \cdot dx \qquad \text{This is a substitution integral.}$$

Let $r = x^2 + 1$ then $\dfrac{1}{2}dr = x \cdot dx$
$$= x \cdot \tan^{-1}(x) - \frac{1}{2} \int \frac{dr}{r} \cdot dx$$
$$= x \cdot \tan^{-1}(x) - \frac{1}{2} \ln(r) + C$$
$$= x \cdot \tan^{-1}(x) - \frac{1}{2} \ln(x^2 + 1) + C$$

Notice that since U and dU are used in integration by parts, we used "r-substitution" instead of u-substitution for the part of the integral that needed substitution.

$$\Diamond$$

Knowledge Box 7.4.

Integral of the arctangent function

$$\int \tan^{-1}(x) \cdot dx = x \cdot \tan^{-1}(x) - \frac{1}{2}\ln(x^2 + 1) + C$$

Example 7.17. Compute $\int x^2 \cdot \ln(x) \cdot dx$.

Solution:

Tactically, we know that $\ln(x)$ turns into a (negative) power of x when we take its derivative. So a natural choice to try is $U = \ln(x)$, $dV = x^2 \cdot dx$. This makes $dU = \dfrac{dx}{x}$ and $V = \dfrac{1}{3}x^3$. Apply the formula.

$$\int U \cdot dV = UV - \int V \cdot dU$$

$$\int x^2 \cdot \ln(x) \cdot dx = \frac{1}{3}x^3 \cdot \ln(x) - \int \frac{1}{3}x^3 \cdot \frac{dx}{x}$$

$$= \frac{1}{3}x^3 \cdot \ln(x) - \frac{1}{3}\int x^2 \cdot dx$$

$$= \frac{1}{3}x^3 \cdot \ln(x) - \frac{1}{9}x^3 + C$$

◇

Sometimes integration by parts must be applied more than once to finish a problem.

Example 7.18. Compute $\int x^2 \cdot e^x \cdot dx$.

Solution:

Since taking the derivative of a power of x at least makes it a smaller power of x, choose $u = x^2$, $dV = e^x \cdot dx$. Then $dU = 2x \cdot dx$ and $V = e^x$. Integrate by parts:

$$\int x^2 \cdot e^x \cdot dx = x^2 \cdot e^x - \int 2x \cdot e^x \cdot dx$$

$$= x^2 \cdot e^x - 2\int x \cdot e^x \cdot dx \qquad\qquad \text{Integrate by parts again.}$$

$$U = x; \ dV = e^x \cdot dx$$
$$dU = dx; \ V = e^x$$

$$= x^2 \cdot e^x - 2\left(x \cdot e^x - \int e^x \cdot dx\right)$$

$$= x^2 \cdot e^x - 2x \cdot e^x + 2e^x + C$$

◇

Notice that the second integration by parts in this example was very similar to the one in Example 7.15. We could have simply plugged that result in. The example was worked in full to show how two integrations by parts are needed.

Let's consider for a minute what happened in the last example. To deal with the x^2 part of the integral we integrated by parts twice – reducing the power by one in each step. That means that, for example,

$$\int x^5 \cdot e^x \cdot dx$$

would require integration by parts *five times*. Fortunately, there is a shortcut. What happens if we integrate

$$\int f(x) \cdot e^x \cdot dx$$

Choose $U = f(x)$ and $dV = e^x \cdot dx$. Then $dU = f'(x)$ and $V = e^x$. Integrate by parts and we get:

$$\int f(x) \cdot e^x \cdot dx = f(x)e^x - \int f'(x)e^x \cdot dx$$

Which, used correctly, is a remarkable shortcut.

Knowledge Box 7.5.

Shortcut for $\int p(x)e^x \cdot dx$

Suppose that $p(x)$ is a polynomial. Then

$$\int p(x)e^x \cdot dx = \left(p(x) - p'(x) + p''(x) - p'''(x) + \cdots\right)e^x + C$$

The formula in Knowledge Box 7.5 comes from applying the formula for $\int p(x)e^x \cdot dx$ many times. Let's use the shortcut in an example.

Example 7.19. Find

$$\int x^5 \cdot e^x \cdot dx.$$

Solution:

Applying the shortcut we get that

$$\int x^5 \cdot e^x \cdot dx = \left(x^5 - 5x^4 + 20x^3 - 60x^2 + 120x - 120\right)e^x + C$$

All we need to do is take alternating signs of x^5 and its derivatives. If the polynomial is not just a power of x this is a little more complicated.

◊

Example 7.20. Find $\int (x^2 + 2x + 2)e^x\,dx$.

Solution:

$$p(x) = x^2 + 2x + 2$$
$$p(x)' = 2x + 2$$
$$p(x)'' = 2$$

So

$$\int (x^2 + 2x + 2)e^x = (x^2 + 2x + 2 - (2x + 2) + 2)e^x + C$$
$$= (x^2 + 2)e^x + C$$

The key fact is that if you keep taking derivatives of a polynomial, then at some point you get to zero. With a non-polynomial function the shortcut can produce an infinite object.

Now we look at another technique, **circular integration by parts**.

Example 7.21. Compute $\int \sin(x)e^x \cdot dx$.

Solution:
There is no natural choice of parts. So pick anything and plunge in. Let $U = \sin(x)$ and $dV = e^x \cdot dx$. Then $dU = \cos(x) \cdot dx$ and $V = e^x$.

$$\int \sin(x)e^x \cdot dx = \sin(x)e^x - \int \cos(x)e^x \cdot dx$$

Let: $U = \cos(x)$ and $dV = e^x \cdot dx$ (Second integration by parts.)

Then: $dU = -\sin(x) \cdot dx$ and $V = e^x$

$$= \sin(x)e^x - \left(\cos(x)e^x - \int -\sin(x)e^x \cdot dx \right)$$

$$\int \sin(x)e^x \cdot dx = (\sin(x) - \cos(x))e^x - \int \sin(x)e^x \cdot dx$$

Notice that the remaining integral equals the left hand side.

This is the circular part.

So, $2\int \sin(x)e^x \cdot dx = (\sin(x) - \cos(x))e^x$

Divide by two and we get:

$$\int \sin(x)e^x \cdot dx = \frac{1}{2}(\sin(x) - \cos(x))e^x + C$$

Notice that we had to separately remember to put in the $+\,C$, because we finessed the final integral where we would normally have generated it.

$\Diamond$

The power of these methods of integration are expanded in the homework problems. We will also see a horrific example of circular integration by parts in Section 7.3.

Problems

Problem 7.13. *Do each of the following integrals using integration by parts.*

(a) $\int x \cdot \sin(2x) \cdot dx$

(b) $\int x \cdot e^{5x} \cdot dx$

(c) $\int \ln(x) \cdot dx$

(d) $\int x \cdot \ln(x) \cdot dx$

(e) $\int x \cdot \sin(x) \cos(x) dx$

(f) $\int \ln(x^2 + 1) \cdot dx$

Problem 7.14. *For Examples 7.15-7.18, verify the formulas by taking an appropriate derivative.*

Problem 7.15. *Compute* $\int x^n \ln(x) \cdot dx$ *for $n \geq 1$.*

Problem 7.16. *In this section we develop a shortcut for $\int p(x)e^x \cdot dx$ where $p(x)$ is a polynomial. Find the corresponding shortcut for*
$$\int p(x)e^{-x} \cdot dx$$

Problem 7.17. *In this section we develop a shortcut for $\int p(x)e^x \cdot dx$ where $p(x)$ is a polynomial. Find the corresponding shortcut for*
$$\int p(x)\sin(x) \cdot dx$$

Problem 7.18. *In this section we develop a shortcut for $\int p(x)e^x \cdot dx$ where $p(x)$ is a polynomial. Find the corresponding shortcut for*
$$\int p(x)\cos(x) \cdot dx$$

Problem 7.19. *Compute the following integrals using integration by parts.*

(a) $\int x^4 e^x \cdot dx$

(b) $\int (x^3 + 2x^2 + 3x + 4)e^x \cdot dx$

(c) $\int x^3 e^{-x} \cdot dx$

(d) $\int \left(x^2 + 1\right) e^{-x} \cdot dx$

(e) $\int x^2 \sin(x) \cdot dx$

(f) $\int x^5 \cos(x) \cdot dx$

Problem 7.20. *Compute the following integrals using integration by parts.*

(a) $\int \cos(x)e^x \cdot dx$

(b) $\int \sin(2x)e^x \cdot dx$

(c) $\int \cos(x)e^{3x} \cdot dx$

(d) $\int \sin(x)e^{-x} \cdot dx$

(e) $\int (\cos(x) + \sin(x)) e^x \cdot dx$

(f) $\int \cos(2x)e^{5x} \cdot dx$

Problem 7.21. *Compute*
$$\int x \cdot \tan^{-1}(x) \cdot dx$$

Problem 7.22. *Compute*
$$\int x^3 \cdot (x^2 + 1)^{10} \cdot dx$$

Problem 7.23. *Compute*
$$\int x \left(\cos(ax) + \sin(bx)\right) \cdot dx$$

Problem 7.24. *Compute*
$$\int \sin(ax)e^{bx} \cdot dx$$

7.3. Integrating Trig Functions

For completeness we start with a Knowledge Box of the integrals of trigonometric functions we have already obtained. Most of the section is about how to deal with products of powers of trig functions, and these integrals are building blocks.

Knowledge Box 7.6.

Trig-related integral forms

- $\int \sin(x) \cdot dx = -\cos(x) + C$

- $\int \cos(x) \cdot dx = \sin(x) + C$

- $\int \tan(x) \cdot dx = \ln|\sec(x)| + C$

- $\int \sec(x) \cdot dx = \ln|\tan(x) + \sec(x)| + C$

- $\int \sec^2(x) \cdot dx = \tan(x) + C$

- $\int \csc^2(x) \cdot dx = -\cot(x) + C$

- $\int \sec(x)\tan(x) \cdot dx = \sec(x) + C$

- $\int \csc(x)\cot(x) \cdot dx = -\csc(x) + C$

Odd powers of sine and cosine

Functions of this sort are integrated by exploiting the Pythagorean identity. We leave one power of the trig function to be dU and transform the remaining even powers via one of

$$\cos^2(x) = 1 - \sin^2(x)$$

or

$$\sin^2 = 1 - \cos^2(x)$$

as appropriate, transforming the problem into an integral of a polynomial function.

Example 7.22. Find $\displaystyle\int \cos^5(x) \cdot dx$.

Solution:

$$\int \cos^5(x) \cdot dx = \int \left(\cos^2(x)\right)^2 \cdot \cos(x) \cdot dx$$

$$= \int \left(1 - \sin^2(x)\right)^2 \cdot \cos(x) \cdot dx$$

Let $u = \sin(x)$, $du = \cos(x) \cdot dx$
$$= \int (1 - u^2)^2 \cdot du$$

$$= \int \left(1 - 2u^2 + u^4\right) \cdot du$$

$$= u - \frac{2}{3}u^3 + \frac{1}{5}u^5 + C$$

$$= \sin(x) - \frac{2}{3}\sin^3(x) + \frac{1}{5}\sin^5(x) + C$$

$$\Diamond$$

Odd powers of sine work the same way, but with the other function in a starring role.

Even powers of sine and cosine

These are much messier and rely on the power reduction identities:

$$\sin^2(x) = \frac{1 - \cos(2x)}{2}$$

$$\cos^2(x) = \frac{1 + \cos(2x)}{2}$$

They can always be used to transform an even power into a bunch of much lower powers, both odd and even. This means that even powers of trig functions often end up as furballs, but they can be done with patience and persistence.

Example 7.23. Compute:

$$\int \sin^4(x) \cdot dx$$

Solution:

$$\int \sin^4(x) \cdot dx = \int \left(\sin(x)^2\right)^2 \cdot dx$$

$$= \int \left(\frac{1 - \cos(2x)}{2}\right)^2 \cdot dx$$

$$= \frac{1}{4} \int \left(1 - 2\cos(2x) + \cos^2(2x)\right) \cdot dx$$

$$= \frac{1}{4} \int \left(1 - 2\cos(2x) + \frac{1 + \cos(4x)}{2}\right) \cdot dx$$

$$= \frac{1}{8} \int \left(2 - 4\cos(2x) + 1 + \cos(4x)\right) \cdot dx$$

$$= \frac{1}{8} \int \left(3 - 4\cos(2x) + \cos(4x)\right) \cdot dx$$

$$= \frac{1}{8} \left(3x - 4\frac{1}{2}\sin(2x) + \frac{1}{4}\sin(4x)\right) + C$$

$$= \frac{3}{8}x - \frac{1}{4}\sin(2x) + \frac{1}{32}\sin(4x) + C$$

So, ick, but possible.

$$\Diamond$$

Even powers of secant

These are pretty easy, again by hitting them with Pythagorean identities. Save one $\sec^2(x)$ to be du and apply

$$\sec^2(x) = 1 + \tan^2(x)$$

to set up the u-substitution $u = \tan(x)$.

Example 7.24. Compute: $\displaystyle\int \sec^8(x) \cdot dx$

Solution:

$$\int \sec^8(x) \cdot dx = \int \left(\sec^2(x)\right)^3 \cdot \sec^2(x) \cdot dx$$

$$= \int \left(1 + \tan^2(x)\right)^3 \cdot \sec^2(x) \cdot dx$$

$$\text{Let } u = \tan(x),\ du = \sec^2(x) \cdot dx$$

$$= \int \left(1 + u^2\right)^3 du$$

$$= \int \left(1 + 3u^2 + 3u^4 + u^6\right) du$$

$$= u + u^3 + \frac{3}{5}u^5 + \frac{1}{7}u^7 + C$$

$$= \tan(x) + \tan^3(x) + \frac{3}{5}\tan^5(x) + \frac{1}{7}\tan^7(x) + C$$

$$\Diamond$$

Odd powers of secant – these are typically a nightmare!

Example 7.25. Compute: $\displaystyle\int \sec^3(x) \cdot dx$

Solution:

$$\int \sec^3(x) \cdot dx = \int \left(1 + \tan^2(x)\right)\sec(x) \cdot dx$$

$$= \int \sec(x) \cdot dx + \int \tan^2(x)\sec(x) \cdot dx$$

$$= \ln|\sec(x) + \tan(x)| + \int \tan(x) \cdot \sec(x)\tan(x) \cdot dx$$

$$U = \tan(x); \quad dV = \sec(x)\tan(x) \cdot dx \qquad\qquad\qquad \text{Use parts}$$

$$du = \sec^2(x) \cdot dx; \quad V = \sec(x)$$

$$= \ln|\sec(x) + \tan(x)| + \sec(x)\tan(x) - \int \sec^3(x) \cdot dx$$

$$2\int \sec^3(x) \cdot dx = \ln|\sec(x) + \tan(x)| + \sec(x)\tan(x) \qquad\qquad \text{Circularize!}$$

$$\int \sec^3(x) \cdot dx = \frac{1}{2}\left(\ln|\sec(x) + \tan(x)| + \sec(x)\tan(x)\right) + C$$

As with all circular integrations by parts, we need to remember to finish with a $+C$. Higher odd powers are worse.

$$\Diamond$$

Powers of tangent.

Even powers of tangent can be transformed into even powers of secant by using $\tan^2(x) = \sec^2(x) - 1$.

Example 7.26. Compute:

$$\int \tan^4(x) \cdot dx$$

Solution:

$$\int \tan^4(x) \cdot dx = \int \left(\sec^2(x) - 1\right)^2 \cdot dx$$

$$= \int \left(\sec^4(x) - 2\sec^2(x) + 1\right) \cdot dx$$

$$= \int \sec^4(x) \cdot dx - 2 \int \sec^2(x) \cdot dx + \int dx$$

$$= \int (\tan^2 + 1)\sec^2 \cdot dx - 2 \int \sec^2(x) \cdot dx + \int dx$$

Let: $u = \tan(x)$, $du = \sec^2(x) \cdot dx$

$$= \int (u^2 + 1) \cdot du - 2 \int \sec^2(x) \cdot dx + \int dx$$

$$= \frac{1}{3}u^3 + u - 2\tan(x) + x + C$$

$$= \frac{1}{3}\tan^3(x) + \tan(x) - 2\tan(x) + x + C$$

$$= \frac{1}{3}\tan^3(x) - \tan(x) + x + C$$

◊

Odd powers of tangent can be reduced to lower odd powers of tangent. Assume $2n + 1 > 1$. Then:

$$\int \tan^{2n+1} \cdot dx = \int (\sec^2(x) - 1) \tan^{2n-1}(x) \cdot dx$$

$$= \int \tan^{2n-1}(x) \sec^2(x) \cdot dx - \int \tan^{2n-1}(x) \cdot dx$$

Let: $u = \tan(x),\ du = \sec^2(x) \cdot dx$

$$= \int u^{2n-1} \cdot du - \int \tan^{2n-1}(x) \cdot dx$$

$$= \frac{1}{2n} u^{2n} - \int \tan^{2n-1}(x) \cdot dx$$

$$= \frac{1}{2n} \tan^{2n}(x) - \int \tan^{2n-1}(x) \cdot dx$$

Since we also know that

$$\int \tan(x) \cdot dx = \ln|\sec(x)| + C$$

we can integrate any odd power of tangent by applying the above formula until the power is down to one.

Example 7.27. Find $\int \tan^5(x) \cdot dx$.

Solution:

$$\int \tan^5(x) \cdot dx = \frac{1}{4} \tan^4(x) - \int \tan^3(x) \cdot dx$$

$$= \frac{1}{4} \tan^4(x) - \left(\frac{1}{2} \tan^2(x) - \int \tan(x) \cdot dx \right)$$

$$= \frac{1}{4} \tan^4(x) - \frac{1}{2} \tan^2(x) + \ln|\sec(x)| + C$$

$\Diamond$

Mixed functions

When you wish to integrate a mixed product of trig functions, one strategy is to turn everything into sines and cosines and look for a good trig substitution. The fact that $u = \cos(x) \implies du = -\sin(x) \cdot dx$ and $u = \sin(x) \implies du = \cos(x) \cdot dx$ permits us to transform the trig integral into the integral of a rational function.

Example 7.28. Find

$$\int \sin(x) \tan^2(x) \cdot dx$$

Solution:

$$\int \sin(x) \tan^2(x) \cdot dx = \int \frac{\sin^3(x)}{\cos^2(x)} \cdot dx$$

$$= \int \frac{1 - \cos^2(x)}{\cos^2(x)} \cdot \sin(x) \cdot dx$$

Let $u = \cos(x)$, $du = -\sin(x) \cdot dx$

$$= \int \frac{1 - u^2}{u^2} \cdot (-du)$$

$$= -\int \left(\frac{1}{u^2} - 1 \right) du$$

$$= -\left(-\frac{1}{u} - u \right) + C$$

$$= \frac{1}{u} + u + C = \frac{1}{\cos(x)} + \cos(x) + C$$

$$= \sec(x) + \cos(x) + C$$

$$\Diamond$$

The mainstay of the techniques for integrating trig functions presented in this section are the application of trigonometric identities. There are literally an infinite number of problems and identities that can be used to solve them. The techniques of u-substitution and integration by parts also appear, arising naturally when we try to integrate trigonometric functions.

Many of the transformations created by u-substitution make the trig function integrals into rational and even polynomial function integrals. What, however, is the point of learning to integrate these functions? In the next section we will learn to transform integrals involving square roots into trigonometric integrals. Integration is one of the best examples of how mathematics builds on itself.

Problems

Problem 7.25. *Perform the following integrals.*

(a) $\int \cos^3(x) \cdot dx$

(b) $\int \sin^5(x) \cdot dx$

(c) $\int \cos^7(x) \cdot dx$

(d) $\int \cos^4(x) \cdot dx$

(e) $\int \sin^6(x) \cdot dx$

(f) $\int \sin^2(x) \cos^2(x) \cdot dx$

Problem 7.26. *Perform the following integrals.*

(a) $\int \sec^6(x) \cdot dx$

(b) $\int \tan^6(x) \cdot dx$

(c) $\int \tan^3(x) \sec^2(x) \cdot dx$

(d) $\int \tan^7(x) \cdot dx$

(e) $\int \cot^6(x) \cdot dx$

(f) $\int \sec^3(x) \tan^3(x) \cdot dx$

Problem 7.27. *Compute:*

$$\int \sec^5(x) \cdot dx$$

Problem 7.28. *Find the power reduction integral for*

$$\int \cot^{2n+1}(x) \cdot dx$$

that is analogous to the odd powers of tangent technique in this section.

Problem 7.29. *Compute*

$$\int \csc^3(x) \cdot dx$$

Problem 7.30. *Perform the following integrals.*

(a) $\int \cos^4(x) \sin^3(x) \cdot dx$

(b) $\int \cos^6(x) \tan^2(x) \sin(x) \cdot dx$

(c) $\int \cot(x) \csc^3(x) \cdot dx$

(d) $\int \sec^3(x) \sin^3(x) \cdot dx$

(e) $\int \dfrac{\tan^2(x)}{\sin(x)} \cdot dx$

(f) $\int \tan^5(x) sec(x) \cdot dx$

Problem 7.31. *Find:*

$$\int (\cos(x) + 1)^5 \cdot dx$$

Problem 7.32. *Find:*

$$\int \sin^2(x) \cdot \cos^5(x) \cdot dx$$

Problem 7.33. *Find:*

$$\int \sin^2(x) \cdot \cos^4(x) \cdot dx$$

Problem 7.34. *Find:*

$$\int \sin^4(x) \cdot \cos^5(x) \cdot dx$$

Problem 7.35. *Find*

$$\int \left(\sin^2(x) + 1\right)^8 \cos^3(x) \cdot dx$$

Problem 7.36. *Find*

$$\int \left(\cos^2(x) + 1\right)^8 \sin^3(x) \cdot dx$$

Problem 7.37. *Find*

$$\int \sin(x) \cos(x) \cdot dx$$

7.4. Trigonometric Substitution

We have the following integrals that use inverse trig functions:

$$\int \frac{dx}{x^2 + 1} \cdot dx = \tan^{-1}(x) + C$$

$$\int \frac{dx}{\sqrt{1 - x^2}} \cdot dx = \sin^{-1}(x) + C$$

and

$$\int \frac{dx}{|x|\sqrt{x^2 - 1}} \cdot dx = \sec^{-1}(x) + C$$

The goal of this section is to enlarge our library of integrals of this type. Let's start with an example that shows how to use **trigonometric substitution**.

Example 7.29. Compute

$$\int \frac{dx}{\sqrt{x^2 + 1}}$$

Solution:

Examine the following picture of a carefully labeled right triangle.

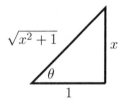

We see that $\dfrac{x}{1} = x = \tan(\theta)$, which means $dx = \sec^2(\theta) \cdot d\theta$. Also, $\dfrac{\sqrt{x^2 + 1}}{1} = \sqrt{x^2 + 1} = \sec(\theta)$. So, we can substitute and integrate.

$$\int \frac{dx}{\sqrt{x^2 + 1}} = \int \frac{\sec^2(\theta)}{\sec(\theta)} \cdot d\theta$$

$$= \int \sec(\theta) \cdot d\theta$$

$$= \ln(|\tan(\theta) + \sec(\theta)|) + C$$

$$= \ln(|x + \sqrt{x^2 + 1}|) + C$$

For the last substitution – back to x – we need to refer back to our substitution triangle at the beginning of the integral.

$$\Diamond$$

When making the substitution triangle remember that it has to obey the Pythagorean theorem and should include all the parts in your integral.

So what just happened? Whenever you have a radical like $\sqrt{ax^2 + b}$ or $\sqrt{b - ax^2}$ in an integral, it is possible to draw a right triangle that structures a substitution for you. This substitution transforms the integral into a trigonometric integral – which goes a long way toward explaining why we worked so hard on integrating trig functions in the last section. Let's do another example that uses a different type of right triangle.

Example 7.30. Compute:
$$\int \sqrt{1 - x^2} \cdot dx$$

Solution:

Here is the natural triangle for this problem:

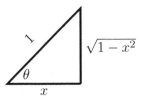

Then the substitutions are $x = \cos(\theta)$ and $\sqrt{1 - x^2} = \sin(\theta)$. Taking a derivative, we see that $dx = -\sin(\theta) \cdot d\theta$. Substituting all this into the problem yields:

$$\int \sqrt{1 - x^2} \cdot dx = \int \sin(\theta)(-\sin(\theta)) \cdot d\theta$$

$$= -\int \sin^2(\theta) \cdot d\theta$$

$$= -\frac{1}{2} \int (1 - \cos(2\theta)) \cdot d\theta$$

$$= \frac{1}{4} \sin(2\theta) - \frac{1}{2}\theta + C$$

But now we need to substitute back. $\sin(2\theta) = 2\sin(\theta)\cos(\theta)$ so that's not too bad. We also know that $x = \cos(\theta)$ so $\cos^{-1}(x) = \theta$. This is enough, and we get that:

$$\frac{1}{4} \sin(2\theta) - \frac{1}{2}\theta + C = \frac{1}{2} \sin(\theta)\cos(\theta) - \frac{1}{2}\theta + C$$

$$= \frac{1}{2} \left(x\sqrt{1 - x^2} - \cos^{-1}(x) \right) + C$$

Done!

◇

Notice that we have a choice of which formula goes on what leg of the right triangle. Sometimes one choice leads to a slightly easier problem than the other, so it is worth thinking about which substitution you will use.

Example 7.31. Compute

$$\int \frac{x^2}{\sqrt{4-x^2}} \cdot dx$$

Solution:

Here is the natural picture:

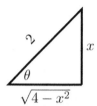

This makes the substitutions $\dfrac{x}{2} = \sin(\theta)$ and $\dfrac{\sqrt{4-x^2}}{2} = \cos(\theta)$ so that

$$\sqrt{4-x^2} = 2\cos(\theta)$$
$$x = 2\sin(\theta)$$
$$dx = 2\cos(\theta)d\theta$$

Plug all this in and integrate:

$$\int \frac{x^2}{\sqrt{4-x^2}} = \int \frac{4\sin^2(\theta)2\cos(\theta)}{2\cos(\theta)}d\theta$$
$$= 4\int \sin^2(\theta) \cdot d\theta$$
$$= 2\int (1-\cos(2\theta)) \cdot d\theta$$
$$= 2\left(\theta - \frac{1}{2}\sin(2\theta)\right) + C$$
$$= 2\theta - \sin(2\theta) + C$$

As before, $\theta = \sin^{-1}\left(\dfrac{x}{2}\right)$ and $\sin(2\theta) = 2\sin(\theta)\cos(\theta) = 2x\sqrt{4-x^2}$ so

$$\int \frac{x^2}{\sqrt{4-x^2}} \cdot dx = 2\sin^{-1}\left(\frac{x}{2}\right) - 2x\sqrt{4-x^2} + C$$

◊

Example 7.32. Directly verify the area formula for a circle by integration using Cartesian coordinates.

Solution:

It would be very easy to just verify the formula using polar coordinates, but for this example we are using Cartesian coordinates. Since a circle of radius R centered at the origin has the form $x^2 + y^2 = R^2$, our strategy is to take the function that graphs as the upper half of the circle

$$y = \sqrt{R^2 - x^2}$$

integrate it and double the result. Since it is an even function, we can actually do the problem

$$\text{Area} = 4 \int_0^R \sqrt{R^2 - x^2} \cdot dx$$

Here is the substitution triangle:

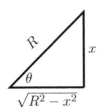

So the substitutions are:

$$\sqrt{R^2 - x^2} = R\cos(\theta)$$
$$x = R\sin(\theta)$$
$$dx = R\cos(\theta) \cdot d\theta$$

When $x = R$, we have that $\theta = \dfrac{\pi}{2}$. When $x = 0$, we have that $\theta = 0$. This gives us the following definite integral.

$$\text{Area} = 4 \int_0^{\pi/2} R\cos(\theta) \cdot R\cos(\theta) \cdot d\theta$$

$$= 4R^2 \int_0^{\pi/2} \cos^2(\theta) \cdot d\theta$$

$$= 2R^2 \int_0^{\pi/2} (1 + \cos(2\theta)) \cdot d\theta$$

$$= 2R^2 \left(\theta + \frac{1}{2}\sin(2\theta) \right)\Big|_0^{\pi/2}$$

$$= 2R^2 \left(\frac{\pi}{2} + \frac{1}{2}\sin(\pi) - 0 - 0 \right)$$

$$= \pi R^2$$

Which verifies the formula.

$$\Diamond$$

The next example will draw on a formula we already have – and paid blood for:

$$\int \sec^3(u) \cdot du = \frac{1}{2}\left(\sec(u)\tan(u) + \ln(|\sec(u) + \tan(u)|) \right) + C$$

Example 7.33. Compute:

$$\int \sqrt{x^2 + 1} \cdot dx$$

Solution:

For this one we get to re-use the first triangle picture:

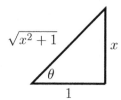

So the substitutions are:

$$\sqrt{x^2 + 1} = \sec(\theta)$$

$$x = \tan(\theta)$$

$$dx = \sec^2(\theta)d\theta$$

Substituting we get:

$$\int \sqrt{x^2 + 1} \cdot dx = \int \sec(\theta) \cdot \sec^2(\theta) \cdot d\theta$$

$$= \int \sec^3(\theta) \cdot d\theta$$

Use the known result

$$= \sec(\theta) \tan(\theta) + \ln(|\sec(\theta) + \tan(\theta)|) + C$$

$$= x\sqrt{x^2 + 1} + \ln(|x + \sqrt{x^2 + 1}|) + C$$

Without the known form, this one is a bear. This integral comes up again in Chapter 12.

◊

Example 7.34. Compute:

$$\int \frac{\sqrt{9 - x^2}}{x^2} \cdot dx$$

Solution:

Use the triangle:

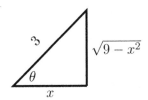

This gives us the following substitutions.

$$\sqrt{9 - x^2} = 3\sin(\theta)$$

$$x = 3\cos(\theta)$$

$$dx = -3\sin(\theta) \cdot d\theta$$

Which permits us to integrate.

$$\int \frac{\sqrt{9 - x^2}}{x^2} \cdot dx = \int \frac{3\sin(\theta) \cdot (-3\sin(\theta))}{9\cos^2(\theta)} \cdot d\theta$$

$$= -\int \frac{\sin^2(\theta)}{\cos^2(\theta)} \cdot d\theta$$

$$= -\int \tan^2(\theta) \cdot d\theta$$

$$= -\int (\sec^2(\theta) - 1) \cdot d\theta$$

$$= -\tan(\theta) + \theta + C$$

We can pull $\tan(\theta)$ directly off the triangle and, by the usual trick, $\theta = \cos^{-1}(x/3)$ making the result of substituting back to x:

$$-\frac{\sqrt{9-x^2}}{x} + \cos^{-1}(x/3) + C$$

$\Diamond$

The next example looks at a power of a radical.

Example 7.35. Compute:

$$\int \left(1-x^2\right)^{3/2} \cdot dx$$

Solution:

We get to re-use one of our earlier triangles:

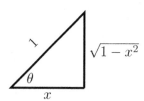

Making the substitutions:

$$\sqrt{1-x^2} = \sin(\theta)$$
$$x = \cos(\theta)$$
$$dx = -\sin(\theta) \cdot d\theta$$

Moving on to the integral,

$$\int \left(1 - x^2\right)^{3/2} \cdot dx = \int \sin^3(\theta) \cdot (-\sin(\theta)) \cdot d\theta$$

$$= -\int \sin^4(\theta) \cdot d\theta$$

Which is example 7.23.

$$= \frac{3}{8}\theta - \frac{1}{4}\sin(2\theta) + \frac{1}{32}\sin(4\theta) + C$$

$$= \frac{3}{8}\theta - \frac{1}{2}\sin(\theta)\cos(\theta) + \frac{1}{16}\sin(2\theta)\cos(2\theta) + C$$

$$= \frac{3}{8}\theta - \frac{1}{2}\sin(\theta)\cos(\theta) + \frac{1}{8}\sin(\theta)\cos(\theta)\left(\cos^2(\theta) - \sin^2(\theta)\right) + C$$

$$= \text{Using } \theta = \cos^{-1}(x)$$

$$= \frac{3}{8}\cos^{-1}(x) - \frac{1}{2}x\sqrt{1-x^2} + \frac{1}{8}x\sqrt{1-x^2}(x^2 - (1-x^2)) + C$$

$$= \frac{3}{8}\cos^{-1}(x) - \frac{1}{8}\left(4x\sqrt{1-x^2} + x\sqrt{1-x^2}(2x^2 - 1)\right) + C$$

$$= \frac{3}{8}\cos^{-1}(x) - \frac{1}{8}\left(4x\sqrt{1-x^2} + \sqrt{1-x^2}(2x^3 - x)\right) + C$$

$$= \frac{3}{8}\cos^{-1}(x) - \frac{1}{8}\left(\sqrt{1-x^2}(4x + 2x^3 - x)\right) + C$$

$$= \frac{3}{8}\cos^{-1}(x) - \frac{1}{8}\sqrt{1-x^2}\left(2x^3 + 3x\right) + C$$

$$= \frac{3}{8}\cos^{-1}(x) - \frac{x}{8}\sqrt{1-x^2}\left(2x^2 + 3\right) + C$$

◇

The really challenging part of this example was the back-substitution. We used two trig identities to get to where we could back-substitute:

$$\sin(2\theta) = 2\sin(\theta)\cos(\theta)$$
$$\cos(2\theta) = \cos^2(\theta) - \sin^2(\theta)$$

Trigonometric substitution is useful for square roots of sums and differences of squares. This is because these fit well onto the three sides of a right triangle. This permits us to translate to the domain of trigonometry where we can use identities to manipulate the form of the integral, often into something doable.

Problems

Problem 7.38. *For each of the following, show a substitution triangle that might work well. Do **not** do the integrals.*

(a) $\int x^5 \sqrt{1 + 4x^2} \cdot dx$

(b) $\int \frac{\sqrt{4 - 9x^2}}{x^3} \cdot dx$

(c) $\int (x^2 + 2x + 1)\sqrt{x^2 + 2x} \cdot dx$

(d) $\int \frac{x^2}{\sqrt{2 - 3x^2}} \cdot dx$

(e) $\int \frac{x}{\sqrt{x^2 + 1}} \cdot dx$

(f) $\int \frac{\sqrt{x^2 + 1}}{x^3} \cdot dx$

Problem 7.39. *Perform the following integrals. Not all require trig substitution, but for those that do, show the substitution triangle.*

(a) $\int x\sqrt{x^2 + 9} \cdot dx$

(b) $\int x^2 \sqrt{x^2 + 1} \cdot dx$

(c) $\int x^3 \sqrt{x^2 + 4} \cdot dx$

(d) $\int \frac{x^3}{\sqrt{9x^2 + 16}} \cdot dx$

(e) $\int \sqrt{1 + 4x^2} \cdot dx$

(f) $\int \frac{x^2}{(x^2 + 1)^{3/2}} \cdot dx$

Problem 7.40. *Compute the integral* $\int \frac{dx}{\sqrt{ax^2 + b^2}}$ *where a, b are non-zero constants.*

Problem 7.41. *Compute the integral:* $\int \sqrt{ax^2 + b^2} \cdot dx$, *where a, b are non-zero constants.*

Problem 7.42. *Perform the following integrals.*

(a) $\int \frac{\sqrt{1 - x^2}}{x} \cdot dx$

(b) $\int \frac{\sqrt{4 - x^2}}{x^2} \cdot dx$

(c) $\int \frac{dx}{\sqrt{9 - 4x^2}} \cdot dx$

(d) $\int x^3 \sqrt{1 - x^2} \cdot dx$

(e) $\int \frac{x^2}{\sqrt{16 - x^2}} \cdot dx$

(f) $\int \frac{x^3}{\sqrt{9 - x^2}} \cdot dx$

Problem 7.43. *Compute the integral*

$$\int \frac{dx}{\sqrt{a^2 - b^2 x^2}} \cdot dx$$

where a, b are non-zero constants.

Problem 7.44. *Compute the integral*

$$\int \sqrt{a^2 - b^2 x^2} \cdot dx$$

where a, b are non-zero constants.

Problem 7.45. *Perform the following integrals.*

(a) $\int \sqrt{\sin^2(x) + 1} \cdot \cos(x) \cdot \cdot dx$

(b) $\int e^{3x} \sqrt{e^{2x} + 4} \cdot dx$

(c) $\int (x + 1)\sqrt{x^2 + 2x + 2} \cdot dx$

(d) $\int \sqrt{1 - e^{2x}} \cdot dx$

(e) $\int \frac{\sqrt{1 - \tan^2(x)}}{\tan(x)} \sec^2(x) \cdot dx$

(f) $\int x \cdot \sqrt{1 - x^2}\sqrt{1 + x^2} \cdot dx$

Problem 7.46. *Compute*

$$\int x\sin^{-1}(x) \cdot dx$$

7.5. Partial Fractions

This section introduces a novel algebraic technique for setting up integrals. The first thing we need is a useful fact about polynomials over the real numbers.

Knowledge Box 7.7.

Factorization of polynomials

The only polynomials that do not factor over the real numbers are quadratics with no roots. Any polynomial of degree 3 or more factors into linear polynomials and quadratics that do not factor.

What can we do with this fact? Let's look at an example to set the stage.

Example 7.36. Compute

$$\int \frac{dx}{x^2 - 3x + 2}$$

Solution:

Notice that

$$\frac{1}{x-2} - \frac{1}{x-1} = \frac{(x-1)-(x-2)}{(x-1)(x-2)} = \frac{1}{x^2 - 3x + 2}$$

This means that:

$$\int \frac{dx}{x^2 - 3x + 2} = \int \frac{dx}{x-2} - \int \frac{dx}{x-1}$$

$$= \ln|x-2| - \ln|x-1| + C$$

The problem with this solution is that "notice" is not a recognized algebra technique. In order to gain the ability to "notice" in this fashion we need one additional fact. **Equal polynomials have equal coefficients**.

Suppose that

$$\frac{A}{x-1} + \frac{B}{x-2} = \frac{1}{(x-1)(x-2)}$$

Then, if we cross-multiply on the left side of the equation, we get

$$\frac{A(x-2) + B(x-1)}{(x-1)(x-2)} = \frac{1}{(x-1)(x-2)}$$

which simplifies to

$$(A+B)x - 2A - B = 1 = 0x + 1$$

Since these polynomials are equal we get the system of equations:

$$A + B = 0$$
$$-2A - B = 1$$
$$A = -1 \qquad \text{Second equation plus first.}$$
$$B = 1 \qquad \text{Substitute A}= -1 \text{ into the first equation.}$$

So $A = -1$ and $B = 1$ tell us the thing we noticed:

$$\frac{1}{x-2} - \frac{1}{x-1} = \frac{1}{x^2 - 3x + 2}$$

$$\Diamond$$

This is the basis of **integration by partial fractions**. The fact that a polynomial factors into linear and unfactorable quadratic terms means that we can always break a rational function into (i) polynomial terms and (ii) rational functions of the form

$$\frac{A}{x-u}$$

or

$$\frac{Bx + C}{x^2 + sx + t}$$

Integration by partial fractions is used to integrate rational functions.

Knowledge Box 7.8.

Steps of integration by partial fractions

1 If the numerator of the rational function is not lower degree than the denominator, divide to transform the problem into a polynomial plus a rational function with a higher degree denominator.
2 Factor the denominator.
3 Set the rational function equal to the sum of partial fraction terms.
4 Clear the denominator by cross multiplication.
5 Solve for the coefficients of the partial fraction terms based on the equal polynomials resulting from cross multiplication.
6 Perform the resulting integrals.

If a factor of the denominator is repeated, it gets one partial fraction term for each power of the denominator. Let's work some examples.

Example 7.37. Find:

$$\int \frac{dx}{x^2 - 4x + 3}$$

Solution:

The numerator is already lower degree. Factor the denominator. $x^2 - 4x + 3 = (x - 3)(x - 1)$ which gives us

$$\frac{1}{x^2 - 4x + 3} = \frac{A}{x - 1} + \frac{B}{x - 3}$$

meaning that

$$\frac{1}{x^2 - 4x + 3} = \frac{A(x - 3) + B(x - 1)}{x^2 - 4x + 3}$$

yielding the polynomial equality $1 = (A + B)x - 3A - B$. We get two equations – one for the coefficient of x and one for the constant term:

$$3A + B = -1 \qquad \text{for constant term}$$
$$A + B = 0 \qquad \text{for coefficient of } x$$
$$A = -B$$
$$-2B = -1$$
$$B = 1/2$$
$$A = -1/2$$

So

$$\frac{1}{x^2 - 4x + 3} = \frac{1}{2}\left(\frac{1}{x - 3} - \frac{1}{x - 1}\right)$$

Setting up the integral:

$$\int \frac{dx}{x^2 - 4x + 3} = \frac{1}{2}\int\left(\frac{1}{x - 3} - \frac{1}{x - 1}\right)dx$$
$$= \frac{1}{2}\left(\ln|x - 3| - \ln|x - 1|\right) + C$$

$\Diamond$

Example 7.38. Compute:

$$\int \frac{2x + 5}{x^2 - x}$$

Solution:

The numerator is already lower degree. Factor the denominator and set up the partial fractions.

$$\frac{2x + 5}{x(x - 1)} = \frac{A}{x} + \frac{B}{x - 1}$$

Cross multiply and extract the polynomials equations.

$$2x + 5 = A(x - 1) + Bx$$

which gives us the linear system

$$A + B = 2$$
$$5 = -A$$
$$A = -5$$
$$B = 7$$

So

$$\int \frac{2x+5}{x^2-x} = \int \frac{7}{x-1}dx - \int \frac{5}{x} \cdot dx$$

$$= 7\ln|x-1| - 5\ln|x| + C$$

$$\Diamond$$

The next example will be our first with a repeated root, letting us showcase the fact that partial fractions have one part for each point of multiplicity of a root.

Example 7.39. Compute

$$\int \frac{dx}{x^3 - x^2}$$

Solution:

We begin, as always, by factoring: $x^3 - x^2 = x^2(x-1)$. The root $x = 0$ has multiplicity two. This means that both x and x^2 are factors. So, the partial fraction decomposition is:

$$\frac{1}{x^3 - x^2} = \frac{A}{x} + \frac{B}{x^2} + \frac{C}{x-1}$$

So, once we clear the denominator ($= x^2(x-1)$) we get

$$1 = Ax(x-1) + B(x-1) + Cx^2$$

or

$$(A+C)x^2 + (B-A)x - B = 0x^2 + 0x + 1$$

The resulting system of equations is:

$$A + C = 0$$
$$B - A = 0$$
$$-B = 1$$

So $B = -1$, $A = -1$, and $C = 1$. Now we can integrate:

$$\int \frac{dx}{x^3 - x^2} = \int \left(-\frac{1}{x} - \frac{1}{x^2} + \frac{1}{x-1} \right) \cdot dx$$

$$= -\ln|x| + \frac{1}{x} + \ln|x-1| + C$$

$$\Diamond$$

Next we look at a shortcut that permits us to, in some cases, avoid having to solve the system of equations for the coefficients of the partial fraction decomposition. Instead, we plug the roots into the equation.

Example 7.40. Compute:

$$\int \frac{dx}{x^3 - 6x + 11x - 6}$$

Solution:

Start by factoring: $x^3 - 6x + 11x - 6 = (x-1)(x-2)(x-3)$ which gives us a partial fractions form:

$$\frac{1}{x^3 - 6x + 11x - 6} = \frac{A}{x-1} + \frac{B}{x-2} + \frac{C}{x-3}$$

Clearing the denominator we get

$$1 = A(x-2)(x-3) + B(x-1)(x-3) + C(x-1)(x-2)$$

which sets us up for the shortcut:

$$1 = A(-1)(-2) = 2A \qquad\qquad \text{When } x = 1$$
$$1 = B(1)(-1) = -B \qquad\qquad \text{When } x = 2$$
$$1 = C(2)(1) = 2C \qquad\qquad \text{When } x = 3$$

So $A = 1/2$, $B = -1$, and $C = 1/2$. Integrate:

$$\int \frac{dx}{x^3 - 6x + 11x - 6} = \int \left(\frac{1/2}{x-1} - \frac{1}{x-2} + \frac{1/2}{x-3} \right) \cdot dx$$

$$= \frac{1}{2} \ln|x-1| - \ln|x-2| + \frac{1}{2} \ln|x-3| + C$$

$$\Diamond$$

What happened? Let's summarize in a Knowledge Box.

Knowledge Box 7.9.

Finding the coefficients of a partial fractions decomposition

- Equal polynomials have equal coefficients for equal power terms.
- Equal polynomials are equal for any specific value of their variable.

The shortcut for finding the partial fraction constants is to plug in values of x that zero out all but one term. Warning: this isn't always possible. Terms with multiplicity above one or non-factorable quadratic terms can mess up the shortcut. The next set of examples involve integrating reciprocals of quadratics that don't factor. These will arise as part of partial fraction decompositions.

Example 7.41. Compute:

$$\int \frac{dx}{x^2 + 4x + 5}$$

Solution:

The denominator doesn't factor. Let's trying completing the square.

$$\int \frac{dx}{x^2 + 4x + 5} = \int \frac{dx}{x^2 + 4x + 4 + 1}$$

$$= \int \frac{dx}{(x+2)^2 + 1}$$

Let $u = x + 2$, so $du = dx$

$$= \int \frac{du}{u^2 + 1}$$

$$= \tan^{-1}(u) + C$$
$$= \tan^{-1}(x + 2) + C$$

$\Diamond$

Knowledge Box 7.10.

Integrating the reciprocal of unfactorable quadratics

1 Complete the square to place the quadratic in the form $a(x - b)^2 + c$

2 $\dfrac{1}{a(x-b)^2 + c} = \dfrac{1}{c\left(\frac{a}{c}(x-b)^2 + 1\right)} = \dfrac{1}{c\left(\left(\sqrt{\frac{a}{c}}(x-b)\right)^2 + 1\right)}$

3 Let $u = \sqrt{\dfrac{a}{c}}(x - b)$ so that $du = \sqrt{\dfrac{a}{c}}dx$ and $dx = \sqrt{\dfrac{c}{a}}du$

4 $\displaystyle\int \dfrac{1}{a(x-b)^2 + c} \cdot dx = \dfrac{1}{c}\sqrt{\dfrac{c}{a}}\tan^{-1}(u) + C = \dfrac{1}{\sqrt{ac}}\tan^{-1}\left(\sqrt{\dfrac{a}{c}}(x - b)\right) + C$

With this particular type of u-substitution in place we can do the next batch of integrations by partial fractions.

Example 7.42. Compute:

$$\int \frac{dx}{x^3 + x}$$

Solution:

Factor and we see that $x^3 + x = x(x^2 + 1)$. So the partial fraction form is:

$$\frac{1}{x^3 + x} = \frac{A}{x} + \frac{Bx + C}{x^2 + 1}$$

We need $Bx + C$ in the numerator of the second part instead of just B because $x^2 + 1$ is quadratic. Clearing the denominator we get the polynomial equation:

$$1 = A(x^2 + 1) + x(Bx + C)$$

Plug in $x = 0$ and we see that $1 = A$. This tells us that

$$1 = (x^2 + 1) + Bx^2 + Cx$$

or

$$-x^2 + 0x = Bx^2 + Cx$$

so $B = -1$ and and $C = 0$. We are now prepared to integrate.

$$\int \frac{dx}{x^3 + x} = \int \left(\frac{1}{x} - \frac{x}{x^2 + 1} \right) \cdot dx$$
$$= \ln|x| - \frac{1}{2} \ln|x^2 + 1| + C$$

The second part of the integral is a u-substitution with $u = x^2 + 1$; we have done it before.

$\Diamond$

Let's try a less neat integral of this sort.

Example 7.43. Compute:

$$\int \frac{3x \cdot dx}{x^3 - 9x^2 + 25x - 25}$$

Solution:

Factor, and we get $x^3 - 9x^2 + 25x - 25 = (x - 5)(x^2 - 4x + 5) = (x - 5)((x - 2)^2 + 1)$. That makes the partial fraction form

$$\frac{3x \cdot dx}{x^3 - 9x^2 + 25x - 25} = \frac{A}{x - 5} + \frac{Bx + C}{x^2 - 4x + 5}$$

Clear the denominator and we get

$$0x^2 + 3x + 0 = A(x^2 - 4x + 5) + (Bx + C)(x - 5) = Ax^2 - 4Ax + 5A + Bx^2 - 5Bx + Cx - 5C$$

yielding the simultaneous equations

$$A + B = 0$$
$$-4A - 5B + C = 3$$
$$5A - 5C = 0$$

So, $A = -B$, $A = C$ and $-4A + 5A + A = 2A = 3$. So, $A = C = 3/2$ and $B = -3/2$. This permits us to integrate:

$$\int \frac{3x \cdot dx}{x^3 - 9x^2 + 25x - 25} = \frac{3}{2} \int \left(\frac{1}{x - 5} - \frac{x - 1}{(x - 2)^2 + 1} \right) \cdot dx$$

$$= \frac{3}{2} \int \left(\frac{1}{x - 5} - \frac{x - 2}{(x - 2)^2 + 1} + \frac{1}{(x - 2)^2 + 1} \right) \cdot dx$$

Let $u = x - 2$ with $du = dx$ for the second and third terms

$$= \frac{3}{2} \left(\ln|x - 5| - \frac{1}{2} \ln(u^2 + 1) + \tan^{-1}(u) \right) + C$$

$$= \frac{3}{2} \left(\ln|x - 5| - \frac{1}{2} \ln(x^2 - 4x + 5) + \tan^{-1}(x - 2) \right) + C$$

◊

Example 7.44. Compute:

$$\int \frac{dx}{x^3 - 1}$$

Solution:

Factor: $x^3 - 1 = (x - 1)(x^2 + x + 1)$. Then

$$\frac{1}{x^3 - 1} = \frac{A}{x - 1} + \frac{Bx + C}{x^2 + x + 1}$$

Clear the denominator and we get

$$1 = A(x^2 + x + 1) + (Bx + C)(x - 1)$$

yielding:

$$A + B = 0$$
$$A - B + C = 0$$
$$A - C = 1$$

Solving we get $A = 1/3$, $B = -1/3$, $C = -2/3$.

$$\int \frac{dx}{x^3-1} = \frac{1}{3}\int \left(\frac{1}{x-1} - \frac{x+2}{x^2+x+1}\right)\cdot dx$$

$$= \frac{1}{3}\int \left(\frac{1}{x-1} - \frac{1}{2}\cdot\frac{2x+4}{x^2+x+1}\right)\cdot dx$$

$$= \frac{1}{3}\int \left(\frac{1}{x-1} - \frac{1}{2}\cdot\frac{2x+1+3}{x^2+x+1}\right)\cdot dx$$

$$= \frac{1}{3}\int \left(\frac{1}{x-1} - \frac{1}{2}\cdot\frac{2x+1}{x^2+x+1} - \frac{1}{2}\cdot\frac{3}{x^2+x+1}\right)\cdot dx$$

The middle term is a u-substitution that leads to a log.

$$= \frac{1}{3}\ln|x-1| - \frac{1}{6}\ln|x^2+x+1| - \frac{1}{2}\int \frac{dx}{x^2+x+1}$$

$$= \frac{1}{3}\ln|x-1| - \frac{1}{6}\ln|x^2+x+1| - \frac{1}{2}\int \frac{dx}{(x+1/2)^2+3/4}$$

$$= \frac{1}{3}\ln|x-1| - \frac{1}{6}\ln|x^2+x+1| - \frac{1}{2}\int \frac{dx}{\dfrac{3}{4}\left(\dfrac{4}{3}(x+1/2)^2+1\right)}$$

$$= \frac{1}{3}\ln|x-1| - \frac{1}{6}\ln|x^2+x+1| - \frac{1}{2}\cdot\frac{4}{3}\int \frac{dx}{\left(\dfrac{4}{3}(x+1/2)^2+1\right)}$$

$$= \frac{1}{3}\ln|x-1| - \frac{1}{6}\ln|x^2+x+1| - \frac{2}{3}\int \frac{dx}{\left(\dfrac{2}{\sqrt3}(x+1/2)\right)^2+1}$$

Let $u = \dfrac{2}{\sqrt3}(x+\tfrac{1}{2})$ and so $du = \dfrac{2}{\sqrt3}dx$

$$= \frac{1}{3}\ln|x-1| - \frac{1}{6}\ln|x^2+x+1| - \frac{2}{3}\cdot\frac{\sqrt3}{2}\int \frac{du}{u^2+1}$$

$$= \frac{1}{3}\ln|x-1| - \frac{1}{6}\ln|x^2+x+1| - \frac{1}{\sqrt3}\tan^{-1}(u) + C$$

$$= \frac{1}{3}\ln|x-1| - \frac{1}{6}\ln|x^2+x+1| - \frac{1}{\sqrt3}\tan^{-1}\left(\frac{2}{\sqrt3}\left(x+\frac{1}{2}\right)\right) + C$$

See? Easy!

◊

Some of the problems develop general formulas for this sort of super messy inverse tangent integral. Like Knowledge Box 7.10, but with more parameters.

Problems

Problem 7.47. *Perform the following integrals.*

(a) $\int \dfrac{dx}{x^2 - x}$

(b) $\int \dfrac{dx}{x^2 - 6x + 8}$

(c) $\int \dfrac{dx}{x^2 + 4x + 3}$

(d) $\int \dfrac{dx}{6x^2 - 5x + 1}$

(e) $\int \dfrac{dx}{x^2 - x - 1}$

(f) $\int \dfrac{dx}{x^2 - 11x + 30}$

Problem 7.48. *Perform each of the following integrals. Partial fractions are not needed.*

(a) $\int \dfrac{1}{x^2 + 2x + 2} \cdot dx$

(b) $\int \dfrac{1}{x^2 + 3x + 4} \cdot dx$

(c) $\int \dfrac{1}{x^2 + x + 2} \cdot dx$

(d) $\int \dfrac{x + 1}{x^2 + 2x + 2} \cdot dx$

(e) $\int \dfrac{2x + 2}{x^2 + 3x + 4} \cdot dx$

(f) $\int \dfrac{x - 1}{x^2 + x + 2} \cdot dx$

Problem 7.49. *Perform each of the following integrals.*

(a) $\int \dfrac{dx}{x^3 + 6x^2 + 11x + 6}$

(b) $\int \dfrac{dx}{x^4 - 16}$

(c) $\int \dfrac{dx}{x^3 - 5x^2 - 2x + 24}$

(d) $\int \dfrac{x^2 + x + 1}{x^3 - 6x^2 + 11x + 6} \cdot dx$

(e) $\int \dfrac{x^2 + 1}{x^3 + 4x} \cdot dx$

(f) $\int \dfrac{3x - 2}{x^3 + 2x^2 - 5x - 6} \cdot dx$

Problem 7.50. *For each of the following, compute the partial fractions form for decomposing the integral, but do not solve for the partial fractions coefficients or perform the integral.*

(a) $\int \dfrac{dx}{x^5 + 4x^3 + x}$

(b) $\int \dfrac{dx}{x^4 + 4x^3 + 7x^2 + 6x + 3}$

(c) $\int \dfrac{dx}{x^5 + x^3}$

(d) $\int \dfrac{dx}{x^6 - 5x^5 + 6x^4 - x^3 + 5x^2 - 5x + 1}$

(e) $\int \dfrac{dx}{x^6 - 3x^5 + 3x^4 + x^3}$

(f) $\int \dfrac{dx}{x^6 + 2x^5 + 5x^3 + 6x^2 + 3x + 2}$

Problem 7.51. *Compute:*

$$\int \frac{dx}{ax^2 + bx + c}$$

There are three cases each of which requires its own method of integration.

Problem 7.52. *Compute:*

$$\int \frac{dx}{x^3 - a^3}$$

where a is a positive constant.

Problem 7.53. *Compute:*

$$\int \frac{dx}{x^4 - a^4}$$

where a is a positive constant.

Problem 7.54. *Perform the integrals in Problem 7.50.*

Problem 7.55. *Compute:*

$$\int \frac{x^5}{x^2 + x + 1} \cdot dx$$

7.6. Practicing Integration

Knowledge Box 7.11.

Which method of integration do I use?

(a) Polynomials are integrated with the power rule, one term at a time.

(b) If the thing you are integrating is the derivative of something you recognize, use the fundamental theorem – the integral is the thing you recognize, plus "C".

(c) Look at the known derivatives, e.g. $Dx \arctan(x) = \frac{dx}{x^2+1}$. If you find one, apply rule "b".

(d) Will an algebraic re-arrangement of the terms of the integral set up an integral you can do?

(e) Look through the examples in this chapter and in Chapter 4. Are any of them similar enough to your problem to help?

(f) Check if there is a u-substitution that makes a simpler integral. Remember that du needs to be in there in an appropriate form.

(g) Are there natural parts for integration by parts? Remember you may need to integrate by parts several times.

(h) Is your integral packed with trig functions? Flip through Section 7.3 which has special methods for many different combinations of trig functions. Also remember that rule "c" includes applying trig identities to set up one of these special trig methods.

(i) Does the integral contain $\sqrt{x^2 \pm a^2}$ or $\sqrt{a^2 \pm x^2}$? Consider trigonometric substitution. This may yield integrals covered by any of rules "b", "c", or "d".

(j) Is the integral a ratio of polynomials? If the numerator is *not* lower degree, divide to get a remainder that does have a lower degree top. Integrate the resulting ratio of polynomials with partial fractions.

(k) If one of these rules seems to make some progress, keep going, you may need many steps to finish an integral.

(l) Try rule "d" again. Really.

(m) There are a lot of integrals that you can't do, there are many that noone can do. If you are really stuck, check with someone who has more experience than you. They may be able to recognize impossible integrals or integrals well above your level.

One of the important skills for performing integrals is figuring out which method of integration to use. This section is just more problems – but, unlike all the other sections, you cannot guess which method is useful by looking at the section the problem appears in. Practice! Practice! Practice!

Problem 7.56. *Pick and state a method of integration and then perform the integration for each of the following problems.*

(a) $\int x^3 \mathrm{e}^x \cdot dx$

(b) $\int x^2 \cdot (x^3 + 8) \cdot dx$

(c) $\int \dfrac{x}{x^2 + 1} \cdot dx$

(d) $\int \dfrac{x}{x^3 - 1} \cdot dx$

(e) $\int \sin(x) \cos(x) \cdot dx$

(f) $\int \sin^4(3x) \cdot dx$

Problem 7.57. *Pick and state a method of integration and then perform the integration for each of the following problems.*

(a) $\int \sin(2x) \mathrm{e}^{3x} \cdot dx$

(b) $\int \sqrt{25x^2 + 1} \cdot dx$

(c) $\int \cos(x) \sin(2x) \cdot dx$

(d) $\int \dfrac{\ln^3(x) + 1}{x} \cdot dx$

(e) $\int \csc^4(x) \cdot dx$

(f) $\int \dfrac{x^2 - 1}{x^2 + 1} \cdot dx$

Problem 7.58. *Pick and state a method of integration and then perform the integration for each of the following problems.*

(a) $\int \dfrac{dx}{\sqrt{1 - 4x^2}}$

(b) $\int \dfrac{x}{\sqrt{1 - 9x^2}} \cdot dx$

(c) $\int x \cdot \mathrm{e}^{\sqrt{x}} \cdot dx$

(d) $\int \dfrac{\mathrm{e}^x}{1 + \mathrm{e}^{2x}} \cdot dx$

(e) $\int \dfrac{\mathrm{e}^x}{\mathrm{e}^{2x} - 3\mathrm{e}^x + 2} \cdot dx$

(f) $\int x \cdot \ln(x) \cdot dx$

Problem 7.59. *Pick and state a method of integration and then perform the integration for each of the following problems.*

(a) $\int \dfrac{x^2}{x^4 - 1} \cdot dx$

(b) $\int x \cdot \mathrm{e}^{\sqrt[3]{x}} \cdot dx$

(c) $\int \dfrac{\cos(x)}{\sin^2(x) - 1} \cdot dx$

(d) $\int \dfrac{\sin(2x)}{\cos^2(x) - 9} \cdot dx$

(e) $\int x^5 (x^2 + 2)^4 \cdot dx$

(f) $\int \left(\cos^2(x) - \sin^2(x) \right) \mathrm{e}^{\sin(2x)} \cdot dx$

Problem 7.60. *Compute*
$$\int x^n \cdot \ln(x) \cdot dx$$

Problem 7.61. *Compute*
$$\int x \mathrm{e}^{\sqrt[n]{x}} \cdot dx$$

Problem 7.62. *Find a problem in this section that can be done in two ways and demonstrate them.*

Chapter 8

Limits and Continuity: the Details

In this chapter we go into the material usually presented at the beginning of calculus courses – the formal definition of limits and continuity. While we've taught you a boat-load of tricks for doing limits, we've largely ignored continuity (other than to say you need it) and the limits, so far, are all rules of thumb. Continuity is based on limits so we do them first.

8.1. Limits and Continuity

If $\lim_{x \to a} f(x) = L$ then, informally, as x gets closer to a, $f(x)$ needs to get closer to L. The absolute value function gives us the ability to figure out if two things are "close". If $|x - a|$ is small, x is close to a; if $|f(x) - L|$ is small, then $f(x)$ is close to L. We have two traditional symbols to represent small values: the Greek letters epsilon, ϵ, and delta ,δ. The closeness of x to a is encoded with δ, while the closeness of $f(x)$ to L is encoded by ϵ. With that preparation, here is the definition:

Knowledge Box 8.1.

The formal definition of a limit

Let $f(x)$ be a function defined on an open interval containing the value $x = a$, except that $f(x)$ might not be defined at a itself. We say that the limit of $f(x)$ at a is L, written

$$\lim_{x \to a} f(x) = L$$

if, for every $\epsilon > 0$, it is possible to find $\delta > 0$ so that

$$\text{if } 0 < |x - a| < \delta, \text{ then we have that } |f(x) - L| < \epsilon.$$

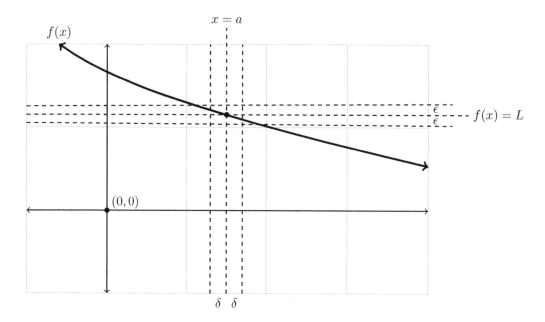

Figure 8.1. The geometry of $\lim\limits_{x \to a} f(x) = L$. As long x is within δ of a we have that $f(x)$ is within ϵ of L.

Example 8.1. Show formally that

$$\lim_{x \to 1} (2x + 1) = 3.$$

Solution:

Since there are no algebraic anomalies, we can just plug in and see that the desired limit is actually 3. But, how do we show it using the formal definition? We have two inequalities to satisfy. For one, we are given a positive number ϵ; for the other we need to find an appropriate positive number δ. Let's start with the inequality where everything is known:

$$|f(x) - L| < \epsilon$$
$$|2x + 1 - 3| < \epsilon$$
$$|2x - 2| < \epsilon$$
$$-\epsilon < 2x - 2 < \epsilon$$
$$-\epsilon < 2(x - 1) < \epsilon$$
$$-\frac{\epsilon}{2} < x - 1 < \frac{\epsilon}{2}$$
$$|x - 1| < \frac{\epsilon}{2}$$

This gives us information about how to choose δ. Since $a = 1$, we need to find a δ such that $|x - 1| < \delta$. So, set $\delta = \dfrac{\epsilon}{2}$. Then we satisfy the formal definition, as shown below.

Set $\delta = \dfrac{\epsilon}{2}$ and suppose that $0 < |x - 1| < \delta$. Then

$$|x - a| < \delta \qquad \text{The start of the formal definition.}$$

$$|x - 1| < \delta \qquad \text{Substitute in actual values.}$$

$$|x - 1| < \frac{\epsilon}{2}$$

$$|2x - 2| < \epsilon \qquad \text{Multiply by 2}$$

$$|2x + 1 - 3| < \epsilon$$

$$|f(x) - L| < \epsilon \qquad \text{Conclusion of the formal definition.}$$

So now we have proven, formally, that the limit is 3.

Lines are the simple case – for any more complex functions, the algebra gets more complicated.

Example 8.2. Show formally that

$$\lim_{x \to 4} x^2 = 16$$

Solution:

Our goal is to have

$$\text{If } 0 < |x - 4| < \delta \text{ then } |x^2 - 16| < \epsilon$$

We will use the fact that when x is near 4 then $x + 4 < 9$. Compute:

$$|x - 4| < \delta$$
$$|x - 4| \cdot |x + 4| < |x + 4|\delta$$
$$|x^2 - 16| < |x + 4|\delta < 9\delta$$
$$|x^2 - 16| < 9\delta$$

So let us set $\delta = \epsilon/9$. Check the formal definition.

$$|x - a| < \delta \qquad \qquad \text{Start of the formal definition.}$$
$$|x - 4| < \delta$$
$$|x - 4||x + 4| < |x + 4|\delta$$
$$|x^2 - 16| < |x + 4|\delta < 9\delta \qquad \qquad \text{Recall: } |x + 4| < 9$$
$$|x^2 - 16| < 9\delta$$
$$|x^2 - 16| < \epsilon$$
$$|f(x) - L| < \epsilon \qquad \qquad \text{Conclusion of the formal definition.}$$

And we have satisfied the formal definition.

$$\Diamond$$

Notice that the formulas we are using that tell us how ϵ and δ are related have the property that, as δ shrinks to zero, so does ϵ.

Like derivatives and integrals, limits have some useful algebraic properties.

Knowledge Box 8.2.

Algebraic properties of limits

Suppose that $\lim_{x \to a} f(x) = L$, $\lim_{x \to a} g(x) = M$, and that c is a constant.
Then:

- $\lim_{x \to a} (f(x) \pm g(x)) = L \pm M$

- $\lim_{x \to a} c \cdot f(x) = c \cdot L$

Since derivatives are based on limits, these algebraic roles underlie the summing and scaling properties of derivatives. If one needs to take a problematic limit, these properties permit you to break it up into pieces.

In Chapter 2 we informally defined limits of a function from above and below a value $x = c$. The formal definitions of these limits are closely parallel to the formal definition of a limit.

Knowledge Box 8.3.

The formal definition of limits from above and below

We say that
$$\lim_{x \to a^-} f(x) = L$$
if, for each $\epsilon > 0$, we can find $\delta > 0$ so that
$$\text{If } a - \delta < x < a, \text{ then } |f(x) - L| < \epsilon.$$

Similarly, we say that
$$\lim_{x \to a^+} f(x) = L$$
if, for each $\epsilon > 0$, we can find $\delta > 0$ so that
$$\text{If } a < x < a + \delta, \text{ then } |f(x) - L| < \epsilon.$$

The following example needs a one sided limit, because the function we are taking the limit of exists only on one side of the limiting point.

Example 8.3. Prove that
$$\lim_{x \to 0^+} \sqrt{x} = 0$$

Solution:

When we have that $0 < x < \delta$ we need for $|\sqrt{x} - 0| < \epsilon$, which is equivalent to $\sqrt{x} < \epsilon$. Squaring both sides we get $x < \epsilon^2$. Choose $\delta = \epsilon^2$ and check the definition.

$a < x < a + \delta$	Beginning of formal definition
$0 < x < \delta$	
$0 < x < \epsilon^2$	
$0 < \sqrt{x} < \epsilon$	
$0 < \sqrt{x} - 0 < \epsilon$	
$|\sqrt{x} - 0| < \epsilon$	Conclusion of the formal definition

◊

At this point, for completeness, we also restate the definition of limits at infinity. We originally encountered these in Chapter 3.

Knowledge Box 8.4.

The formal definition of a limit at infinity

If we can find a number L so that, for every number $\epsilon > 0$, we can find a value X_ϵ such that, when $x \geq X_\epsilon$, we have that $|f(x) - L| < \epsilon$, then L is the limit of $f(x)$ as $x \to \infty$. Written:

$$\lim_{x \to \infty} f(x) = L$$

Example 8.4. Using the definition of the limit, show that

$$\lim_{x \to \infty} \frac{1}{x^2 + 3} = 0$$

Solution:

The trick here is to find an appropriate X_ϵ, given an ϵ. Remember that when you take the reciprocal of both sides of an inequality, it reverses.

$$\frac{1}{3 + X_\epsilon^2} < \epsilon$$

$$3 + X_\epsilon^2 > \frac{1}{\epsilon}$$

$$X_\epsilon^2 > \frac{1}{\epsilon} - 3$$

$$X_\epsilon > \sqrt{\frac{1}{\epsilon} - 3}$$

So the value

$$X_\epsilon = \sqrt{\frac{1}{\epsilon} - 3}$$

meets the requirements if $\epsilon \leq 1/3$. If $\epsilon > 1/3$, then this value is undefined. So, in that case, just pick $X_\epsilon = 0$. $f(0) = 1/3$ and f is decreasing. So, when $x \geq 0, f(x) \leq 1/3 < \epsilon$ for $\epsilon > 1/3$.

◊

We also remind the reader of the formal definition of diverging to infinity from Chapter 3.

Knowledge Box 8.5.

Knowledge Box 8.5.

The formal definition of divergence to ∞

If, for any constant c, we can find X_c so that $x > X_c$ means $f(x) \geq c$, then

$$\lim_{x \to \infty} f(x)$$

diverges to infinity.

Example 8.5. Show formally that

$$\lim_{x \to \infty} x^4 = \infty.$$

Solution:

For $f(x) = x^4$ to be bigger than c we need $x > \sqrt[4]{c}$. So the choice of X_c is pretty simple. Let $X_c = \sqrt[4]{c}$ and check the formal definition:

Suppose that $x > X_c = \sqrt[4]{c}$, then $f(x) \geq (\sqrt[4]{c})^4 = c$, and we meet the requirements of the formal definition.

$$\Diamond$$

Once we have limits, we can define continuity. Suppose that a function $f(x)$ exists on an interval $a < x < b$. Then, informally, $f(x)$ is continuous on an interval if you can draw the graph of the function on $a < x < b$ *without lifting your pencil.* (Mathematics should not, of course, be done in pen unless required for some formality of testing.) The formal definition is more complex.

Knowledge Box 8.6.

The definition of continuity at a point

A function $f(x)$ is **continuous at x=c** if three things are true.

1 $\lim_{x \to c^-} f(x) = v$,

2 $\lim_{x \to c^+} f(x) = v$, and

3 $f(c) = v$,

In other words, $f(x)$ is continuous at $x = c$ if the limits from above and below exist and are equal to the value of the function at $x = c$.

The definition of continuity at a point can fail in a number of ways. A function could lack one or both of the required limits; the limits could disagree with one another or the value of the function;

or the limits might be fine but the function might not exist as with the somewhat contrived example

$$f(x) = \frac{x^2 - 4}{x - 2}$$

from Chapter 3.

Example 8.6. Is the function

$$f(x) = \begin{cases} x^2 & x < 2 \\ 3x - 2 & x \geq 2 \end{cases}$$

continuous at $x = 2$?

Solution:

First compute $f(2) = 3 \cdot 2 - 2 = 4$. The limit from below takes place where the rule of the function is x^2. So

$$\lim_{x \to 2^-} f(x) = 4$$

by substituting into x^2. The limit from above takes place where the rule of the function is $3x - 2$. So

$$\lim_{x \to 2^+} f(x) = 4$$

by substituting into $3x - 2$. The function exists with a value of 4 at $x = 2$ which agrees with the limits from above and below. So the function is continuous at $x = 2$.

$$\Diamond$$

Example 8.7. Is

$$g(x) = \sqrt{x}$$

continuous at $x = 0$?

Solution:

The function is not continuous at $x = 0$ because the limit from below is in an area where the function does not exist.

$$\Diamond$$

The informal definition of continuity said that a function is continuous on an interval if you can draw the graph without lifting your pencil. The formal definition, so far, talks only of continuity at a point. Knowledge Box 8.7 provides the formal definition of continuity on an interval.

Knowledge Box 8.7.

The definition of continuity on an interval

A function is continuous on an interval if it is continuous at every point in that interval.

Which is a bit of an anticlimax. Mathematicians have already checked the continuity of a large number of functions and worked out some rules that will permit you to check if functions are continuous without laborious reference to the formal definitions. The first group are simply continuous everywhere.

Knowledge Box 8.8.

Functions that are continuous on $-\infty < x < \infty$

- polynomials
- $\sin(x)$, $\cos(x)$, e^x, $\tan^{-1}(x)$
- $\sqrt[n]{x}$, when n is odd

The next list is not quite *as* nice because the functions have vertical asymptotes or fail to exist for some ranges of x.

Knowledge Box 8.9.

Functions that are continuous on intervals where they exist

- rational functions,
- $\tan(x)$, $\cot(x)$, $\sec(x)$, $\csc(x)$, $\ln(x)$, $\sin^{-1}(x)$, $\sec^{-1}(x)$
- $\sqrt[n]{x}$, when n is even.

As you would expect, we have algebraic rules for making more continuous functions. In essence, doing standard arithmetic on continuous functions yields a continuous function unless you do something impossible like divide by zero or take an even root of a negative number. If we know the domain and range of two functions, then we may also be able to compose continuous functions to obtain a continuous function.

Knowledge Box 8.10.

Operations that preserve continuity

Suppose that $f(x)$ and $g(x)$ are functions that are continuous on the interval $a < x < b$ and that c is a constant. Then the following are also continuous on the interval $a < x < b$.

- $f(x) \pm g(x)$
- $f(x) \cdot g(x)$
- $c \cdot f(x)$
- $f(x)/g(x)$, if $g(x)$ has no roots r with $a < r < b$

Example 8.8. Show that $f(x) = \dfrac{\sin(x)}{x^2 + 1}$ is continuous everywhere.

Solution:

We have $\sin(x)$ and $x^2 + 1$ on our list of functions that are continuous everywhere. Since $x^2 \geq 0$, we know that $x^2 + 1 > 0$ so the rule for dividing functions says that $f(x)$ is continuous everywhere, as $x^2 + 1$ has no roots on $-\infty < x < \infty$.

$\Diamond$

Example 8.9. Determine the largest intervals on which

$$g(x) = \frac{1}{x}$$

is continuous.

Solution:

The function $g(x)$ is a rational function. It exists when $x \neq 0$ so it is continuous on any interval not containing zero. This means the largest intervals where it is continuous are

$$-\infty < x < 0 \text{ and } 0 < x < \infty$$

$\Diamond$

Knowledge Box 8.11.

Composition of continuous functions

Suppose that $g(x)$ is continuous on $a < x < b$, that on the interval $a < x < b$ all values of $g(x)$ are in the interval $c < x < d$, and that $f(x)$ is continuous on $c < x < d$.

Then, $f(g(x))$ is continuous on $a < x < b$.

Let's look at how this might work in practice to show a complicated function is continuous.

Example 8.10. Show that

$$h(x) = \sqrt{x^2 + 1}$$

is continuous everywhere.

Solution:

Set $g(x) = x^2 + 1$ and $f(x) = \sqrt{x}$. Then $h(x) = f(g(x))$. We know $g(x)$ is continuous everywhere because it is a polynomial. We also have already shown that $g(x) > 0$ everywhere, and so the values it takes on are all positive. This means that $f(x)$ is continuous for every value that $g(x)$ can produce – which permits us to use the composition rule. We conclude that $h(x)$ is continuous everywhere.

$\Diamond$

Example 8.11. Find the largest interval(s) on which

$$h(x) = \sqrt{x^2 - 1}$$

is continuous.

Solution:
The danger here is taking square roots of negative numbers. The composition rule may be applied, taking $g(x) = x^2 - 1$ and $f(x) = \sqrt{x}$ when

$$x^2 - 1 > 0.$$

This happens when x is at least one. So the maximal intervals of continuity of $h(x)$ are $(-\infty, -1)$ and $(1, \infty)$.

Example 8.12. Find the largest interval(s) on which

$$q(x) = \sqrt{x^2 - 4} \cdot \sqrt{\ln(x + 1)}$$

is continuous.

Solution:
The danger here is also taking square roots of negative numbers. Multiplication preserves continuity but we need both halves of the product to be continuous for this to work. That means we need $x^2 - 4 \geq 0$ and $x + 1 \geq 1$. The solution to the first inequality is $(-\infty, -2) \cup (2, \infty)$ while the solution to the second is $(0, \infty)$. Since *both* must be true to set up the product to yield a continuous function we get that the negative range of the first solution is disallowed by the second and $(0, 2)$ is excluded from the second solution by the first leaving us with the range $(2, \infty)$ where this function is continuous.

Example 8.13. Determine where $f(x) = \tan(x)$ is continuous.

Solution:
The tangent function is

$$\tan(x) = \frac{\sin(x)}{\cos(x)}$$

and we know that both sine and cosine are continuous everywhere. This means that the danger is division by zero so the answer is $\cos(x) \neq 0$. The cosine is zero at odd multiples of $\pi/2$ so this means the tangent function is continuous on

$$\left\{ x : x \neq (2n + 1)\frac{\pi}{2} \right\}$$

where n is any whole number.

Problems

Problem 8.1. *Use the formal definition of limits to prove the following limits.*

(a) $\lim_{x \to 1} x + 3 = 4$

(b) $\lim_{x \to 2} 2x + 1 = 5$

(c) $\lim_{x \to -3} 5x + 3 = -12$

(d) $\lim_{x \to 5} x^2 = 25$

(e) $\lim_{x \to 1} x^2 = 1$

(f) $\lim_{x \to 2} x^3 = 8$

Problem 8.2. *Suppose that*
$\lim_{x \to 2} f(x) = 3$, $\lim_{x \to 2} g(x) = -2$, *and* $\lim_{x \to 2} h(x) = 7$.

Compute the following limits.

(a) $\lim_{x \to 2} f(x) + g(x)$

(b) $\lim_{x \to 2} 2.1 f(x) - 3.5 g(x)$

(c) $\lim_{x \to 2} f(x) + g(x) - h(x)$

(d) $\lim_{x \to 2} 3h(x) + 2f(x) + g(x)$

(e) $\lim_{x \to 2} \pi f(x) + e \cdot g(x)$

(f) $\lim_{x \to 2} h(x) - \pi g(x)$

Problem 8.3. *Using the formal definition, show*
$$\lim_{x \to \infty} \frac{1}{x^3} = 0.$$

Problem 8.4. *Using the formal definition, show*
$$\lim_{x \to \infty} \frac{1}{e^x} = 0.$$

Problem 8.5. *Using the formal definition, show*
$$\lim_{x \to \infty} x^7 = \infty.$$

Problem 8.6. *Using the formal definition, show*
$$\lim_{x \to \infty} 5x + 6 = \infty.$$

Problem 8.7. *For each of the following functions, determine if the function is continuous at the value given. If the function is not continuous, say where the definition of continuity fails, e.g. "Not continuous because the function does not exist there".*

(a) $f(x) = 2x - 4$ at $x = 1$

(b) $g(x) = \begin{cases} x^2 & x < 0 \\ x^3 & x \geq 0 \end{cases}$ at $x = 0$

(c) $h(x) = \dfrac{2x}{x^2 - 16}$ at $x = 4$

(d) $r(x) = \ln(x)$ at $x = -1$

(e) $s(x) = \begin{cases} 5x + 11 & x < 3 \\ x^3 + x^2 - 10 & x \geq 3 \end{cases}$ at $x = 3$

(f) $q(x) = \begin{cases} 5x + 11 & x < 2 \\ x^3 + x^2 - 10 & x \geq 2 \end{cases}$ at $x = 1$

Problem 8.8. *Find all the largest possible intervals on which the following functions are continuous. Justify your results with a sentence or two.*

(a) $f(x) = \dfrac{x^2}{x^2 - 5}$

(b) $g(x) = \sin^2(x) + 4\sin(x) + 5$

(c) $h(x) = \dfrac{\sin(x)}{\cos^2(x) + 1}$

(d) $r(x) = \sqrt{x^3 - x^2}$

(e) $s(x) = \sqrt[3]{x^3 - x^2}$

(f) $q(x) = \ln(\cos(x))$

Problem 8.9. *If $p(x)$ is a polynomial, where is the function*
$$f(x) = p(\cos(x))$$
continuous?

8.2. Applications: Improper Integrals

One of the odd facts about integration is that it is possible to find a finite area under a curve that has one of its sides infinitely long. These areas can be finite or infinite. *Improper integrals* are the tool we use to sort these out.

Knowledge Box 8.12.

Definition of improper integrals at infinity

$$\int_a^\infty f(x) \cdot dx = \lim_{b \to \infty} \int_a^b f(x) \cdot dx$$

Example 8.14. Find

$$\int_1^\infty \frac{dx}{x^2}$$

Solution:

$$\int_1^\infty \frac{dx}{x^2} = \lim_{b \to \infty} \int_1^b \frac{dx}{x^2}$$

$$= \lim_{b \to \infty} \left. \frac{-1}{x} \right|_1^b$$

$$= \lim_{b \to \infty} \left(\frac{-1}{b} - \frac{-1}{1} \right)$$

$$= \lim_{b \to \infty} \left(1 - \frac{1}{b} \right)$$

$$= 1 - \lim_{b \to \infty} \frac{1}{b}$$

$$= 1 - 0 = 1$$

Which shows that the area under $f(x) = \dfrac{1}{x^2}$ from $x = 1$ all the way to infinity is just one square unit.

$\Diamond$

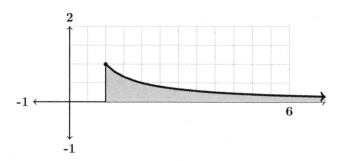

Figure 8.2. $\int_{1}^{\infty} \dfrac{dx}{x^2}$ from Example 8.14.

Figure 8.2 shows the improper integral from Example 8.14. The shaded area continues off to the right infinitely far.

Let's do an example where the area is infinite. This example is very similar to the last one.

Example 8.15. Find
$$\int_{1}^{\infty} \frac{dx}{x}$$

Solution:

$$\int_{1}^{\infty} \frac{dx}{x} = \lim_{b \to \infty} \int_{1}^{b} \frac{dx}{x}$$

$$= \lim_{b \to \infty} \ln(x) \Big|_{1}^{b}$$

$$= \lim_{b \to \infty} (\ln(b) - \ln(1))$$

$$= \lim_{b \to \infty} (\ln(b) - 0))$$

$$= \lim_{b \to \infty} \ln(b)$$

$$= \infty$$

So this integral does not exist.

◊

The same technique may be used at any point where the function that you are integrating does not exist. The most common application of this kind of improper integral is when you are integrating up to or away from a vertical asymptote. Knowledge Box 8.13 formalizes this.

Knowledge Box 8.13.

Improper integrals at a discontinuity

Suppose that $f(x)$ has a discontinuity at $x = c$. Then

$$\int_a^c f(x) \cdot dx = \lim_{b \to c^-} \int_a^b f(x) \cdot dx$$
and
$$\int_c^a f(x) \cdot dx = \lim_{b \to c^+} \int_b^a f(x) \cdot dx$$

Example 8.16. Find

$$\int_0^1 \sqrt{x} \cdot dx$$

Solution:

The discontinuity is at $c = 0$ so the improper integral works like this:

$$\int_0^1 \sqrt{x} \cdot dx = \lim_{b \to 0^+} \int_b^1 x^{1/2} \cdot dx$$

$$= \lim_{b \to 0^+} \int_b^1 x^{1/2} \cdot dx$$

$$= \lim_{b \to 0^+} \frac{2}{3} x^{3/2} \Big|_b^1$$

$$= \lim_{b \to 0^+} \left(\frac{2}{3} - \frac{2}{3} b^{3/2} \right)$$

$$= \frac{2}{3} - \lim_{b \to 0^+} \frac{2}{3} b^{3/2}$$

$$= \frac{2}{3} - 0 = \frac{2}{3}$$

$\Diamond$

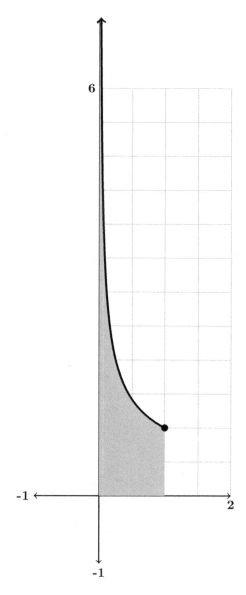

Figure 8.3. $\displaystyle\int_0^1 \sqrt{x} \cdot dx$ from Example 8.16.

Figure 8.3 shows the improper integral from Example 8.16. The shaded area continues upward infinitely far.

In general you replace the problematic (integral) limit of your integral with an unknown constant and take the (calculus) limit as that variable approaches the problematic value.

Example 8.17. Find $\int_0^\infty e^{-x} \cdot dx$.

Solution:

For this example the infinite limit is the problematic one so,

$$\int_0^\infty e^{-x} \cdot dx = \lim_{b \to \infty} \int_0^b e^{-x} \cdot dx$$

$$= \lim_{b \to \infty} -e^{-x} \Big|_0^b$$

$$= \lim_{b \to \infty} -e^{-b} - (-1)$$

$$= 1 - \lim_{b \to \infty} \frac{-1}{e^b} = 1 - 0 = 1$$

◇

So we have another area with an infinite side length but a finite value.

Recall that e^x grows faster than any polynomial. That explains why the next example is also finite.

Example 8.18. Find

$$\int_0^\infty x^2 \cdot e^{-x} \cdot dx$$

Solution:

In Problem 7.16 you developed a shortcut for this integral similar to the shortcut in Knowledge Box 7.5: $\int p(x)e^{-x}dx = -(p(x) + p'(x) + \ldots)e^{-x} + C$.

Here the infinite limit is the problematic one so,

$$\int_0^\infty x^2 e^{-x} \cdot dx = \lim_{b \to \infty} \int_0^b x^2 e^{-x} \cdot dx$$

$$= \lim_{b \to \infty} -(x^2 + 2x + 2)e^{-x} \Big|_0^b \qquad \text{shortcut}$$

$$= \lim_{b \to \infty} -(b^2 + 2b + 2)e^{-b} + (0 + 0 + 2)e^0$$

$$= 2 - \lim_{b \to \infty} -\frac{b^2 + 2b + 2}{e^b}$$

$$= 2 - 0 = 2$$

◇

8.2.1. Limits that don't exist because of really bad behavior. In the course of doing improper integrals there are limits that don't exist because the behavior of the functions involved is pathological. Classifying all the pathological behaviors is way too hard, so we will give a couple of examples that will help you develop intuition about another type of discontinuity – one that cannot be patched by taking a limit.

Example 8.19. Consider $\lim\limits_{x \to 0} \sin\left(\dfrac{1}{x}\right)$. Does this exist?

Solution:

A good place to begin with this problem is to look at the graph.

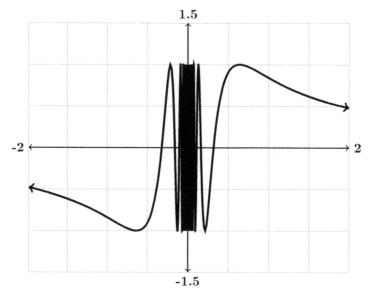

The function undergoes an infinite number of oscillations between $y = -1$ and $y = 1$ in any interval containing zero. This is such an extreme behavior that the limit does not exist.

The cognate behavior to this is much easier to understand.

Example 8.20. Consider $\lim\limits_{x \to \infty} \sin(x)$. Does this exist?

Solution:

On the way to ∞, the function $\sin(x)$ never approaches a particular value and so has no limit.

Problems

Problem 8.10. *Perform each of the following improper integrals.*

(a) $\int_1^\infty \dfrac{1}{x^3} \cdot dx$

(b) $\int_2^\infty \dfrac{1}{x^{3/2}} \cdot dx$

(c) $\int_1^\infty \dfrac{1}{\sqrt{x}} \cdot dx$

(d) $\int_0^\infty \dfrac{1}{x^2 + 1} \cdot dx$

(e) $\int_{-\infty}^\infty \dfrac{1}{x^2 + 1} \cdot dx$

(f) $\int_0^\infty \dfrac{x}{x^2 + 1}$

Problem 8.11. *Perform each of the following improper integrals.*

(a) $\int_0^1 \dfrac{1}{\sqrt[3]{x}}$,

(b) $\int_0^1 \dfrac{1}{x^2}$,

(c) $\int_0^2 \dfrac{5}{x^{3/2}}$

(d) $\int_1^2 \dfrac{1}{x^2 - 1}$

(e) $\int_0^4 \dfrac{3}{x^{1.25}}$

(f) $\int_0^1 \dfrac{1}{\sqrt[3]{x^4}}$

Problem 8.12. *Suppose that all the roots of a polynomial $p(x)$ are negative and that $q(x)$ is a polynomial with a degree at least two smaller than the degree of $p(x)$. Do your best to explain if the following exists:*

$$\int_0^\infty \frac{q(x)}{p(x)} \, dx$$

Problem 8.13. *Perform each of the following improper integrals.*

(a) $\int_0^\infty x \, e^{-x} \cdot dx$

(b) $\int_0^\infty x^3 e^{-x} \cdot dx$

(c) $\int_0^\infty e^{-3x+1} \cdot dx$

(d) $\int_1^\infty (x^2 + x + 1) e^{-x} \cdot dx$

(e) $\int_0^\infty \sin(x) \cdot e^{-x} \cdot dx$

(f) $\int_0^\infty \cos(2x) e^{-x} \cdot dx$

Problem 8.14. *Demonstrate that if $p(x)$ is a polynomial then*

$$\int_0^\infty p(x) e^{-x} \cdot dx$$

always exists.

Problem 8.15. *For which values of c does*

$$\int_1^\infty \frac{dx}{x^c}$$

exist?

Problem 8.16. *For which values of c does*

$$\int_0^1 \frac{dx}{x^c}$$

exist?

Problem 8.17. *Compute*

$$\int_0^1 \ln(x) \cdot dx$$

Problem 8.18. *Compute*

$$\int_0^1 x \cdot \ln(x) \cdot dx$$

Problem 8.19. *Compute*

$$\int_0^\infty e^{-ax+b} \cdot dx$$

where $a > 0$ and b are constants.

8.3. The Squeeze Theorem, The Mean Value Theorem

Sometimes it is possible to compute a limit by bounding it by much easier limits. Consider trying to compute

$$\lim_{x \to \infty} \frac{\cos(x)}{x}$$

We already know that the infinite limits of sine and cosine do not exist, but this is different. Consider the graph of $\dfrac{\cos(x)}{x}$ shown in Figure 8.4.

The cosine function is never farther from 0 than ± 1 but x grows without limit, making it look like the limit is zero. This lets us use a trick called the **squeeze theorem**.

Example 8.21. Compute:

$$\lim_{x \to \infty} \frac{\cos(x)}{x}$$

Solution:

Bound the limit and proceed.

$$-1 \le \cos(x) \le 1$$

$$\frac{-1}{x} \le \frac{\cos(x)}{x} \le \frac{1}{x} \qquad\qquad x > 0$$

$$\lim_{x \to \infty} \frac{-1}{x} \le \lim_{x \to \infty} \frac{\cos(x)}{x} \le \lim_{x \to \infty} \frac{1}{x}$$

$$0 \le \lim_{x \to \infty} \frac{\cos(x)}{x} \le 0$$

Since the limit we are interested in is *squeezed* between zero and zero, it follows that its value is also zero.

$$\Diamond$$

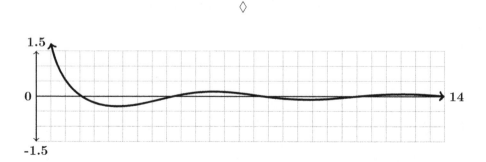

Figure 8.4. Graph of $\dfrac{\cos(x)}{x}$.

The trick, in general, is to squeeze the limit of interest between two manageable limits. This trick is only useful when you can trap a limit in the fashion shown above. This most often happens when you have a ratio of functions that goes to zero. Let's formally state the result.

Knowledge Box 8.14.

The Squeeze Theorem

Suppose that
$$f(x) \le g(x) \le h(x)$$
on an interval containing a. Then if
$$\lim_{x \to a} f(x) = L = \lim_{x \to a} h(x)$$
we may deduce that
$$\lim_{x \to a} g(x) = L$$
This works if the interval is of the form (c, ∞) and the limit is as $x \to \infty$, as well.

Example 8.22. Compute $\displaystyle\lim_{x \to \infty} \frac{\sin(x) + \cos(x)}{3x + 2}$.

Solution:

We know that $-1 \le \sin(x), \cos(x) \le 1$.
This means that:
$$-2 \le \sin(x) + \cos(x) \le 2$$

$$\frac{-2}{3x + 2} \le \frac{\sin(x) + \cos(x)}{3x + 2} \le \frac{2}{3x + 2} \qquad\qquad x > 0$$

$$\lim_{x \to \infty} \frac{-2}{3x + 2} \le \lim_{x \to \infty} \frac{\sin(x) + \cos(x)}{3x + 2} \le \lim_{x \to \infty} \frac{2}{3x + 2}$$

$$0 \le \lim_{x \to \infty} \frac{\sin(x) + \cos(x)}{3x + 2} \le 0$$

Since the upper and lower bounding limits are 0 we may deduce
$$\lim_{x \to \infty} \frac{\sin(x) + \cos(x)}{3x + 2} = 0$$

$\Diamond$

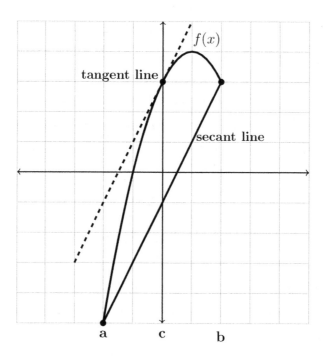

Figure 8.5. Illustration of the Mean Value Theorem.

Knowledge Box 8.15.

The Mean Value Theorem

Suppose that $f(x)$ is continuous and differentiable on the interval $a \leq x \leq b$. Then there is a value c with $a \leq c \leq b$ so that

$$\frac{f(b) - f(a)}{b - a} = f'(c).$$

This theorem seems a bit arcane when first read, but in fact it is a simple upgrade of Rolle's theorem (Knowledge Box 6.3). Look at Figure 8.5.

Notice that

$$\frac{f(b) - f(a)}{b - a}$$

is the slope of the secant line to $f(x)$ on the interval $[a, b]$. What the mean value theorem says is that (at least) one point c in the interval $[a, b]$ has a tangent line to $f(x)$ that is parallel to the secant line – since $f'(c)$ is the slope of the tangent line at $x = c$. All that is happening is that we are applying Rolle's theorem to $f(x)$ after subtracting the secant line.

Example 8.23. Apply the mean value theorem to the function $f(x) = x^2$ on the interval $[-1, 2]$.

Solution:

The secant slope is

$$m = \frac{2^2 - (-1)^2}{2 - (-1)} = \frac{3}{3} = 1$$

Solve:

$$f'(x) = 1$$
$$2x = 1$$
$$x = \frac{1}{2}$$

So the value c predicted by the mean value theorem is $c = 1/2$.

$$\Diamond$$

Example 8.24. Apply the mean value theorem to the function $f(x) = 3 - x^2$ on the interval $[0, 2]$ and draw the situation.

Solution:

The secant slope is $m = \dfrac{-1 - 3}{2 - 0} = \dfrac{-4}{2} = -2$. Solve:

$$f'(x) = -2$$
$$-2x = -2$$
$$x = 1$$

So the value c predicted by the mean value theorem is $c = 1$.

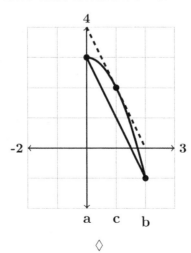

$$\Diamond$$

At this point the mean value theorem is a technical theorem. If you continue in your study of mathematics, it will turn out to be useful for proving facts about the nature of functions. For now, consider it a source of skill-building exercises.

Problems

Problem 8.20. *Use the Squeeze Theorem to compute the following limits.*

(a) $\displaystyle\lim_{x\to\infty} \frac{\cos(x)}{\ln(x)}$

(b) $\displaystyle\lim_{x\to\infty} \frac{\cos(x^2)}{x^2}$

(c) $\displaystyle\lim_{x\to\infty} \frac{\sin(x^2+1)}{x^3}$

(d) $\displaystyle\lim_{x\to\infty} \frac{\tan^{-1}(x)}{\sqrt{x}}$

(e) $\displaystyle\lim_{x\to\infty} \frac{\cos(2x)}{x+1}$

(f) $\displaystyle\lim_{x\to\infty} \frac{\cos(\ln(x))}{x^2+1}$

(g) $\displaystyle\lim_{x\to\infty} \frac{\tan^{-1}(x)}{x^3+x^2+x+1}$

(h) $\displaystyle\lim_{x\to 0} x\sin\left(\frac{1}{x}\right)$

(i) $\displaystyle\lim_{x\to 0} x\tan^{-1}\left(\frac{1}{x}\right)$

Problem 8.21. *Find the value of c for the Mean Value Theorem for the following functions and intervals* $[a,b]$.

(a) $f(x) = 4 - x^2$ on [-2,1]

(b) $g(x) = x^3$ on [-1,1]

(c) $h(x) = \ln(x)$ on [1,4]

(d) $r(x) = \sqrt{x}$ on [0,9]

(e) $s(x) = \sin(x)$ on $[0, \frac{\pi}{2}]$

(f) $q(x) = \tan^{-1}(x)$ on [0,1]

Problem 8.22. *For each part of Problem 8.21, carefully sketch the situation as in the examples in this section.*

Problem 8.23. *True or false:*

$$\lim_{x\to 0} x\cdot\sin\left(\frac{1}{x}\right)$$

is zero. Explain your answer.

Problem 8.24. *True or false:*

$$\lim_{x\to 0} \sqrt[3]{x}\cdot\cos\left(\frac{1}{x}\right)$$

is zero. Explain your answer.

Problem 8.25. *True or false:*

$$\lim_{x\to 0} \sqrt{x}\cdot\sin\left(\frac{1}{x}\right)$$

exists. Explain your answer.

Problem 8.26. *Construct an example of* $f(x)$, a *and* b *so that the point* c *satisfying the Mean Value Theorem is not unique.*

Problem 8.27. *Using the Mean Value Theorem, prove that*

$$|\sin(u) - \sin(v)| \le |u - v|$$

for any u, v.

Problem 8.28. *Find two values* a, b *such that the secant line of* $y = e^x$ *has a slope of* $m = 4$. *Having done this, find the formula of the tangent line with the same slope.*

Problem 8.29. *Find two values* a, b *such that the secant line of* $y = \ln(x)$ *has a slope of* $m = 2$. *Having done this, find the formula of the tangent line with the same slope.*

Problem 8.30. *Find the maximum slope of any secant line of*

$$y = \cos(x)$$

Problem 8.31. *Suppose we take a car trip and have a digital logger record our velocity throughout the trip. Since the car is a mechanical device in the real world, we know that velocity as a function of time is a continuous function. If the trip took three hours and we went 140 kilometers, what is the slope at the Mean Value Theorem point* c *for the logged function?*

Problem 8.32. *Show that there is a value* $x = a$ *so that the tangent line to* $y = \cos(x)$ *at* $(a, \cos(a))$ *goes through the point* $(4, 3)$.

Chapter 9

Simple Differential Equations

A **differential equation** is an equation that has one or more derivatives in it. The good news is that you already know how to solve many differential equations – by integrating.

Example 9.1. Solve:

$$y' = \cos(3x)$$

Solution:

Notice that this is a differential equation – we have a derivative that is equal to a relatively simple expression. This is the structurally simplest type of differential equation, one that we can solve simply by getting the x and y variables (and their differentials) on opposite sides of the equation and then integrating.

$$y' = \cos(3x)$$
$$\frac{dy}{dx} = \cos(3x)$$
$$dy = \cos(3x) \cdot dx$$
$$\int dy = \int \cos(3x)dx$$
$$y = \frac{1}{3}\sin(3x) + C$$

$\Diamond$

In this chapter we will look at simple differential equations like the one in Example 9.1 as well as more complex ones which require more sophisticated methods. You may want to review Chapters 4 and 7 for integration methods and Chapter 6, since polynomials are used in solving differential equations.

9.1. Separable Equations

A differential equation is **separable** if it is possible to get the the x and y terms of the equation on different sides of the equals sign. A very useful and relatively simple separable equation is that for exponential growth and decay.

Example 9.2. Suppose we know that the rate at which the amount $A(t)$ of something at a time t is changing is proportional to the amount of that substance present. If the original amount of the substance is $A_0 = A(0)$, then can we find an expression for $A(t)$?

Solution:

Derivatives are rates of change. So, the statement about $A(t)$ describes the differential equation

$$A'(t) = rA(t)$$

where the number r, the **constant of proportionality**, gives us the quantitative measurement of how $A'(t)$ depends on $A(t)$. This permits us to write and solve a differential equation:

$$A'(t) = rA(t)$$

$$\frac{dA}{dt} = r \cdot A$$

$$\frac{dA}{A} = r \cdot dt$$

$$\int \frac{dA}{A} = \int r \cdot dt$$

$$\ln(A) = rt + C$$

$$e^{\ln(A)} = e^{rt+C}$$

$$A = e^{rt} \cdot e^C$$

$$A(t) = Q \cdot e^{rt} \qquad\qquad \text{Set } Q = e^C, \text{ neaten up.}$$

$$A(0) = Q \cdot 1$$

$$A_0 = Q \qquad\qquad \text{We know } A(0) = A_0, \text{ solve for } Q$$

This gives us the exponential growth/decay equation:

$$A(t) = A_0 e^{rt}$$

◇

Knowledge Box 9.1.

The exponential growth and decay formula

If we have A_0 as the original amount of a substance and

$$A'(t) = r \cdot A(t)$$

then the closed form solution is:

$$A(t) = A_0 \cdot e^{rt}$$

The formula in Knowledge Box 9.1 is for exponential growth and decay – which is a little confusing. When $r > 0$ we get exponential growth like one might observe in a biological population. When $r < 0$, exponential decay occurs like with a radioactive substance. We call r the **constant of proportionality**. It is also called the **growth constant** (when $r > 0$) or the **decay constant** when $(r < 0)$.

In both these cases the behavior is intuitive: new creatures require parents, so the rate of growth of a biological population is proportional to its size. Similarly, radioactive decay is caused by particles emitted by individual decay events and so is faster when there is more of the substance present. Again, rate proportional to amount. Examples will help.

Example 9.3. Suppose that a population of 2000 bacteria increase to 8000 after an hour. If we assume they are undergoing exponential growth, then what is the formula of that growth?

Solution:

Assume that 2000 is the amount of bacteria at time $t = 0$ giving us that $T_0 = 2000$. Measure time in hours and get that

$$8000 = A(1) = A_0 \cdot e^{r \cdot 1} = 2000 \cdot e^r$$

or $e^r = 4$ and so $r = \ln(4) \cong 1.386$, meaning

$$A(t) = 2000e^{\ln(4)t} \cong 2000e^{1.329t}$$

◊

Example 9.4. Suppose that we have 6.4 grams of a radioisotope to begin with and 6.2 grams after five years. Find the decay constant r and compute the amount of the substance left after a century.

Solution:
Plug in $A_0 = 6.4$ and $t = 5$ and solve for r:

$$6.2 = 6.4 \cdot e^{5r}$$

$$\frac{6.2}{6.4} = e^{5r}$$

$$\ln\left(\frac{6.2}{6.4}\right) = 5r$$

$$r = \frac{1}{5}\ln\left(\frac{6.2}{6.4}\right)$$

$$r \cong -0.00635$$

So we have $A(t) = 6.4e^{-0.00635t}$. Plug in $t = 100$ years and we predict that the amount of the radioisotope left after a century is $A(100) \cong 3.4$ grams. This stuff decays fairly slowly.

$$\Diamond$$

Let's do some examples of separable equations with more than one nontrivial integral.

Example 9.5. Solve: $y' = yx^2$

Solution:
The strategy is to separate and solve.

$$y' = yx^2$$

$$\frac{dy}{dx} = yx^2$$

$$\frac{dy}{y} = x^2 \cdot dx$$

$$\int \frac{dy}{y} = \int x^2 \cdot dx$$

$$\ln(y) = \frac{1}{3}x^3 + C$$

$$e^{\ln(y)} = e^{\frac{x^3}{3} + C}$$

$$y = e^C \cdot e^{\frac{x^3}{3}}$$

$$y = A \cdot e^{\frac{x^3}{3}}$$

Notice that we put a 'x '+ C" only on one side. This is because the sum or difference of two unknown constants is just another unknown constant. Also the constant re-naming $A = e^C$ is a common practice. It's not required, but e to an unknown constant power is still an unknown constant, and the resulting expression is simpler to deal with.

$$\Diamond$$

Example 9.6. Solve: $y' = \sqrt{x}\sec(y)$

Solution:
Same strategy as the previous example:

$$y' = \sqrt{x}\sec(y)$$

$$\frac{dy}{dx} = \sqrt{x}\sec(y)$$

$$\frac{dy}{\sec(y)} = x^{1/2} \cdot dx$$

$$\cos(y) \cdot dy = x^{1/2} \cdot dx$$

$$\int \cos(y) \cdot dy = \int x^{1/2} \cdot dx$$

$$\sin(y) = \frac{2}{3}x^{3/2} + C$$

$$y = \sin^{-1}\left(\frac{2}{3}x^{3/2} + C\right)$$

After separation this is all pretty standard, except that the $+ C$ is inside an inverse sine. This kind of thing is common with separable differential equations.

$\Diamond$

Example 9.7. Solve:

$$y' = \frac{(x+1)}{y-1}$$

Solution:

$$\frac{dy}{dx} = \frac{(x+1)}{y-1}$$

$$\int (y-1)dy = \int (x+1)dx$$

$$\frac{1}{2}y^2 - y = \frac{1}{2}x^2 + x + C$$

$$y^2 - 2y - x^2 - 2x - 2C = 0 \qquad\qquad \text{Apply the quadratic formula.}$$

$$y = \frac{2 \pm \sqrt{4 + 4(x^2 + 2x + 2C)}}{2}$$

$$y = 1 \pm \sqrt{x^2 + 2x + 2C + 1} = 1 \pm \sqrt{x^2 + 2x + A}$$

Double check the quadratic: $a = 1$, $b = -2$, $c = -x^2 - 2x - 2C$.

$\Diamond$

We now return to models of growth with a famous differential equation that defines **logistic growth**. Exponential growth is a common model of biological growth, but it's not a plausible one. When we are studying the biophysics of water systems, bacteria that grow mats are very important. They can absorb (or create) pollutants. In this situation, however, the bacteria have severely limited growth potential. They may be inside a pipe or a small stream. Let's look at a separable differential equation that deals with self-limiting growth due to crowding.

The parameter K in Example 9.8 represents the maximum population the environment can support – the **carrying capacity** of the environment. The term $(K - P(t))$ makes the growth rate drop to zero as the population approaches the value K.

Example 9.8. Consider the differential equation:

$$P'(t) = rP(t) \cdot (K - P(t))$$

Solve this equation for $P(t)$ assuming that the population starts with P_0 members.

Solution:

The first move in solving this equation is to separate the variables. This means we need to use the fact that $P'(t) = \dfrac{dP}{dt}$. Here are the calculations:

$$P'(t) = rP(t) \cdot (K - P(t))$$

$$\frac{dP}{dt} = r \cdot P \cdot (K - P)$$

$$\frac{dP}{P(K - P)} = r \cdot dt$$

$$\int \frac{dP}{P(K - P)} = \int r \cdot dt \qquad\qquad \text{Separation complete.}$$

Integrate both sides.
We will need partial fractions.

$$\int \left(\frac{A}{P} + \frac{B}{K - P} \right) dP = rt + C$$

Do the partial fractions decomposition.

$$\frac{1}{P(K - P)} = \frac{A}{P} + \frac{B}{K - P}$$

$$1 = A(K - P) + BP$$

$$P = K \text{ gives us } B = \frac{1}{K}$$

$$P = 0 \text{ gives us } A = \frac{1}{K}$$

Using the partial fractions decomposition we get:

$$\frac{1}{K} \int \left(\frac{1}{K-P} + \frac{1}{P} \right) dP = rt + C$$

$$\frac{1}{K} \left(-\ln(K-P) + \ln(P) \right) = rt + C$$

$$\ln \left(\frac{P}{K-P} \right) = (rK)t + CK$$

$$\frac{P}{K-P} = e^{(rK)t + CK}$$

$$\frac{P}{K-P} = e^{At+B} \qquad \text{Rename the constants for clarity.}$$

$$\frac{P}{K-P} = e^{At} e^{B}$$

$$\frac{P}{K-P} = Qe^{At} \qquad \text{Rename } e^{B} = Q \text{ for clarity.}$$

$$P = KQe^{At} - PQe^{At}$$

$$P \left(1 + Qe^{At} \right) = KQe^{At}$$

$$P(t) = K \frac{Qe^{At}}{1 + Qe^{At}} \qquad \text{Recall: } K \text{ is the carrying capacity.}$$

So what did we get? We know from the original formulation of the equations that K is the maximum population the environment can support. The constant A is, roughly, the growth rate when the population is very small and controls the speed of the transition from a low to a high population, relative to K. The constant Q controls the population at time $t = 0$.

$\Diamond$

Example 9.9. Assume $K = 1$, meaning we are looking at "fraction of carry capacity," and graph the solution from Example 9.8 for $A = 1$ varying the Q parameter across $Q = 1, 2$ and 4.

Solution:

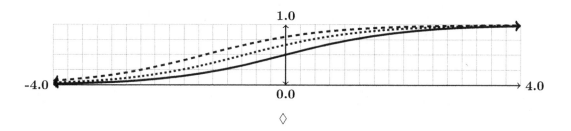

It is worth noting that the logistic solution

$$P(t) = K \cdot \frac{Qe^{At}}{1 + Qe^{At}}$$

joins $y = \tan^{-1}(x)$ as an example of a function that has two horizontal asymptotes, something you are asked to prove in the problems.

9.1.1. Verifying Solutions to Differential Equations. In Chapter 4 we learned that the solution to an integral can be checked by taking a derivative. If we claim that $y(t)$ solves a differential equation, then we can check the claim by plugging $y(t)$ into the differential equation to see if it satisfies it.

Example 9.10. Verify that $y = \cos(2x)$ solves the differential equation $y'' + 4y = 0$.

Solution:

Compute two derivatives, $y' = -2\sin(2x)$ and $y'' = -4\cos(x)$ and plug in:

$$y'' + y4 = -4\cos(2x) + 4\cos(2x) = 0$$

So we verify the solution.

$\Diamond$

The differential equation $y'' + cy = 0$ is the one that describes **simple periodic motion**. This is the situation that arises when the force felt by an object is proportional to the negative of the distance it is displaced from a starting position – as with a mass on a spring.

Knowledge Box 9.2.

Simple periodic motion

In a situation where $F = -mx$, where x is the distance from rest that an object is, the object undergoes simple periodic motion satisfying the differential equation

$$y'' + cy = 0$$

with solutions of the form

$$y = A \cdot \sin(\sqrt{c}\, x) + B \cdot \cos(\sqrt{c}\, x).$$

The constants A and B determine the particular solution of the differential equation and are computed from observations.

We include the differential equation for simple periodic motion at this point because it is quite useful in physics. In Section 9.3 we will learn how to find the solution to this type of differential equation.

Example 9.11. Verify that $y = \dfrac{-3}{x^3 + 1}$ solves $y' = x^2 y^2$.

Solution:

Compute that

$$y' = \frac{9x^2}{(x^3 + 1)^2}$$

with the reciprocal rule. Plug in and we get

$$x^2 y^2 = x^2 \cdot \left(\frac{-3}{x^3 + 1} \right)^2$$

$$= x^2 \cdot \frac{9}{(x^3 + 1)^2} = y'$$

which verifies the solution as desired.

Solving differential equations shares with integration the property that there are many methods, and it is often not clear which is the best for a given problem. Differential equations also share with integrals the annoying property that many of them do not have a closed form solution.

Problems

Problem 9.1. *Find the exponential growth or decay model for each of the following situations.*

(a) Deer population with $P(0) = 12$ and $P(8) = 3072$

(b) Radioisotope decay with $A(0) = 3g$ and $A(6) = 0.375g$

(c) Bacterial growth with $B(0) = 2500$ and $B(2) = 3160$

(d) Exotic particles with $E(0) = 100$ and $E(0.05) = 40$

(e) Fish population with $F(20) = 1000$ and $F(30) = 1600m$

(f) A mixture of heavy metals with $M(4) = 4.56$ and $M(7) = 4.03$.

Problem 9.2. *For each of the growth or decay situations in Problem 9.1, find the population at the specified time. The problem letters correspond.*

(a) Find $P(10)$

(b) find $A(9)$

(c) find $B(5)$

(d) find $E(0.03)$

(e) find $F(0)$

(f) find $M(20)$

Problem 9.3. *Solve the following separable equations.*

(a) $(1 + x^2)y' = x^3$

(b) $\frac{y'}{x} = \tan^{-1}(x)$

(c) $y' = \frac{x^3}{y^2}$

(d) $yy' = x^2 + x + 1$

(e) $y' = x^2 y + xy + y$

(f) $\cos(x)y' = y$

Problem 9.4. *Verify each of the following are solutions to the given differential equation.*

(a) $y = e^{3x}$ solves $y' - 3y = 0$

(b) $y = e^{2x}$ solves $y'' - 6y' + 8y = 0$

(c) $y = e^{4x}$ solves $y'' - 6y' + 8y = 0$

(d) $y = x^2 + 2x$ solves $y'' - y' + y = x^2$

(e) $y = \sin(0.5x)$ solves $4y'' + y = 0$

(f) $y = \sqrt{x^2 + 1}$ solves $y' = x/y$

Problem 9.5. *Prove that the solution to the logistic differential equation:*

$$P(t) = K \cdot \frac{Qe^{At}}{1 + Qe^{At}}$$

has horizontal asymptotes at $y = 0, K$.

Problem 9.6. *Show that*

$$y = A\sin(\sqrt{c}\,x) \text{ and } y = B\cos(\sqrt{c}\,x)$$

are both solutions to the differential equation $y'' + cy = 0$.

Problem 9.7. *Carefully graph the following simple periodic motion solutions.*

(a) $y = \sin(t)$

(b) $y = \cos(t)$

(c) $t = \frac{1}{2}(\sin(x) + \cos(t))$

(d) $t = \frac{1}{3}(\sin(x) + 2\cos(x))$

Problem 9.8. *If*

$$P(t) = \frac{120e^{2t}}{1 + e^{2t}}$$

at what time is the population $P(t)$ growing the most rapidly?

Problem 9.9. *Which of the following solve the differential equation $y'' - 3y' + 2y = 0$?*

(a) $y = e^x$

(b) $y = e^{-x}$

(c) $y = e^{2x}$

(d) $y = e^{-2x}$

(e) $y = 3e^x - e^{2x}$

(f) $y = e^{-x} + e^{-2x}$

9.2. Integrating Factors

In this section we will develop a technique called **integrating factors** that will let us solve some differential equations of the form:

$$y' + P(t) \cdot y = Q(t)$$

This method is frankly a trick that sets up a product rule. The trick works when the integrals that make it possible can be performed. Let's work through the derivation. Set

$$I(t) = e^{\int P(t)dt}$$

Then $I(t)' = e^{\int P(t)dt} \cdot P(t) = I(t) \cdot P(t)$. Now compute:

$$y' + P(t) \cdot y = Q(t)$$

$$I(t) \cdot y' + I(t) \cdot P(t) \cdot y = I(t) \cdot Q(t)$$

$$I(t) \cdot y' + I(t)' \cdot y = I(t) \cdot Q(t) \qquad \text{Left side is a product rule!}$$

$$(I(t) \cdot y)' = I(t) \cdot Q(t)$$

$$I(t) \cdot y = \int I(t) \cdot Q(t)dt$$

Knowledge Box 9.3.

Solving differential equations with an integrating factor

If $y' + yP(t) = Q(t)$ and we set

$$I(t) = e^{\int P(t) \cdot dt}$$

then a solution of this differential equation has the form

$$y = \frac{1}{I(t)} \int I(t) \cdot Q(t)dt$$

This is a fairly special purpose technique for solving differential equations – but this is something a specialist in differential equations must get used to. A large proportion of the techniques for solving differential equations are fairly special purpose. Let's practice this technique with some examples.

Example 9.12. Solve

$$y' + \frac{y}{t} = e^t$$

Solution:

First we compute $I(t)$. Looking at the equation we have $P(t) = 1/t$ and $Q(t) = e^t$.

$$I(t) = e^{\int dt/t}$$
$$= e^{\ln(t)} \qquad\qquad\qquad = t$$

So, using the integrating factor formula we see that

$$y = \frac{1}{t} \int t \cdot e^t dt$$

$$= \frac{1}{t} \cdot \left((t-1)e^t + C\right) \qquad\qquad \text{shortcut in Knowledge Box 7.5}$$

$$= \frac{t-1}{t}e^t + \frac{C}{t}$$

$$\Diamond$$

As before, when we are solving differential equations, the constants get more mixed into the final expression than they do when we are doing integration.

Example 9.13. Verify the solution in Example 9.12.

Solution:

We have $y = \dfrac{t-1}{t}e^t + \dfrac{C}{t}$.

So, $y' = \dfrac{1}{t^2}e^t + \dfrac{t-1}{t}e^t - \dfrac{C}{t^2}$.

$$y' + \frac{y}{t} = \underbrace{\frac{1}{t^2}e^t + \frac{t-1}{t}e^t - \frac{C}{t^2}}_{y'} + \underbrace{\frac{t-1}{t^2}e^t + \frac{C}{t^2}}_{y/t}$$

$$= \frac{1}{t^2}e^t + e^t - \frac{1}{t}e^t + \frac{1}{t}e^t - \frac{1}{t^2}e^t$$

$$= e^t$$

verifying the solution.

$$\Diamond$$

Example 9.14. Solve the differential equation:

$$y' + \frac{y}{t} = \frac{\cos(t)}{t}$$

Solution:

Examining the problem we see $P(t) = \frac{1}{t}$, and so $I(t) = t$ as in Example 9.12. This gives us that

$$y = \frac{1}{t} \int t \cdot \frac{\cos(t)}{t} \cdot dt$$

$$= \frac{1}{t} \int \cos(t) dt$$

$$= \frac{1}{t} \left(\sin(t) + C \right)$$

$\Diamond$

Example 9.15. Solve

$$\cos(t)y' + \sin(t)y = 1$$

Solution:

First we divide through by $\cos(t)$ to put the problem in the standard form:

$$y' + \tan(t)y = \sec(t)$$

This means that

$$I(t) = e^{\int \tan(t) \cdot dt} = e^{\ln|\sec(t)|} = \sec(t)$$

and we are ready to apply the integrating factor method.

$$y = \frac{1}{\sec(t)} \int \sec(t) \cdot \sec(t) \cdot dt$$

$$= \cos(t) \int \sec^2(t) \cdot dt$$

$$= \cos(t) \left(\tan(t) + C \right)$$

$$= \sin(t) + C \cdot \cos(t)$$

$\Diamond$

This example makes the important point that we may need algebra to set up the integrating factors.

Another possibility for integrating factor problems is that we may be able to solve for C if one additional piece of information is available.

Example 9.16. Suppose $(t+1)y' + y = (t+1)e^t$ and $y(0) = 4$. Find an expression for y.

Solution:

Divide through to place the problem in standard form:

$$y' + \frac{y}{t+1} = e^t$$

and see that $P(t) = \dfrac{1}{t+1}$ so

$$I(t) = e^{\int \frac{dt}{t+1}} = e^{\ln(t+1)} = t+1$$

Now apply the method of integrating factors and we obtain:

$$y = \frac{1}{t+1} \int (t+1)e^t dt$$

$$= \frac{1}{t+1} \left((t+1-1)e^t + C \right)$$

$$= \frac{1}{t+1} \left(te^t + C \right)$$

Plug in $y(0) = 4$.

$$4 = \frac{1}{1+0} (0 + C)$$

$$4 = C$$

Which makes the complete solution

$$y = \frac{te^t + 4}{t+1}$$

$\Diamond$

The method of integrating factors shows up in problems related to physics – something that makes it valuable in spite of its special purpose nature. It also permits us to use our other calculus skills in a complex, integrated fashion. Algebra, integration, and more algebra are all part of performing the method of integrating factors.

Problems

Problem 9.10. *Solve each of the following using the method of integrating factors.*

(a) $y' + \dfrac{3y}{t} = e^t$

(b) $y' + \dfrac{y}{t^2} = e^{1/t}$

(c) $y' + \dfrac{y}{3t + 2} = e^t$

(d) $y' + \dfrac{2y}{t} = \dfrac{\cos(t)}{t^2}$

(e) $y' - \dfrac{y}{t} = -te^{-t}$

(f) $y' + \dfrac{5y}{t} = t^3$

Problem 9.11. *Solve the general problem*
$$y' + \frac{qy}{at + b} = e^t$$

Problem 9.12. *Solve the general problem*
$$y' + \frac{qy}{at + b} = e^{-t}$$

Problem 9.13. *Solve the general problem*
$$y' + \frac{qy}{at + b} = \sin(t)$$

Problem 9.14. *Solve the general problem*
$$y' + \frac{qy}{at + b} = \cos(t)$$

Problem 9.15. *For each of the parts of Problem 9.10, verify your solution is correct by plugging it into the original problem and simplifying.*

Problem 9.16. *Solve:*
$$ty' + ky = t^2 + t + 1$$

Problem 9.17. *Solve:*
$$ty' + ky = t^3 + t^2 + t + 1$$

Problem 9.18. *Solve:*
$$ty' + ky = t^4 + 4$$

Problem 9.19. *Solve each of the following using the method of integrating factors.*

(a) $ty' + 4y = e^t$,

(b) $\cos(t)y' + y = \cos^2(t)$

(c) $ty' + 3y = \dfrac{t + 1}{t}$

(d) $ty' + 4y = \cos(t)$

(e) $t^2y' + ty = t^4 e^t$

(f) $\cos(t)y' + \sin(t)y = 4$

Problem 9.20. *Completely solve:*
$$y' + \frac{3y}{t + 2} = e^t$$
if $y(0) = 3$.

Problem 9.21. *Completely solve:*
$$y' + \frac{2y}{t + 4} = e^t$$
if $y(0) = 0.75$.

Problem 9.22. *Completely solve:*
$$y' + \frac{y}{t} = \frac{\cos(t)}{t}$$
if $y(\pi) = 0.5$.

Problem 9.23. *Completely solve:*
$$\cos(t)y' + \sin(t)y = 2$$
if $y(0) = 4$

Problem 9.24. *Completely solve:*
$$\sin(t)y' + \cos(t)y = 1$$
if $y(0) = 0$

Problem 9.25. *Completely solve:*
$$y't + y = \cos(t)$$
if $y(\pi/2) = 0.5$

Problem 9.26. *Completely solve:*
$$y't + y = \sin(t)$$
if $y(2) = 0.15$

9.3. Linear Second Order Homogeneous Equations

The name of this section is a little intimidating, so let's start by defining what these things are. A **linear second order homogeneous differential equation** is a differential equation of the form

$$ay'' + by' + cy = 0$$

which isn't so bad. **Second order** means that the highest derivative that appears is the second; **linear** means, roughly, that the the sums of multiples of solutions to these differential equations are also solutions to these differential equations, and **homogeneous** means that the right hand side – without y's – is zero.

Solving this type of differential equation sets the stage for more complex types of differential equations later; nonlinear, non-homogeneous, and higher order. These equations are solved by assuming that the solution has the form

$$y = Ve^{rt}$$

with $V \neq 0$ and then looking at how that solution plays out.

Example 9.17. Suppose that $y = Ve^{rt}$; determine when it solves

$$ay'' + by' + cy = 0$$

Solution:

Start by computing the derivatives: $y = Ve^{rt}$ so $y' = Vre^{rt}$ and $y'' = Vr^2e^{rt}$. Plug in and simplify:

$$ay'' + by' + cy = 0$$

$$aVr^2e^{rt} + bVre^{rt} + cVe^{rt} = 0$$

$$Ve^{rt}\left(ar^2 + br + c\right) = 0$$

Since the quantity Ve^{rt} cannot be zero, the function $y = Ve^{rt}$ solves the differential equation when

$$ar^2 + br + c = 0.$$

In other words, the solutions are of the form $y = Ve^{rt}$ where r is a root of the quadratic equation $ax^2 + bx + c = 0$.

$$\diamond$$

This has the interesting effect that there is more than one type of solution – one for each root. There is more, related to linearity, that we need.

- A constant multiple of a solution to a linear homogeneous differential equation is also a solution.

- The sum of solutions to a linear homogeneous differential equation is also a solution.

Example 9.18. Find the general form of the solution to

$$y'' - 3y' + 2y = 0.$$

Solution:

According to the work we have already done, the solutions should be constant multiples of e^{rt} where r is a root of $r^2 - 3r + 3 = (r - 2)(r - 1) = 0$. This means $r = 1, 2$, and we get that Ae^t and Be^{2t} are both solutions. Using the sum rule the general solution is

$$y = Ae^t + Be^{2t}$$

$\Diamond$

Definition 9.1. For the differential equation $ay'' + by' + cy = 0$ we call

$$p(r) = ar^2 + br + c$$

the **characteristic polynomial** of the differential equation.

The characteristic polynomial of a differential equation is exactly the object whose roots we need to get the solution. Abstracting it out permits us to understand the solution of differential equations in terms of the roots of polynomials. There are three classes of solutions, shown in Knowledge Box 9.4, that depend on how the roots come out. We have derived the first class, and there is a hint in Knowledge Box 9.4 as to how to derive the third class, but deriving the second class of solutions involves mathematics that is beyond the scope of this course.

Knowledge Box 9.4.

The general solution to $ay'' + by' + cy = 0$

Suppose that we have the differential equation $ay'' + by' + cy = 0$ with characteristic polynomial $p(r) = ar^2 + br + c$, then the solution to the differential equation depends on the roots.

- If $p(r)$ has two distinct real roots r_1, r_2, then the general solution is:
$$y(t) = Ae^{r_1 t} + Be^{r_2 t}$$

- If $p(r)$ has one real root r, then the general solution is:
$$y(t) = Ae^{rt} + Bte^{rt}$$

- If $p(r)$ has no real roots, then it has two complex roots of the form $u \pm vi$, and by using Euler's identity we can find that the general solution has the form:
$$y(t) = e^{ut}\left(A\sin(vt) + B\cos(vt)\right)$$

Let's try this out.

Example 9.19. Find the general solution to

$$y'' - 4y'' + 4y = 0.$$

Solution:

We begin by extracting the characteristic polynomial $p(r) = r^2 - 4r + 4 = (r - 2)^2$. Using the general solution rule we get that:

$$y(t) = Ae^{2t} + Bte^{2t}$$

$$\Diamond$$

The constants A and B play the same role as the $+\,C$ we get when performing integration. Since the differential equation is second order, solving it creates two unknown constants, rather than one, and so we will need two added pieces of information to find the value of these constants.

Example 9.20. Solve $y'' - 6y' + 8y = 0$ if $y(0) = 4$ and $y'(0) = 14$.

Solution:

The characteristic polynomial is $p(r) = r^2 - 6r + 8 = (r - 2)(r - 4)$. So we get that the general solution is $y(t) = Ae^{2t} + Be^{4t}$. This also tells us that $y'(t) = 2Ae^{2t} + 4Be^{4t}$. Plugging in the facts that $y(0) = 4$ and $y'(0) = 14$, we get that:

$$
\begin{aligned}
4 &= A + B \\
14 &= 2A + 4B \\
8 &= 2A + 2B && \text{Double the first equation} \\
6 &= 2B && \text{Subtract the second and third lines} \\
B &= 3 \\
A &= 1 && \text{Substitute into the first line.}
\end{aligned}
$$

So:

$$y(t) = e^{2t} + 3e^{4t}$$

$$\Diamond$$

Example 9.21. Find the general solution to

$$y'' - 2y' + 5y = 0$$

Solution:

The characteristic polynomial is $p(r) = r^2 - 2r + 5$. Use the quadratic equation to get that $r = 1 \pm 2i$. So the general solution is

$$y(t) = e^t \left(A\sin(2t) + B\cos(2t) \right)$$

$$\Diamond$$

Example 9.22. Solve

$$y'' + 4y' + 4y = 0$$

if $y(0) = 3$ and $y'(0) = -4$.

Solution:

The characteristic polynomial is $r^2 + 4r + 4$. So we have a repeated real root $r = -2$. This means the general solution is

$$y(t) = Ae^{-2t} + Bte^{-2t}$$

and that $y'(t) = -2Ae^{-2t} + Be^{-2t} - 2Bte^{-2t}$. Plugging in the known values for $y(0)$ and $y'(0)$ we get:

$$3 = A$$
$$-4 = -2A + B$$

so

$$A = 3$$
$$B = 2$$

Giving us a solution of

$$y(t) = 3e^{-2t} + 2te^{-2t}$$

$$\Diamond$$

At this point we can look again at the equation for simple periodic motion from Knowledge Box 9.2:

$$y'' + cy = 0$$

This equation has characteristic polynomial $p(r) = r^2 + c$, which has roots

$$r = 0 \pm \sqrt{c}\, i$$

Plugging into the general solution, we get

$$y(t) = e^0 \left(A\sin(\sqrt{c}\, t) + B\cos(\sqrt{c}\, t) \right) = A \cdot \sin(\sqrt{c}\, x) + B \cdot \cos(\sqrt{c}\, x)$$

which is exactly the solution given and verified in Section 9.1.

The generalization of this material to more complex differential equations typically happens in a course called **Differential Equations**. In this section we guess at the solution and manage to verify that it works. Once we have the general solution to $ay'' + by' + cy = 0$, most of what we do is plugging in – with possibly the additional work of solving a system of linear equations for the unknown constant.

Problems

Problem 9.27. *For each of the following differential equations, find the characteristic polynomial.*

(a) $y'' - y' - 2y = 0$

(b) $y'' - 2y' + y = 0$

(c) $y'' - 6y' + 8y = 0$

(d) $y'' + y' + y = 0$

(e) $y'' + 9y = 0$

(f) $y'' + 2y' + 3y = 0$

(g) $y''' - 8y = 0$

(h) $y''' + 4y'' - 2y' + 4$

Problem 9.28. *For the differential equations in parts a-f of Problem 9.27 find the general solution.*

Problem 9.29. *Suppose $y = f(t)$ and $y = g(t)$ are both solutions to*

$$ay'' + by' + cy = 0.$$

Show that $y = A \cdot f(t) + B \cdot g(t)$ is also a solution.

Problem 9.30. *Find the general solution to*

$$y'' + y' + y = 0.$$

Problem 9.31. *Find the general solution to*

$$y'' + 2y' + 2y = 0.$$

Problem 9.32. *Find the general solution to*

$$y'' - y' + 2y = 0.$$

Problem 9.33. *Given a differential equation with general solution:*

$$y(t) = e^{at}\left(A\sin(\sqrt{c}\,t) + B\cos(\sqrt{c}\,t)\right)$$

Discuss the possible behaviors of the solution when $a > 0$, $a = 0$, and $a < 0$.

Problem 9.34. *Find a solution to each of the following differential equations.*

(a) $y'' - 3y' + 2y = 0$
   if $y(0) = 6$ and $y'(0) = 8$

(b) $y'' - 4y' + 3y = 0$
   if $y(0) = 3$ and $y'(0) = 1$

(c) $y'' + 3y' + 2y = 0$
   if $y(0) = 3$ and $y'(0) = -4$

(d) $y'' - 2y' + y = 0$
   if $y(0) = 8$ and $y'(0) = 11$

(e) $y'' + 4y' + 4y = 0$
   if $y(0) = -1$ and $y'(0) = 2$

(f) $y'' + y' + y = 0$
   if $y(0) = 1$ and $y'(0) = \sqrt{3}$

(g) $y'' - 5y' + 6y = 0$
   if $y(0) = 3$ and $y'(0) = 7$

(h) $y'' + 5y' + 6y = 0$
   if $y(0) = 1$ and $y'(0) = 7$

Problem 9.35. *Which of the types of solutions in Problem 9.33 have limits at infinity?*

Problem 9.36. *Find a general solution to*

$$y''' - 6y'' + 11y' + 6y = 0$$

Problem 9.37. *Find the general solution to*

$$y'''' - 16y = 0$$

Problem 9.38. *Find a second order homogeneous equation that has each of the following as a solution.*

(a) $y = 2e^t - e^{-t}$

(b) $y = 5e^{2t} + e^{-3t}$

(c) $y = 4e^{-3t} - e^{5t}$

(d) $y = e^{2t} + e^t$

(e) $y = 24e^{-2t} - e^{12t}$

(f) $y = 17e^{4t} + e^t$

(g) $y = 5e^{4t} + 3te^{4t}$

(h) $y = e^{3t} - te^{3t}$

Chapter 10

Advanced Derivatives

This chapter deals with the issues of derivatives of functions that have multiple independent variables. In Chapter 2 we learned to take derivatives of functions of the form $y = f(x)$. Now we will learn to work with functions of the form $z = f(x, y)$, functions that graph as surfaces in a space like the one shown in Figure 10.1.

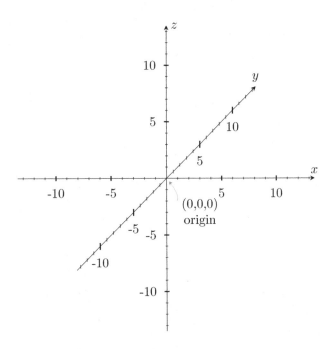

Figure 10.1. Three dimensional coordinate system.

Where before we had points (x, y), we now have points (x, y, z). We've looked at the graph of functions like $y = x^2$ over and over. Now let's look at the three-dimensional analog:

$$z = x^2 + y^2$$

for $-4 \leq x, y \leq 4$. A graph of the function is shown in Figure 10.2.

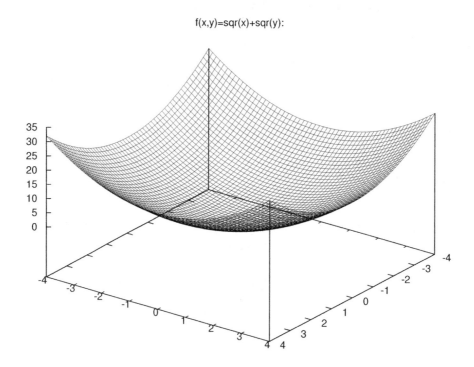

Figure 10.2. A graph of $f(x, y) = x^2 + y^2$ for $-4 \leq x, y \leq 4$.

10.1. Partial Derivatives

So far in this text we have had one independent and one dependent variable. Now we have two independent variables, which has a huge effect on derivatives. The number of directions on the number line is two – left and right or plus and minus. As we learned in Chapter 5, there are an infinite number of directions when there are two variables to choose directions amongst. Each unit vector starting at the origin points in a different direction. Every single vector can be written as:

$$\vec{v} = (a, b) = a \cdot (1, 0) + b \cdot (0, 1)$$

In other words, each vector is a combination of the two fundamental vectors $\vec{e}_1 = (1, 0)$ and $\vec{e}_2 = (0, 1)$.

On the surface of a function $z = f(x,y)$ the function has a rate of change in every direction at every point. Pick a direction – the slope in that direction is the rate of change in that direction. The basis for all of these rates of change are the *partial derivatives* – derivatives in the direction of x and y.

Knowledge Box 10.1.

Partial Derivatives

If $z = f(x,y)$ is a function of two variables, then there are two fundamental derivatives

$$z_x = \frac{\partial f}{\partial x} \text{ and } z_y = \frac{\partial f}{\partial y}$$

These are called the **partial derivatives** of z (or f) with respect to x and y, respectively.

In order to take the partial derivative of a function with respect to one variable, all other variables are treated as constants. This is most easily understood through examples.

Example 10.1. Suppose that $z = x^2 + 3xy + y^2$. Find z_x and z_y.

Solution:

$$z_x = 2x + 3y \text{ and } z_y = 3x + 2y$$

To see this, notice that when we are taking the derivative with respect to x the derivative of $3xy$ is $3y$ *because y is treated as a constant*. Similarly, the derivative of $3xy$ with respect to y is $3x$. The derivatives for x^2 and y^2 are $2x$ and $2y$ when they are the active variable and zero when the other variable is active, because the derivative of a constant is zero.

It is important to remember that every derivative rule we have learned so far applies when we are taking partial derivatives – the product, quotient, and chain rules and all the individual formulas for functions.

Example 10.2. Find $\dfrac{\partial f}{\partial x}$ if

$$f(x,y) = \frac{x}{x^2 + y^2}$$

Solution:

This problem requires the quotient rule. Since a "prime" hash-mark doesn't carry its identity (with respect to x or with respect to y) we used the symbol $\dfrac{\partial}{\partial x}$ to mean "derivative with respect

to x". This means that

$$\frac{\partial}{\partial x}\left(\frac{x}{x^2+y^2}\right) = \frac{\frac{\partial}{\partial x}x \cdot (x^2+y^2) - \frac{\partial}{\partial x}(x^2+y^2) \cdot x}{(x^2+y^2)^2}$$
$$= \frac{1 \cdot (x^2+y^2) - 2x \cdot x}{(x^2+y^2)^2}$$
$$= \frac{y^2 - x^2}{(x^2+y^2)^2}$$

$$\Diamond$$

The next example uses the chain rule. It also uses the additional notation "f_y" as another way of saying "the partial derivative of $f(x,y)$ with respect to y."

Example 10.3. Find $f_y(x,y)$ if

$$f(x,y) = \sin(2xy + 1)$$

Solution:

$$f_y(x,y) = \cos(2xy + 1) \cdot 2x$$

$$\Diamond$$

As long as we remember that y is the active variable and x is treated as a constant, this is not difficult.

Example 10.4. Find the partial derivatives with respect to x and y of

$$f(x,y) = (3x + 4y)^5$$

Solution:

$$f_x(x,y) = 5(3x + 4y)^4 \cdot 3$$

$$f_y(x,y) = 5(3x + 4y)^4 \cdot 4$$

The only point where things are different is the way the chain rule acts depending on if x or y is the active variable.

Example 10.5. Find the partial derivatives with respect to x and y for

$$f(x,y) = \frac{x}{y}$$

Solution:

$$f_x(x,y) = \frac{1}{y} \qquad f_y(x,y) = \frac{-x}{y^2}$$

For f_x, $\frac{1}{y}$ is effectively a constant, while, for f_y, we employ the reciprocal rule while x plays the part of a constant.

◊

10.1.1. Implicit Partial Derivatives. Since natural laws are often stated in the form of equations that are not in functional form, implicit derivatives are very useful in physics. It turns out that implicit partial derivatives are a lot like standard partial derivatives – as long as you remember which variable is active.

Example 10.6. Find z_x and z_y if

$$x^2 + y^2 + z^2 = 16$$

Solution:

z is the dependent variable and so gets a z_x or a z_y each time we take a derivative of it, while x and y take turns being the active variable and a constant respectively. So,

$$2x + 2z \cdot z_x = 0 \text{ and } 2y + 2z \cdot z_y = 0$$

Simplifying we get that

$$z_x = -\frac{x}{z} \text{ and } z_y = -\frac{y}{z}$$

◊

Example 10.7. Find z_y if

$$(xyz + 1)^3 = 6y$$

Solution:

In this example z is the dependent variable, y is the active variable, and x is acting like a constant. So:

$$3(xyz + 1)^2(xz + xy \cdot z_y) = 6$$

$$xz + xy \cdot z_y = \frac{2}{(xyz + 1)^2}$$

$$z_y = \frac{2}{xy \cdot (xyz + 1)^2} - \frac{z}{y}$$

◊

10.1.2. Higher-order Partial Derivatives. In earlier chapters, when we wanted the second derivative of a function we just took the derivative again. The fact that we have multiple choices of which derivative to take complicates this. If the first derivative is with respect to x or y and so is the second, then we get four possible nominal second derivatives:

$$f_{xx} \qquad\qquad f_{xy} \qquad\qquad f_{yx} \qquad\qquad f_{yy}$$

A useful fact saves us from having the number of higher order partial derivatives explode.

Knowledge Box 10.2.

When working with $f(x, y)$,

$$f_{xy}(x, y) = f_{yx}(x, y)$$

and in general the order in which partial derivatives are taken does not affect the result of taking them.

Example 10.8. If

$$f(x, y) = x^3 + 3xy + y^2$$

find f_{xx}, f_{xy}, f_{yx} and f_{yy}, and verify that $f_{xy} = f_{yx}$.

Solution:

Start by finding f_x and f_y and then keep going.

$$f_x = 3x^2 + 3y$$
$$f_y = 3x + 2y$$
$$f_{xx} = 6x$$
$$f_{xy} = 3$$
$$f_{yx} = 3$$
$$f_{yy} = 2$$

And we see that $f_{xy} = 3 = f_{yx}$, achieving the desired verification. From this point on we will only compute one of the two mixed partials f_{xy} and f_{yx}.

◇

The "order doesn't matter" rule means, for example, that $f_{xxy} = f_{xyx} = f_{yxx}$. So the degree to which this rule reduces the number of higher order derivatives increases with the order of the derivative.

Example 10.9. Find f_{xx}, f_{xy}, and f_{yy} for

$$f(x, y) = x^2 \sin(y).$$

Solution:

Compute the first partials first and keep going.

$$f_x = 2x \sin(y)$$
$$f_y = x^2 \cos(y)$$
$$f_{xx} = 2 \sin(y)$$
$$f_{xy} = 2x \cos(y)$$
$$f_{yy} = -x^2 \sin(y)$$

$\Diamond$

Example 10.10. Find f_{xy} for

$$f(x, y) = \frac{xy}{x^2 + 1}.$$

Solution:

The fact that we get the same result by computing the partials with respect to x and y in either order means that we may choose the order to minimize our work. The order y then x is easier, because y vanishes.

$$f(x, y) = y \cdot \frac{x}{x^2 + 1}$$

$$f_y = \frac{x}{x^2 + 1}$$

$$f_{xy} = \frac{(x^2 + 1)(1) - x(2x)}{(x^2 + 1)^2} \qquad \text{Quotient rule.}$$

$$= \frac{1 - x^2}{(x^2 + 1)^2}$$

$\Diamond$

Problems

Problem 10.1. *For each of the following functions find f_x and f_y.*

(a) $f(x,y) = 2x^2 + 3xy + y^2 + 4x + 2y + 7$

(b) $g(x,y) = \sin(xy)$

(c) $h(x,y) = x^3 y^3$

(d) $r(x,y) = \ln(x^2 + y^2 + 1)$

(e) $s(x,y) = \dfrac{1}{x^2 + y^2 + 1}$

(f) $q(x,y) = \dfrac{x-y}{x+y}$

(g) $a(x,y) = e^x \sin(y)$

(h) $b(x,y) = (x^2 + xy + 1)^6$

Problem 10.2. *For each of the following functions find f_{xx}, f_{xy}, and f_{yy}.*

(a) $f(x,y) = x^2 - 5xy + 2y^2 + 3x - 6y + 11$

(b) $g(x,y) = \tan^{-1}(xy)$

(c) $h(x,y) = (x^3 + 1)(y^3 + 1)$

(d) $r(x,y) = e^{x^2 + y^2 - 5}$

(e) $s(x,y) = \dfrac{1}{x^2 + y^2 + 1}$

(f) $q(x,y) = \dfrac{2x - 3y}{5x + y}$

(g) $a(x,y) = \ln(x \sin(y))$

(h) $b(x,y) = (2x^2 - xy)^4$

Problem 10.3. *Find z_x and z_y if*
$$(xyz + 2)^3 = 4$$

Problem 10.4. *Find z_x and z_y if*
$$\frac{x}{yz} = 1$$

Problem 10.5. *Find z_x and z_y if*
$$\frac{x+z}{y+z} = 4$$

Problem 10.6. *Find z_x and z_y if*
$$\frac{3x + z}{2z} = 4xy$$

Problem 10.7. *Find z_x and z_y if*
$$\cos(x + y + z) = \frac{\sqrt{2}}{2}$$

Problem 10.8. *Find z_x and z_y if*
$$\tan^{-1}(z - xy) = 1$$

Problem 10.9. *Find z_x and z_y if*
$$(xy + xz + yz)^3 = 16$$

Problem 10.10. *For each of the following functions find f_{xy}.*

(a) $f(x,y) = x^2 y + x^3 y + y \sin(x)$

(b) $g(x,y) = x \sin(y) + x \tan^{-1}(x)$

(c) $h(x,y) = xy^2 + x^2 y + xy$

(d) $r(x,y) = \left(x(y+1)^5 + 1 \right)^2$

(e) $s(x,y) = x \cdot \sin(xy)$

(f) $q(x,y) = y \cdot \cos(xy)$

(g) $a(x,y) = \dfrac{x}{y} + \dfrac{y}{x} + y^3$

(h) $b(x,y) = (1 + x + y + xy)^4$

Problem 10.11. *Find f_x, f_y, f_{xx}, f_{xy}, and f_{yy} if*
$$f(x,y) = \frac{x^2}{x^2 + y^2}$$

Problem 10.12. *Find g_x, g_y, g_{xx}, g_{xy}, and g_{yy} if*
$$g(x,y) = \tan^{-1}(xy + 1)$$

Problem 10.13. *If*
$$h(x,y) = \sin(xy)$$
find f_{xx}, f_{xxy}, and f_{xxyy}.

Problem 10.14. *Find z_x and z_y if*
$$z = (x^2 + 1)^y$$

Problem 10.15. *Find z_x and z_y if*
$$z^{xy} = 2$$

Problem 10.16. *Find z_x and z_y if*
$$z = \frac{x^3 (x+1)^2 (x-1)^3}{y^2 (y-1)^3 (y+1)^5}$$

10.2. The Gradient and Directional Derivatives

We are now ready to look at some of the opportunities that are available once we understand partial derivatives. At any point on the surface that forms the graph of $z = f(x, y)$ there are an infinite number of directions and so an infinite number of rates at which the function is changing. Pick a direction, and the function has a rate of change *in that direction*.

It turns out that there is some order to this richness of directions and rates of growth, in the form of a simple formula for the direction in which the function is growing fastest.

Knowledge Box 10.3.

If $z = f(x, y)$ is a function of two variables, then
$$\nabla f(x, y) = (f_x, f_y)$$
is called the **gradient** of $f(x, y)$. The gradient of a function points in the direction it is growing most quickly; the rate of growth is the magnitude of the gradient.

Example 10.11. Find the gradient of the function $f(x, y) = x^2 + y^2 + 3xy$.

Solution:

Using the formula given, $\nabla f(x, y) = (2x + 3y, 3x + 2y)$.
$\Diamond$

We can ask much more complex questions about the gradient than simply computing its value.

Example 10.12. At what points is the function
$$g(x, y) = \sin(x) + \cos(y)$$
changing the fastest in its direction of maximum increase?

Solution:

This question wants us to maximize the magnitude of the gradient. First compute the gradient:
$$\nabla g(x, y) = (\cos(x), -\sin(y))$$
The magnitude of this is
$$\sqrt{\cos^2(x) + \sin^2(y)}$$
Since x and y vary independently, the answer is simply those points that make $\cos^2(x)$ and $\sin^2(y)$ both one, and so the answer is those points (x, y) such that
$$x = n\pi \text{ and } y = \frac{2m+1}{2}\pi$$
where n and m are whole numbers.
$\Diamond$

The graph in Figure 10.3 might help you understand Example 10.12.

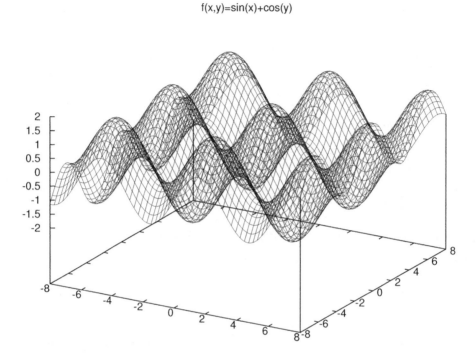

Figure 10.3. A graph of $z = \sin(x) + \cos(y)$ for $-8 \le x, y \le 8$.

Example 10.13. Find the gradient of

$$z = x^y.$$

Solution:

The partial derivative with respect to x is not hard because y is treated as a constant – so $z_x = yx^{y-1}$. The partial derivative with respect to y is trickier. It uses the formula for the derivative of a constant to a variable power (Knowledge Box 2.7) which gives $z_y = x^y \cdot \ln(x)$. This makes the gradient

$$\nabla z = \left(yx^{y-1}, x^y \cdot \ln(x) \right)$$

$\Diamond$

What is the physical meaning of the gradient? We have already noted that it points in the direction in which a surface grows fastest away from the point where the gradient is computed – the steepest uphill slope away from the point. This means that the negative of the function

$$-\nabla f(x, y)$$

is the steepest downhill slope away from the point. In other words, the negative of the gradient is the direction that a ball, starting at rest, will roll.

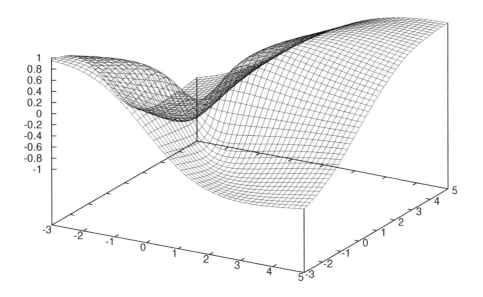

f(x,y)=2xy/(x^2+y^2+1)

Figure 10.4. A graph of $H(x, y) = \dfrac{2x + y}{x^2 + y^2 + 1}$ for $-3 \le x, y \le 5$

Example 10.14. Suppose that the function $H(x, y) = \dfrac{2x + y}{x^2 + y^2 + 1}$ (Figure 10.4) describes the height of a surface. Which direction will a ball placed at the point $(1, 2)$ roll?

Solution:

We need to compute the negative of the gradient of $H(x, y)$.

$$H_x(x, y) = \frac{(x^2 + y^2 + 1)(2) - (2x + y)(2x)}{(x^2 + y^2 + 1)^2}$$

$$H_x(1, 2) = \frac{12 - 8}{36} = 1/9$$

$$H_y(x, y) = \frac{(x^2 + y^2 + 1)(1) - (2x + y)(2y)}{(x^2 + y^2 + 1)^2}$$

$$H_y(1, 2) = \frac{6 - 8}{36} = -1/18$$

So the ball rolls in the direction of the vector

$$-\nabla H(1, 2) = \left(\frac{-1}{9}, \frac{1}{18} \right)$$

$\Diamond$

Notice that, if we were designing a game, then we could use a well-chosen equation to give us a height map for the surface, and the gradient could be used to tell which ways balls would roll and water would flow.

Visualization helps us understand functions. This leads to the question: what does a gradient look like? The gradient of a function $f(x, y)$ assigns a vector to each point in space. That means that we could get an idea of what a gradient looks like by plotting the vectors of the gradient on a grid of points.

Example 10.15. Let $f(x, y) = x^2 + y^2$. For all points (x, y) with coordinates in the set $\{\pm 2, \pm 1.5, \pm 1, \pm 0.5, 0\}$, plot the point and the vector starting at that point in the direction $\nabla f(x, y)$.

Solution:

Since $\nabla f(x, y) = (2x, 2y)$ the vectors are:

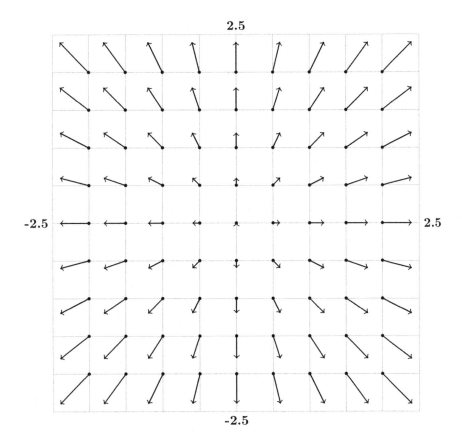

Notice that all the gradient vectors point directly away from the origin. This corresponds to our notion that the gradient is the direction of fastest growth.

$\Diamond$

The function $\dfrac{2x + y}{x^2 + y^2 + 1}$ from Example 10.14, the rolling ball question, will probably have a more interesting gradient than the simple paraboloid in Example 10.15.

Example 10.16. Let $H(x,y) = \dfrac{2x + y}{x^2 + y^2 + 1}$. For all points (x,y) with coordinates in the set $\{\pm 3, \pm 2.5, \pm 2, \pm 1.5, \pm 1, \pm 0.5, 0\}$, plot the point and the vector starting at that point in the direction $\nabla H(x,y)$.

Solution:

We are plotting the vectors drawn from the gradient

$$\nabla H(x,y) = \left(\frac{2y^2 - 2x^2 - 2xy + 2}{(x^2 + y^2 + 1)^2}, \frac{x^2 - y^2 - 4xy + 1}{(x^2 + y^2 + 1)^2} \right)$$

which yields the picture:

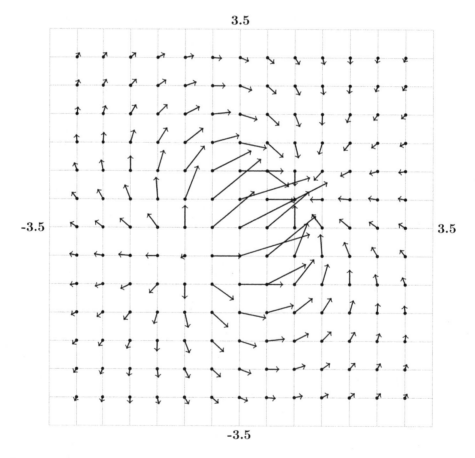

The gradient vectors point in many directions. A ball rolling on this surface could potentially have a very complex path.

$\Diamond$

Example 10.17. Find a reasonable sketch of the vector field associated with the gradient of $g(x, y) = \sin(x) + \cos(y)$ from Example 10.12. Use grid points with coordinates $\pm n$ for $n = 0, 1, \ldots 8$.

Solution:

We are plotting the vectors drawn from the gradient

$$\nabla g(x, y) = (\cos(x), -\sin(y))$$

which yields the picture:

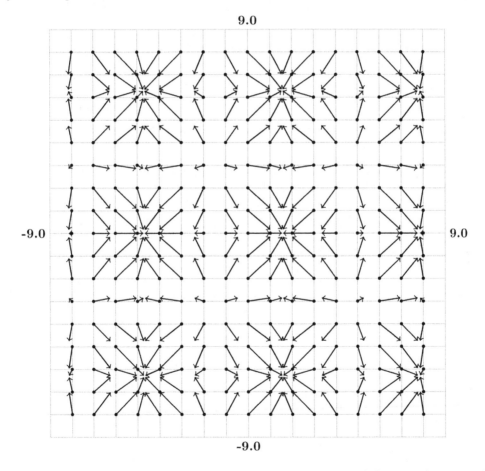

The periodicity of the vector field is easy to see. The different cells of the periodicity are slightly different because the periods are multiples of π, and the grid used to display the vectors is sized in multiples of one.

◇

Now that we have the gradient, it is possible to define the derivative in a particular direction.

Knowledge Box 10.4.

If $z = f(x, y)$ is a differentiable function of two variables, and $\vec{u} = (a, b)$ is a unit vector, then the **derivative of $f(x, y)$ in the direction of $\vec{u}$ is:**

$$\nabla_{\vec{u}} f(x, y) = \vec{u} \cdot \nabla f(x, y) = a f_x + b f_y$$

If $\vec{v} = (r, s)$ is any vector, then the **derivative of $f(x, y)$ in the direction of $\vec{v}$ is:**

$$\nabla_{\vec{v}} f(x, y) = \frac{\vec{v}}{|\vec{v}|} \cdot \nabla f(x, y)$$

Notice that we are continuing the practice of using unit vectors to designate directions – even when computing the derivative of a function in the direction of a general vector, we first coerce it to be a unit vector.

It is worth mentioning that, when computing directional derivatives, we start with a scalar quantity – the function $f(x, y)$. When we compute the gradient, we get the vector quantity $\nabla f(x, y) = (f_x, f_y)$ but then return to a scalar function of two variables $\nabla_{\vec{u}} f(x, y)$. It is important to keep track of the type of object – scalar or vector – that you are working with.

Example 10.18. Find the derivative of

$$f(x, y) = x^2 + y^2$$

in the direction of $\vec{u} = (1/2, \sqrt{3}/2)$.

Solution:

The vector $\vec{u}$ is a unit vector so, starting with $f_x = 2x$, $f_y = 2y$, we get:

$$\nabla_{\vec{u}} f(x, y) = \frac{1}{2} 2x + \frac{\sqrt{3}}{2} 2y = x + \sqrt{3} y$$

$\diamond$

The directional derivative occasionally comes up in the natural course of trying to solve a problem. There is one very natural application: finding level curves. First let's define level curves.

Knowledge Box 10.5.

Level Curves

If $z = f(x, y)$ defines a surface, then the **level curve of height c** of $f(x, y)$ is the set of points that solve the equation

$$f(x, y) = c.$$

Example 10.19. Plot the level curves for

$$f(x, y) = 2x^2 + y^2$$

for $c \in \{1, 2, 3, 4, 5, 6\}$.

Solution:

The equation $2x^2 + y^2 = c$ is an ellipse that is $\sqrt{2}$ times as far across in the x direction as the y directions. Solving for the points where $x = 0$ or $y = 0$ gives us the extreme points of the ellipse for each value of c, and we get the following picture.

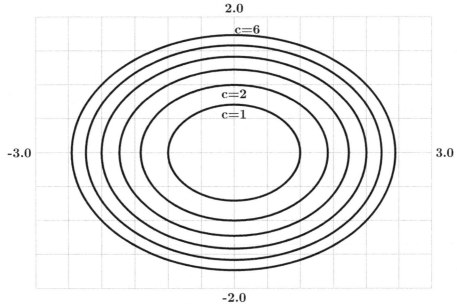

Level curves for $f(x, y) = 2x^2 + y^2$.

Since the values of c used to derive the level curves are equally spaced, the fact that the curves are getting closer together gives a sense of how steeply the graph of $f(x, y) = 2x^2 + y^2$ is sloped.

◇

In Example 10.19 we can plot the level curves because we recognize them as ellipses. In general that's not going to work. If we supply a graphics system with information about the gradient of a curve, then it can sketch level curves by following the correct directional derivative.

Knowledge Box 10.6.

If $z = f(x, y)$ defines a surface, then the level curve at the point (a, b) proceeds in the direction of the unit vector $\vec{u}$ such that
$$|\nabla_{\vec{u}} f(a, b)| = 0.$$
In other words, it proceeds in the direction such that the height of the graph of $f(x, y)$ is not changing.

Example 10.20. Find, in general, the direction of the level curves of
$$f(x, y) = x^2 + 3y^2.$$

Solution:

To find the general direction of level curves, we need to solve
$$|\nabla_{\vec{u}} f(a, b)| = 0$$
for $\vec{u}$.

Let $\vec{u} = (a, b)$.

$$\nabla \left(x^2 + 3y^2\right) \cdot (a, b) = 0$$
$$(2x, 6y) \cdot (a, b) = 0$$
$$2xa + 6yb = 0$$
$$2xa = -6yb$$
$$a = -3\left(\frac{y}{x}\right)b$$

Giving us the direction in which the level curves go. If we set $b = 1$, then, at a given point (x, y) in space, the level curve at that point is in the direction $\left(-\dfrac{3y}{x}, 1\right)$. This encodes the direction.

To be thorough, let's turn this into a unit vector.
$$\left|\left(-\frac{3y}{x}, 1\right)\right| = \sqrt{\left(-\frac{3y}{x}\right)^2 + 1^2} = \sqrt{\frac{9y^2}{x^2} + 1} = \frac{1}{x}\sqrt{9y^2 + x^2}$$

So, the direction, as a unit vector is:
$$\left(-\frac{3y}{\sqrt{9y^2 + x^2}}, \frac{x}{\sqrt{9y^2 + x^2}}\right)$$

Notice that if we had chosen $b = -1$, we would have gotten

$$\left(\frac{3y}{\sqrt{9y^2 + x^2}}, -\frac{x}{\sqrt{9y^2 + x^2}} \right)$$

which is an equally valid solution. The level curve points in two opposite directions.

$\Diamond$

A natural question at this point is: what do the vectors we found in Example 10.20 look like?

Example 10.21. Using the solution to Example 10.20 plot the directions of the level curves as a vector field.

Solution:

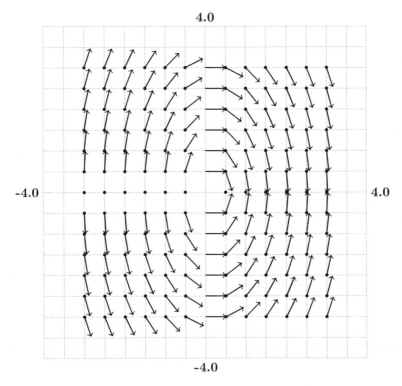

At each point, there are actually two vectors that point in the direction of the level curve. If $\vec{u}$ points in the direction of the level curve, then so does $-\vec{u}$. The figure above sometimes shows one and sometimes the other. The curves that these vectors are following track the vertically long ellipses that one would expect as level curves of:

$$f(x, y) = x^2 + 3y^2$$

$\Diamond$

Problems

Problem 10.17. *For each of the following functions, find their gradient.*

(a) $f(x,y) = x^2 + 4xy + 7y^2 + 2x - 5y + 1$

(b) $g(x,y) = \sin(xy)$

(c) $h(x,y) = \dfrac{x^2}{y^2 + 1}$

(d) $r(x,y) = \left(x^2 + y^2\right)^{3/2}$

(e) $s(x,y) = \sin(x) + \cos(y)$

(f) $q(x,y) = x^5 y + xy^5 + x^3 y^3 + 1$

(g) $a(x,y) = \dfrac{x}{x^2 + y^2}$

(h) $b(x,y) = x^y$

Problem 10.18. *For the function*

$$g(x,y) = x^2 + 3xy + y^2,$$

remembering that directions should be reported as unit vectors, find:

(a) The greatest rate of growth of the curve in any direction at the point $(1,1)$

(b) The direction in which the greatest growth happens at $(1,1)$

(c) The greatest rate of growth of the curve in any direction at the point $(-1,2)$

(d) The direction in which the greatest growth happens at $(-1,2)$

(e) The greatest rate of growth of the curve in any direction at the point $(0,3)$

(f) The direction in which the greatest growth happens at $(0,3)$

Problem 10.19. *Suppose we are trying to find level curves. Is it possible to find points where there are more than two directions in which the surface does not grow? Either explain wy this cannot happen or give an example where it does.*

Problem 10.20. *For each of the following functions, sketch the gradient vector field of the function for the points (x,y) with*

$$x, y \in \{0, \pm1, \pm2, \pm3\}.$$

Exclude points, if any, where the gradient does not exist. This is a problem where a spreadsheet may be useful for performing routine computation.

(a) $f(x,y) = x^2 + y^2 + 1$

(b) $g(x,y) = \sin(xy)$

(c) $h(x,y) = \dfrac{x}{x + y}$

(d) $r(x,y) = \sqrt{x^2 + y^2}$

(e) $s(x,y) = \sin(x) + \cos(y)$

(f) $q(x,y) = x^5 y + xy^5 + x^3 y^3 + 1$

Problem 10.21. *Find the directional derivative of*

$$f(x,y) = x^2 + y^3$$

in the direction of (1,1) at (0,0).

Problem 10.22. *Find the directional derivative of*

$$g(x,y) = x^2 - y^2$$

in the direction of $(\sqrt{3}/2, 1/2)$ at (2,1).

Problem 10.23. *Find the directional derivative of*

$$h(x,y) = (y - 2x)^3$$

in the direction of (0,-1) at (-1,1).

Problem 10.24. *As in Example 10.20 find the general direction for level curves at (x,y) for the function*

$$H(x,y) = \dfrac{1}{x^2 + y^2}$$

and plot the vector field for all points with whole number coordinates in the range $-3 \le x, y \le 3$.

Problem 10.25. *Which way will a ball placed at (1,1) on the surface given by the function in Problem 10.21 roll?*

10.3. Tangent Planes

One of the first things we built after developing skill with the derivative was tangent lines to a curve. With a function $z = f(x, y)$ the analogous object is a tangent *plane*. There are several ways to specify a plane.

Knowledge Box 10.7.

Formulas for planes

If a, b, c, and d are constants, all of the following formulas specify planes.

$$z = ax + by + c$$

$$ax + by + cz = d$$

$$(a, b, c) \cdot (x, y, z) = d$$

Note that the third formula, while phrased in terms of a dot product, is actually the same as the second.

Let's get some practice with converting between the different possible forms of a plane.

Example 10.22. If

$$(3, -2, 5) \cdot (x, y, z) = 2$$

find the other two forms of the plane.

Solution:

$$
\begin{aligned}
(3, -2, 5) \cdot (x, y, z) &= 2 \\
3x - 2y + 5z &= 2 && \text{Second form} \\
5z &= -3x + 2y + 2 \\
z &= -0.6x + 0.4y + 0.4 && \text{First form}
\end{aligned}
$$

◊

The most common way to find a tangent line for a function is to take the point of tangency – which must be on the line – together with the slope of the line found by computing the derivative and use the point slope form to find the formula for the line. It turns out that there is a way of specifying planes that is similar to the point slope form of a line. Remember that if $\vec{v}$ and $\vec{w}$ are vectors that are at right angles to one another, then $\vec{v} \cdot \vec{w} = 0$.

Knowledge Box 10.8.

Formula for a plane at right angles to a vector

Suppose that $\vec{v}$ is a vector at right angles to a plane in three dimensions and that (a, b, c) is a point on that plane. Then a formula for the plane is

$$\vec{v} \cdot (x - a, y - b, z - c) = 0.$$

Notice that (a, b, c) is constructively on the plane and that our knowledge of the dot product tells us it is at right angles to $\vec{v}$.

Example 10.23. Find the plane at right angles to $\vec{v} = (1, 2, 1)$ through the point $(2, -1, 5)$.

Solution:

Simply substitute into the formula given in Knowledge Box 10.8.

$$(1, 2, 1) \cdot (x - 2, y + 1, z - 5) = 0$$
$$1(x - 2) + 2(y + 1) + 1(z - 5) = 0$$
$$x + 2y + z - 2 + 2 - 5 = 0$$
$$x + 2y + z = 5$$

$\Diamond$

Knowledge Box 10.8 seems very special purpose, but it turns out to be very useful in light of another fact. Suppose that

$$f(x, y, z) = c$$

specifies a surface. We need to expand our definition of the gradient just a bit to

$$\nabla f(x, y, z) = (f_x(x, y, z), f_y(x, y, z), f_z(x, y, z)).$$

In this case the vector $\nabla f(a, b, c)$ points directly outward from the surface $f(x, y, z) = c$ at (a, b, c) and *it is at right angles to the tangent plane.*

Knowledge Box 10.9.

Formula for the tangent plane to a surface

If $f(x, y, z) = c$ defines a surface, and (a, b, c) is a point on the surface, then a formula for the tangent plane to that surface at (a, b, c) is:

$$(f_x(a, b, c), f_y(a, b, c), f_z(a, b, c)) \cdot (x - a, y - b, z - c) = 0$$

Defining a surface in the form $f(x, y, z) = c$ is a little bit new – but in fact this is another version of level curves, just one dimension higher. Let's practice.

Example 10.24. Find the tangent plane to the surface $x^2 + y^2 + z^2 = 3$ at the point $(1, -1, 1)$.

Solution:

Check that the point is on the surface: $(1)^2 + (-1)^2 + (1)^2 = 3$ – so it is. Next find the gradient.

$$\nabla x^2 + y^2 + z^2 = (2x, 2y, 2z).$$

This means that the gradient at $(1, -1, 1)$ is $\vec{v} = (2, -2, 2)$. This makes the plane

$$(2, -2, 2) \cdot (x - 1, y + 1, z - 1) = 0$$

or

$$2x - 2y + 2z = 6 \text{ or } x - y + z = 3$$

Notice that we simplified the form of the plane; this is not required but it does make for neater answers.

◊

The problem with finding the tangent plane to a surface $f(x, y, z) = c$ is that it does not solve the original problem – finding tangent planes to $z = f(x, y)$. A modest amount of algebra solves this problem. If $z = f(x, y)$ then $g(x, y, z) = z - f(x, y) = 0$ is in the correct form for our surface techniques. This gives us a new way of finding tangent planes to a function that defines a surface in 3-space.

Knowledge Box 10.10.

Formula for the tangent plane to a functional surface

If $z = f(x, y)$ defines a surface, then the tangent plane to the surface at (a, b) may be obtained as

$$(-f_x(a, b), -f_y(a, b), 1) \cdot (x - a, y - b, z - f(a, b)) = 0$$

This is the result of applying the gradient-of-a-surface formula to the surface $z - f(x, y) = 0$ at the point $(a, b, f(a, b))$.

Example 10.25. Find the tangent plane to $f(x, y) = x^2 - y^3$ at the point $(2, -1)$.

Solution:
Assemble the pieces and plug into Knowledge Box 10.10.
$$f_x(x, y) = 2x$$
$$f_x(2, -1) = 4$$

$$f_y(x, y) = -3y^2$$
$$f_y(2, -1) = -3$$

$$f(2, -1) = 4 - (-1) = 5$$
Put the plane together
$$(-4, 3, 1) \cdot (x - 2, y + 1, z - 5) = 0$$
$$-4x + 8 + 3y + 3 + z - 5 = 0$$
$$-4x + 3y + z = -6$$
Which is the tangent plane desired.

Example 10.26. Find the tangent plane to
$$g(x, y) = xy^2$$
at $(3,1)$.

Solution:
Assemble the pieces and plug into Knowledge Box 10.10.
$$f_x(x, y) = y^2$$
$$f_x(3, 1) = 1$$

$$f_y(x, y) = 2xy$$
$$f_y(3, 1) = 6$$

$$f(3, 1) = 3$$
Put the plane together
$$(-1, -6, 1) \cdot (x - 3, y - 1, z - 3) = 0$$
$$-x + 3 - 6y + 6 + z - 3 = 0$$
$$-x - 6y + z = -6$$
$$\Diamond$$

Problems

Problem 10.26. *Find the plane through (1,-1,1) at right angles to $\vec{v} = (2, 2, -1)$.*

Problem 10.27. *Find the plane through (2,0,5) at right angles to $\vec{v} = (1, 1, 1)$.*

Problem 10.28. *Find the plane through (3,2,1) at right angles to $\vec{v} = (1, -1, 2)$.*

Problem 10.29. *Find, in the form*

$$ax + by + cz = d,$$

the tangent planes to the following curves at the indicated points.

(a) $f(x, y) = x^2 + y^2 - 1$ at $(2, 2)$

(b) $g(x, y) = 3x^2 + 2xy + 4^2 - 1$ at $(-1, 1)$

(c) $h(x, y) = (x + y + 1)^2$ at $(0, 4)$

(d) $r(x, y) = y \cdot \ln(x^2 + 1)$ at $(3, -1)$

(e) $s(x, y) = e^{x^2 + y^2}$ at $(0, 1)$

(f) $q(x, y) = \sin(x) \cos(y)$ at $(\pi/3, \pi/6)$

Problem 10.30. *Suppose that*

$$x^2 + y^2 + z^2 = 12.$$

Find the tangent plane at each of the following points.

(a) $p = (2, 2, 2)$

(b) $q = (2, -2, 2)$

(c) $r = (-2, -2, -2)$

(d) $u = (1, 1, \sqrt{10})$

(e) $v = (-\sqrt{7}, 1, 2)$

(f) $w = (-1/2, 2.5, \sqrt{22}/2)$

Problem 10.31. *Find all tangent planes to*

$$x^2 + y^2 + z^2 = 75$$

that are at right angles to the vector $(2, 4, 6)$.

Problem 10.32. *Find the tangent plane to*

$$g(x, y) = x \cos(y)$$

at $(-1, \pi/3)$.

Problem 10.33. *Find the tangent plane to each of the following surfaces at the indicated point.*

(a) Surface $x + y + z^2 = 6$ at $(1,1,2)$

(b) Surface $x^2 - y^3 + 5z = 19$ at $(4,-2,-1)$

(c) Surface $3x + y^2 + z^2 = 8$ at $(2,1,1)$

(d) Surface $(x - y)^3 + 2z = 14$ at $(2,0,3)$

(e) Surface $xyz + x + y + z = 4$ at $(1,1,1)$

(f) Surface $xy + yz + xz = 0$ at $(-1,1,-2)$

(g) Surface $x^2 + y^2 + z^3 = 6$ at $(2,1,1)$

(h) Surface $xy + xz + yz = 7$ at $(1,1,3)$

Problem 10.34. *Find the tangent plane to*

$$g(x, y) = e^{-(x^2 + y^2)}$$

at $(0, 0)$.

Problem 10.35. *Find the tangent plane to*

$$\cos(xyz) = 0$$

at $(\pi/4, \pi/4, \pi/4)$.

Problem 10.36. *If*

$$x^2 + y^2 + z^2 = r^2$$

is a sphere of radius r centered at the origin $(0,0,0)$ (it is), show that a sphere has a tangent plane at right angles to any nonzero vector.

Problem 10.37. *If you want to find the tangent plane to a point on a sphere, what is the simplest method? Explain.*

Problem 10.38. *If P and Q are planes that are both at right angles to a vector $\vec{v}$ and they are not equal, what can be said about the intersection of P and Q?*

Problem 10.39. *Suppose that $\vec{u}$ and $\vec{v}$ are vectors so that $\vec{v} \cdot \vec{u} = 0$. If P is a plane at right angles to $\vec{u}$ and Q is a plane at right angles to $\vec{v}$ in three-dimensional space then what is the most that can be said about the intersection of the planes?*

Chapter 11

Multivariate and Constrained Optimization

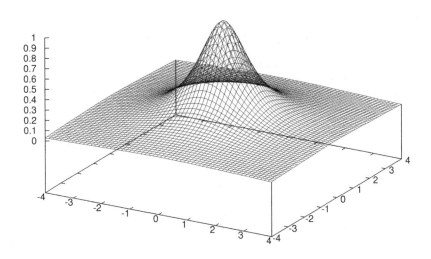

Figure 11.1. A graph of $f(x,y) = \dfrac{1}{x^2 + y^2 + 1}$ on $-4 \leq x, y \leq 4$.

A big application of derivatives in Chapter 3 was optimizing functions. In this chapter we learn to do this for multiple independent variables. In some ways we will be doing pretty much the same thing – but as always, more variables makes the process more complicated.

11.1. Optimization with Partial Derivatives

Examine the hill-shaped graph in Figure 11.1. This function has a single optimum at $(x, y) = (0, 0)$. It is also positive everywhere. It is a good, simple function for demonstration purposes. The level curves are circles centered on the origin – except at the origin on top of the optimum, where we get a "level curve" consisting of a single point. This sort of point – where the directional derivatives are all zero – is the three dimensional equivalent of a critical point.

Knowledge Box 11.1.

Critical points for surfaces

If $z = f(x, y)$ defines a surface, then the **critical points** of the function are at the points (x, y) that solve the equations:

$$f_x(x, y) = 0 \text{ and } f_y(x, y) = 0$$

As with our original optimization techniques, local and global optima occur at critical points or at the boundaries of the domain of optimization.

The information in Knowledge Box 11.1 gives us a good start on finding optima of multivariate functions – but it also contains a land mine. The phrase *the boundaries of the domain of optimization* is more fearsome when there are more dimensions. The next section of this chapter is about dealing with some of the kinds of boundaries that arise when optimizing surfaces.

Example 11.1. Demonstrate that the function in Figure 11.1, $f(x, y) = \dfrac{1}{x^2 + y^2 + 1}$, in fact has a critical point at (0,0) by solving for the partial derivatives equal to zero.

Solution:
The partial derivatives are :

$$f_x(x, y) = \frac{2x}{\left(x^2 + y^2 + 1\right)^2} \text{ and } f_y(x, y) = \frac{2y}{\left(x^2 + y^2 + 1\right)^2}$$

Remembering that a fraction is zero only when its numerator is zero – and noting the denominators of these partial derivative are never zero – we see we are solving the very difficult system of simultaneous equations:

$$2x = 0 \text{ and } 2y = 0$$

So we verify a single critical point at (0,0).

◇

Example 11.2. Find the critical point(s) of

$$g(x, y) = x^2 + 2y^2 + 4xy - 6x - 8y + 2$$

Solution:

Start by computing the partials.

$$g_x(x, y) = 2x + 4y - 6$$
$$g_y(x, y) = 4y + 4x - 8$$

This gives us the simultaneous system:

$$2x + 4y - 6 = 0$$
$$4x + 4y - 8 = 0$$

$$\begin{aligned} 2x + 4y &= 6 \\ 4x + 4y &= 8 \\ 2x &= 2 \qquad \text{Second line minus first.} \\ x &= 1 \\ 2 + 4y &= 6 \qquad \text{Plug x=1 into first line.} \\ 4y &= 4 \\ y &= 1 \end{aligned}$$

So we find a single critical point at $(x, y) = (1, 1)$.

$$\Diamond$$

Now that we can locate critical points of surfaces, we have to deal with figuring out if the point is a local maximum, a local minimum, or something else. Let's begin by understanding the option for "something else." Examine the function in Figure 11.2. The partial derivatives are $f_x = 2x$ and $f_y = 2y$. So, it has a critical point at $(0,0)$ in the center of the graph, but it *does not* have an optimum. This is a type of critical point called a **saddle point**.

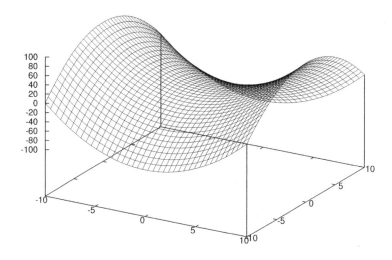

Figure 11.2. A graph of $f(x, y) = x^2 - y^2$ on $-10 \leq x, y \leq 10$. This graph exemplifies a saddle point.

Knowledge Box 11.2.

Types of critical points for surfaces

When we find critical points by solving $f_x = 0$ and $f_y = 0$, there are three possible outcomes:

- if all nearby points are lower, the point is a **local maximum**,

- if all nearby points are higher, the point is a **local minimum**

- if there are nearby points that are both higher and lower, the point is a **saddle point**

If we have a good graph of a function, then we can often look at a critical point to determine its character. This is not always practical. The sign chart technique from Chapter 3 is not possible – it only makes sense in a one-dimensional setting. There *is* an analog to the second derivative test. Look at the functions in Figures 11.1 and 11.2. If we slice them along the x and y axes, we see that the hill in Figure 11.1 is concave down along both axes, while the saddle in Figure 11.2 is concave down along the y axis and concave up along the x axis. This type of observation can be generalized into a type of second derivative test.

Knowledge Box 11.3.

Classifying critical points for surfaces

For a function $z = f(x, y)$ that defines a surface, define

$$D(x, y) = f_{xx}(x, y)f_{yy}(x, y) - f_{xy}^2(x, y)$$

The quantity D is called the **discriminant** of the system. Then for a critical point (a, b)

- If $D(a, b) > 0$ and $f_{xx}(a, b) > 0$ or $f_{yy}(a, b) > 0$, then the point is a local minimum.

- If $D(a, b) > 0$ and $f_{xx}(a, b) < 0$ or $f_{yy}(a, b) < 0$, then the point is a local maximum.

- If $D(a, b) < 0$, then the point is a saddle point.

- If $D(a, b) = 0$, then the test yields no information about the type of the critical point.

If $D(a, b) > 0$ the signs of f_{xx} and f_{yy} will agree – so the test cannot fail at that point.

The fourth outcome – no information – forces you to examine the function in other ways. This situation is usually the result of having functions with repeated roots or high powers of both variables. The function

$$q(x, y) = x^3 + y^3,$$

for example, has a critical point at $(0, 0)$, but all of f_{xx}, f_{yy}, and $f_{x,y}$ are zero. The critical point is a saddle point, but that conclusion follows from careful examination of the graph of the function, not from the second derivative test.

A part of the structure of this multivariate version of the second derivative test is that *either* f_{xx} or f_{yy} may be used to check if a critical point is a minimum or maximum. A frequent question by students is "which should I use?" The answer is "whichever is easier". The reason for this is that the outcome of the discriminant test that tells you a critical point is an optimum of some sort *forces* the signs of f_{xx} and f_{yy} to agree.

Examine the formula for $D(a, b)$ and see if you can tell why this is true.

Example 11.3. Find and classify the critical point of

$$h(x, y) = 4 + 3x + 5y - 3x^2 + xy - 2y^2$$

Solution:

The partials are $f_x = 3 - 6x + y$ and $f_y = 5 + x - 4y$. Setting these equal to zero we obtain the simultaneous system of equations:

$$6x - y = 3$$
$$-x + 4y = 5$$
$$\text{Solve:}$$
$$-6x + 24y = 30$$
$$23y = 33$$
$$y = 33/23$$
$$-x + 132/23 = 115/23$$
$$-x = -14/23$$
$$x = 17/23$$

Now compute $D(x, y)$. We see $f_{xx} = -6$, $f_{yy} = -4$ and $f_{xy} = 1$. So,

$$D = (-6)(-4) - 1^2 = 23.$$

Since $D > 0$, the critical point is an optimum of some sort. The fact that $f_{xx} < 0$ or the fact that $f_{yy} < 0$ suffices to tell us this point is a maximum.

Notice that in Example 11.3 the equation is a quadratic with large negative squared terms and a small mixed term (xy). The fact that its sole critical point is a maximum means that it is a paraboloid opening downward. In the next example we look at a quadratic with a large mixed term.

Example 11.4. Find and classify the critical points of

$$q(x, y) = x^2 + 6xy + y^2 - 14x - 10y + 3.$$

Solution:

Start by computing the needed partials and the discriminant.

$$f_x(x, y) = 2x + 6y - 14$$
$$f_y(x, y) = 6x + 2y - 10$$
$$f_{xx} = 2$$
$$f_{yy} = 2$$
$$f_{xy} = 6$$
$$D(x, y) = 2 \cdot 2 - 6 = -2$$

These calculations show that any critical point we find will be a saddle point since $D < 0$. To find the critical point we solve the first partials equal to zero obtaining the system

$$2x + 6y = 14$$
$$6x + 2y = 10$$
$$6x + 18y = 42 \qquad \text{3x line one}$$
$$16y = 32 \qquad \text{Line 3 minus line two}$$
$$y = 2$$
$$2x + 12 = 14 \qquad \text{Substitute}$$
$$2x = 2$$
$$x = 1$$

So we see that the function has, as its sole critical point, a saddle point at (1,2).

◇

The last two examples have been bivariate quadratic equations with a constant discriminant. This is a feature of all bi-variate quadratics. All functions of this kind have a single critical point and a constant discriminant. For the next example we will tackle a more challenging example.

Example 11.5. Find and classify the critical points of $f(x, y) = x^4 + y^4 - 16xy + 6$.

Solution:

Start by computing the needed partials and the discriminant.

$$f_x(x, y) = 4x^3 - 16y$$
$$f_y(x, y) = 4y^3 - 16x$$
$$f_{xx}(x, y) = 12x^2$$
$$f_{yy}(x, y) = 12y^2$$
$$f_{xy}(x, y) = -16$$
$$D(x, y) = f_{xx}f_{yy} - f_{xy}^2$$
$$D(x, y) = 144x^2y^2 - 256$$

Next we need to find the critical points by solving the system

$$4x^3 = 16y$$
$$4y^3 = 16x$$

or

$$x^3 = 4y$$
$$y^3 = 4x$$

Solve the second equation to get $y = \sqrt[3]{4x}$ and plug into the first, obtaining:

$$x^3 = \sqrt[3]{4x}$$
$$x^9 = 4x$$
$$x^9 - 4x = 0$$
$$x(x^8 - 4) = 0$$
$$x = 0, \pm\sqrt[8]{4}$$
$$x = 0, \pm\sqrt[4]{2}$$

Since the equation system is symmetric in x and y we may deduce that $y = 0, \pm\sqrt[4]{2}$ as well. Referring back to the original equations, it is not hard to see that, when $x = 0$, $y = 0$; when $x = \sqrt[4]{2}$ so must y; when $x = -\sqrt[4]{2}$ so must y. This gives us three critical points: $(-\sqrt[4]{2}, -\sqrt[4]{2}), (0, 0)$, and $(\sqrt[4]{2}, \sqrt[4]{2})$. Next we check the discriminant at the critical points.

$$D(-\sqrt[4]{2}, -\sqrt[4]{2}) = 288 - 256 = 32 > 0$$

Since $f_{xx}(-\sqrt[4]{2}, -\sqrt[4]{2}) > 0$, this point is a minimum.

$$D(\sqrt[4]{2}, \sqrt[4]{2}) = 288 - 256 = 32 > 0$$

Since $f_{xx}(-\sqrt[4]{2}, -\sqrt[4]{2}) > 0$, this point is a minimum.

$$D(0, 0) = -256,$$

so this point is a saddle point.

◇

One application of multivariate optimization is to minimize distances. For that we should state, or re-state, the definition of distance.

Knowledge Box 11.4.

The definition of distance

If $p = (x_0, y_0)$ and $q = (x_1, y_1)$ are points in the plane, then the distance between p and q is

$$d(p, q) = \sqrt{(x_0 - x_1)^2 + (y_0 - y_1)^2}.$$

If $r = (x_0, y_0, z_0)$ and $s = (x_1, y_1, z_1)$ are points in space, then the distance between r and s is

$$d(r, s) = \sqrt{(x_0 - x_1)^2 + (y_0 - y_1)^2 + (z_0 - z_1)^2}.$$

The definition in Knowledge Box 11.4 can be extended to any number of dimensions – the distance between two points is the square root of the sum of the squared differences of the individual coordinates of the point. For this text only two and three dimensional distances are required. With the definition of distance in place, we can now pose a standard type of problem.

Example 11.6. What point on the plane $z = 3x + 2y + 4$ is closest to the origin?

Solution:

The origin is the point $(0, 0, 0)$, while a general point (x, y, z) on the plane has the form $(x, y, 3x + 2y + 4)$. That means that the distance we are trying to minimize is given by

$$d = \sqrt{(x - 0)^2 + (y - 0)^2 + (3x + 2y + 4 - 0)^2}$$

$$= \sqrt{10x^2 + 12xy + 5y^2 + 24x + 16y + 16}$$

The partial derivatives of this function are:

$$d_x = \frac{20x + 12y + 24}{2\sqrt{10x^2 + 12xy + 5y^2 + 24x + 16y + 16}}$$

$$d_y = \frac{12x + 10y + 16}{2\sqrt{10x^2 + 12xy + 5y^2 + 24x + 16y + 16}}$$

We need to solve the simultaneous equation in which each of these partials is zero. Remember a fraction is zero only where its numerator is zero. This maxim gives us the simultaneous equations:

$$20x + 12y + 24 = 0$$
$$12x + 10y + 16 = 0$$
$$\text{Simplify}$$
$$5x + 3y = -6$$
$$6x + 5y = -8$$
$$\text{Solve}$$
$$30x + 18y = -36$$
$$30x + 25y = -40$$
$$7y = -4$$
$$y = -4/7$$
$$5x - 12/7 = -42/7$$
$$5x = -30/7$$
$$x = -6/7$$

So we find a single critical point $(-6/7, -4/7)$ Considering the geometry of a plane, we see that it has a unique closest approach to the origin. So, computing $z = 3x + 2y + 4 = -18/7 - 8/7 + 28/7 = 2/7$, gives us that the point on the plane closest to the origin is

$$(-6/7, -4/7, 2/7).$$

$$\Diamond$$

Clever students will have noticed that when we were minimizing

$$\sqrt{10x^2 + 12xy + 5y^2 + 24x + 128 + 16}$$

the numerators of the partial derivatives were exactly the *partial derivatives* of $10x^2 + 12xy + 5y^2 + 24x + 128 + 16$. This observation is an instance of a more general shortcut.

Knowledge Box 11.5.

Minimization of distance - a shortcut

Suppose that we are optimizing a function $g(x, y) = \sqrt{f(x, y)}$ or $g(x, y, z) = \sqrt{f(x, y, z)}$. Then the critical points for optimization of g and f are the same, and the second derivative tests agree. The actual values of the function are not – which means care is required.

Why does this shortcut work? It is because if $0 < a < b$, then $0 < \sqrt{a} < \sqrt{b}$. So optimizing the value finds where the optimum of the square root of the value is as well.

Example 11.7. Minimize:

$$g(x, y) = \sqrt{x^2 + y^2 + 3}$$

Solution:

In this case $g(x, y) = \sqrt{f(x, y)}$ where $f(x, y) = x^2 + y^2 + 3$, so the shortcut applies. Finding the relevant partials we see that

$$f_x(x, y) = 2x$$
$$f_y(x, y) = 2y$$

which is easy to see has a critical point at $(0,0)$. The second derivative test shows that:

$$f_{xx}f_{yy} - f_{xy}^2 = 2 \cdot 2 - 0 = 4 > 0$$

So the critical point is an optimum. Since $f_{xx} = 2 > 0$, it is a minimum. This means the minimum value of $g(x, y)$ is at the point $(0, 0, \sqrt{3})$.

$\Diamond$

Definition 11.1. A function $m(x)$ is **monotone increasing** if, whenever $a < b$ and $m(x)$ exists on $[a, b]$, then $m(a) < m(b)$.

Notice that $m(x) = \sqrt{x}$ is monotone increasing. In fact the shortcut in Knowledge Box 11.5 works for any monotone increasing function. These functions include: e^x, $\ln(x)$, $\tan^{-1}(x)$, x^n when n is odd, and $\sqrt[n]{x}$.

Knowledge Box 11.6.

A test for a function being monotone increasing

We know that a function is increasing if its first derivative is positive. If a function exists and has a positive derivative on the interval [a,b] it is a monotone increasing function on [a,b].

Example 11.8. Show that $y = \ln(x)$ is monotone increasing where it exists.

Solution:

According to Knowledge Box 11.6, a function is increasing when its first derivative is positive. The function $y = \ln(x)$ only exists on the interval $(0, \infty)$. Its derivative is $y' = 1/x$ which is positive for any positive x. Thus $\ln(x)$ is increasing everywhere that it exists and so is a monotone increasing function.

$\Diamond$

Using the extension of the shortcut from Knowledge Box 11.5 with these other monotone functions will make the homework problems much easier. Let's practice.

Example 11.9. Find the point of closest approach to the origin on the plane
$$3x - 4y - z = 4.$$

Solution:

First find the distance between the origin $(0, 0, 0)$ and a generic point $(x, y, 3x - 4y - 4)$ on the plane. We get that
$$d(x, y) = \sqrt{x^2 + y^2 + 9x^2 - 24xy + 16y^2 - 32y - 24x + 16}$$
$$d(x, y) = \sqrt{10x^2 - 24xy + 17y^2 - 32y - 24x + 16}$$
Using the distance minimization shortcut, find the critical points of
$$d^2(x, y) = 10x^2 - 24xy + 17y^2 - 32y - 24x + 16$$
which is the relatively simple quadratic case.
$$f_x(x, y) = 20x - 24y - 24$$
$$f_y(x, y) = -24x + 34y - 32$$
$$f_{xx}(x, y) = 20$$
$$f_{yy}(x, y) = 34$$
$$f_{x,y}(x, y) = -24$$
$$D(x, y) = 20 \cdot 34 - (-24)^2 = 104$$
So we see that $D > 0$ and $f_{xx} > 0$ meaning we will find a minimum distance. Now we must solve for the critical point.
$$20x - 24y = 24$$
$$-24x + 34y = 32$$

$$5x - 6y = 6$$
$$-12x + 17y = 16$$
$$60x - 72y = 72$$
$$-60x + 85y = 80$$
$$13y = 152$$
$$y = 152/13$$
$$5x - 912/13 = 78/13$$
$$5x = 990/13$$
$$x = 198/13$$

So the critical point is roughly at $x = 15.23$, $y = 11.69$, making the point of closest approach $(15.23, 11.69, 2.93)$. Not a lot easier – but we avoided all sorts of square roots.

◊

Problems

Problem 11.1. *Find the critical points for each of the following functions, and use the second derivative test to classify them as maxima, minima, or saddle points. Also find the value of the function at the critical point.*

(a) $f(x,y) = x^2 + y^2 - 6x - 4y + 4$

(b) $g(x,y) = x^2 + xy + y^2 + 4x - 5y + 2$

(c) $h(x,y) = 12 - x^2 + 3xy - y^2 + 2x - y + 1$

(d) $r(x,y) = x^4 + y^4 - 4xy + 1$

(e) $s(x,y) = x^3 + 2y^3 - 4xy$

(f) $q(x,y) = 3x^2y + y^3 - 3x^2 - 3y^2 + 2$

Problem 11.2. *Show that*

$$f(x,y) = x^4 + y^4$$

is an example of a function where the second derivative test yields no useful information.

Problem 11.3. *Suppose that*

$$f(x,y) = ax^2 + bxy + cy^2 + dx + ey + f.$$

Show that this function has at most one critical point. When it does have a critical point, derive rules based on the constants a-f for classifying that critical point.

Problem 11.4. *Find the point on the plane*

$$x + y + z = 4$$

closest to the origin.

Problem 11.5. *Find the point on the plane*

$$z = -x/3 + y/4 + 1$$

closest to the origin.

Problem 11.6. *Find the point on the tangent plane of*

$$f(x,y) = x^2 + y^2 + 1$$

at (2,1,6) that is closest to the origin.

Problem 11.7. *Find the critical points for each of the following functions, and use the second derivative test to classify them as maxima, minima, or saddle points. Also find the value of the function at the critical point.*

(a) $f(x,y) = \ln(x^2 + y^2 + 4)$

(b) $g(x,y) = e^{x^4 + y^4 - 36xy + 6}$

(c) $h(x,y) = \tan^{-1}(x^3 + 2y^3 - 4xy)$

(d) $r(x,y) = (x^2 + xy + 3y^2 - 4x + 3y + 1)^5$

(e) $s(x,y) = \sqrt{x^2 + y^2 + 14}$

(f) $q(x,y) = $
$\tan^{-1}\left(e^{x^2 + xy + y^2 - 2x - 3y + 1}\right)$

Problem 11.8. *For each of the following functions either demonstrate it is not monotone increasing on the given interval or show that it is.*

(a) $y = e^x$ on $(-\infty, \infty)$

(b) $y = x^3 - x$ on $(-\infty, \infty)$

(c) $y = \dfrac{x}{x^2 + 1}$ on $(-1, 1)$

(d) $y = \tan^{-1}(x)$ on $(-\infty, \infty)$

(e) $y = \dfrac{x^2}{x^2 + 1}$ on $(0, \infty)$

(f) $y = \dfrac{2x}{x^2 + 1}$ on $(-\infty, \infty)$

(g) $y = \dfrac{e^x}{e^x + 1}$ on $(-\infty, \infty)$

Problem 11.9. *Suppose that we are going to cut a line of length L into three pieces. Find the division into pieces that maximizes the sum of the squares of the lengths. This can be phrased as a multivariate optimization problem.*

Problem 11.10. *Find the largest interval on which the function*

$$\frac{x}{x^2 + 16}$$

is monotone increasing.

11.2. The Extreme Value Theorem Redux

The extreme value theorem, from Knowledge Box 3.12, says that *The global maximum and minimum of a continuous, differentiable function must occur at critical points or at the boundaries of the domain where optimization is taking place.* This is just as true for optimizing a function $z = f(x, y)$. So what's changed? The largest change is that the boundary can now have a very complex shape.

In the initial section of this chapter we carefully avoided optimizing functions with boundaries, thus avoiding the issue with boundaries. In the next couple of examples we will demonstrate techniques for dealing with boundaries.

Example 11.10. Find the global maximum of

$$f(x, y) = (xy + 1)e^{-x-y}$$

for $x, y \geq 0$.

Solution:

This function is to be optimized only over the first quadrant. We begin by finding critical points in the usual fashion.

$$f_x(x, y) = ye^{-x-y} + (xy + 1)e^{-x-y}(-1) = (-xy + y - 1)e^{-x-y}$$
$$f_y(x, y) = xe^{-x-y} + (xy + 1)e^{-x-y}(-1) = (-xy + x - 1)e^{-x-y}$$

Since the extreme value theorem tells us that the optima occur at boundaries or critical points, we won't need a second derivative test – we just compare values. This means out next step is to solve for any critical points. Remember that powers of e cannot be zero, giving us the system of equations:

$$-xy + y - 1 = 0$$
$$-xy + x - 1 = 0$$
$$xy = x - 1 = y - 1 \qquad\qquad \text{so } x = y$$
$$-x^2 + x - 1 = 0$$
$$x^2 - x + 1 = 0$$
$$x = \frac{1 \pm \sqrt{1 - 4}}{2} \qquad\qquad \text{There are no critical points!}$$

This means that the extreme value occurs on the boundaries – the positive x and y axes where either $x = 0$ or $y = 0$. This means we need the largest value of $f(0, y) = e^{-y}$ for $y \geq 0$ and $f(x, 0) = e^{-x}$ for $x \geq 0$. Since e^{-x} is largest (for non-negative x) at $x = 0$, we get that the global maximum is $f(0, 0) = 1$.

◊

While the extreme value theorem lets us make decisions without the second derivative test, it forces us to examine the boundaries, which can be hard. It is sometimes possible to solve the problem by adopting a different point of view. The next example will use a transformation to a parametric curve to solve the problem.

Example 11.11. Maximize

$$g(x, y) = x^4 + y^4$$

for those points $\{(x, y) : x^2 + y^2 \le 4\}$.

Solution:

This curve has a single critical point at $(x, y) = (0, 0)$ which is extremely easy to find. The boundary for the optimization domain is the circle $x^2 + y^2 = 4$, a circle of radius 2 centered at the origin. This boundary is also the parametric curve $(2\cos(t), 2\sin(t))$. This means that *on the boundary* the function is

$$g(2\cos(t), 2\sin(t)) = 16\cos^4(t) + 16\sin^4(t).$$

This means we can treat the location of optima on the boundary as a single-variable optimization task of the function $g(t) = 16\cos^4(t) + 16\sin^4(t)$. We see that

$$g'(t) = 64\cos^3(t) \cdot (-\sin(t)) + 64\sin^3(t)\cos(t) = 64\sin(t)\cos(t)(\sin^2(t) - \cos^2(t))$$

Solve:

$$\sin(t) = 0 \qquad\qquad\qquad t = (2n+1)\frac{\pi}{2}$$

$$\cos(t) = 0 \qquad\qquad\qquad t = n\pi$$

$$\sin^2(t) - \cos^2(t) = 0$$

$$\sin^2(t) = \cos^2(t) \qquad\qquad\qquad t = \pm(2n+1)\frac{\pi}{4}$$

If we plug these values of t back into the parametric curve, the points on the boundary that may be optima are: $(0, \pm 2)$, $(\pm 2, 0)$, and $(\pm\sqrt{2}, \pm\sqrt{2})$. Plugging the first four of these into the curve we get that $g(x, y) = 16$. Plugging the last four into the function we get $g(x, y) = 2 \cdot (\sqrt{2})^4 = 2 \cdot 4 = 8$. At the critical point $(0,0)$ we see $g(x, y) = 0$. This means that the maximum value of $g(x, y)$ on the optimization domain is 16 at any of $(0, \pm 2), (\pm 2, 0)$.

$$\diamondsuit$$

Notice that the solutions to the parametric version of the boundary gave us an infinite number of solutions. But, when we returned to the (x, y, z) domain, this infinite collection of solutions were just the eight candidate points repeated over and over. This means that our failure to put a bound on the parameter t was not a problem; bounding t was not necessary.

A problem with the techniques developed in this section is that all of them are special purpose. We will develop general purpose techniques in Section 11.3 – the understanding of which is substantially aided by the practice we got in this section.

Example 11.12. If

$$h(x, y) = x^2 + y^2$$

find the minimum value of $h(x, y)$ among those points (x, y) on the line $y = 2x - 4$.

Solution:

The function $h(x, y)$ is a paraboloid that opens upward. If we look at the points (x, y, z) on $h(x, y)$ that happen to lie on a line, that line will slice a parabolic shape out of the surface defined by $h(x, y)$. First note that $h(x, y)$ has a single critical point at $(0, 0)$ which is *not in* the domain of optimization. This means we may consider only those points on $y = 2x - 4$, in other words on the function

$$h(x, 2x - 4) = x^2 + (2x - 4)^2 = 5x^2 - 16x + 16$$

This means that the problem consists of finding the minimum of $f(x) = 5x^2 - 16x + 16$.

$$f'(x) = 10x - 16 = 0$$
$$10x = 16$$
$$x = 8/5$$
$$y = 16/5 - 20/5 = -4/5$$

Since $f(x)$ opens upward, we see that this point is a minimum; there are no boundaries to the line that is constraining the values of (x, y) so the minimum value of $h(x, y)$ on the line is

$$h(8/5, -4/5) = \frac{64}{25} + \frac{16}{25} = \frac{80}{25} = \frac{16}{5}$$

$$\diamond$$

Example 11.12 did not really use the extreme value theorem. For that to happen we would need to optimize over a line segment instead of a full line.

Example 11.13. If

$$h(x, y) = x^2 + y^2$$

find the minimum and maximum value of $h(x, y)$ among those points (x, y) on the line $y = x + 1$ for $-4 \le x \le 4$.

Solution:

This problem is very similar to Example 11.12. The ends of the line segment are (-4,-3) and (4,5), found by substituting into the formula for the line. On the line $h(x, y)$ becomes

$$h(x, x + 1) = x^2 + (x + 1)^2 = 2x^2 + 2x + 1$$

So we get a critical point at $4x + 2 = 0$ or $x = -1/2$ which is the point $(-1/2, 1/2)$. Plug in and we get

$$h(-4, -3) = 25$$
$$h(-1/2, 1/2) = 1/2$$
$$h(4, 5) = 41$$

This means the minimum value is $1/2$ at the critical point and that the maximum value is 41 at one of the endpoints of the domain of optimization.

$$\diamond$$

Making the domain a line segment is one of the simplest possible options. Let's look at another example with a more complex domain of optimization.

Example 11.14. If

$$s(x, y) = 2x + y - 4$$

find the minimum and maximum value of $s(x, y)$ among those points (x, y) on the curve $y = x^2 - 3$ for $-2 \le x \le 2$.

Solution:

In this problem we are cutting a parabolic segment out of the plane $s(x, y) = 2x + y - 4$. We get that the ends of the domain of optimization are $(2, 1)$ and $(-2, 1)$ by plugging in the ends of the interval in x to the formula for the parabolic segment. Substituting the parabolic segment into the plane yields

$$s(x, x^2 - 3) = 2x + x^2 - 3 - 4 = x^2 + 2x - 7$$

This means our critical point appears at $2x + 2 = 0$ or $x = -1$, making the candidate point $(-1, -2)$. Plugging the candidate points into $s(x, y)$ yields:

$$s(-2, 1) = -7$$
$$s(-1, -2) = -8$$
$$s(2, 1) = 1$$

This means that the maximum value is 1 at $(2, 1)$, and the minimum is -8 at $(-1, -2)$.

$$\diamond$$

Some sets of boundaries are simple enough that we can use geometric reasoning to avoid needing to use calculus on the boundaries.

Example 11.15. Find the maximum value of

$$h(x, y) = x^2 + y^2$$

for $-2 \leq x, y \leq 3$.

Solution:

First of all, we know that this surface has a single critical point at $(0, 0)$ – this surface is an old friend (see graph in Figure 10.2). The domain of optimization is a square with corners $(-2, 2)$ and $(3, 3)$. Along each side of the square, we see that the boundary is a line – and so has extreme values at its ends. This means that we need only check the corners of the square; the interior of the edges – viewed as line segments – cannot attain maximum or minimum values *by the extreme value theorem*. This means we need only add the points $(-2, -2)$, $(-2, 3)$, $(3, -2)$, and $(3, 3)$ to our candidates. Plugging in the candidate points we obtain:

$$h(-2 - 2) = 8$$
$$h(0, 0) = 0$$
$$h(-2, 3) = 13$$
$$h(3, -2) = 13$$
$$h(3, 3) = 18$$

So the maximum is 18 at $(3, 3)$, and the minimum is 0 at $(0, 0)$.

The goal for this section was to set up LaGrange Multipliers – the topic of Section 11.3. The take-home message from this section is that the extreme value theorem implies that optimizing on a boundary is the difficult added portion of optimizing on a bounded domain.

The techniques that we will develop in the next section require that the boundary itself be a differentiable curve. Some of the boundaries in this section are made of several differentiable curves, meaning that the techniques in this section may be easier for those problems. If we have a boundary that is not differential, then the techniques in this section are the only option.

Problems

Problem 11.11. *If we look at the points on*
$$h(x, y) = x^2 + y^2 + 4x + 4$$
that fall on a line, then there is a minimum somewhere on the line. Find that minimum value for the following lines.

(a) $y = 3x + 1$

(b) $y = 4 - x$

(c) $x + y = 6$

(d) $2x - 7y = 24$

(e) $x + y = \sqrt{3}/2$

(f) $-3x + 5x = 7$

Problem 11.12. *Find the maximum of*
$$f(x, y) = (x^2 + y^2)e^{-xy}$$
with (x, y) in the first quadrant where $0 < x, y$. Hint: this function is symmetric. Use this fact.

Problem 11.13. *Find the maximum and minimum of*
$$q(x, y) = \frac{1}{x^2 + y^2 - 2x - 4y + 6}$$
in the first quadrant: $0 < x, y$.

Problem 11.14. *Find the maximum of*
$$g(x, y) = (x^4 + y^4)$$
on the set of points
$$\{(x, y) : x^2 + y^2 \le 25\}.$$

Problem 11.15. *Maximize*
$$h(x) = x^4 + y^4$$
on the set of points $\{(x, y) : x^2 + y^2 \le 25\}$.

Problem 11.16. *If*
$$g(x, y) = 2x^2 + 3y^2,$$
find the minimum of $g(x, y)$ for those points (x, y) on the line $y = x - 1$.

Problem 11.17. *For the function*
$$s(x, y) = x^2 + y^2 - 2x + 4y + 1$$
with (x, y) on the following line segments, find the minimum and maximum values.

(a) $y = 2x - 2$ on $-2 \le x \le 2$

(b) $y = x + 7$ on $-1 \le x \le 4$

(c) $y = 6 - 5x$ on $0 \le x \le 6$

(d) $x + y = 10$ on $4 \le x \le 10$

(e) $2x - 7y = 8$ on $-10 \le x \le 10$

(f) $2x - y = 13$ on $1 \le x \le 5$

Problem 11.18. *Suppose that*
$$p(x, y) = 3x - 5y + 2$$
Find the maximum and minimum values on the following parametric curves.

(a) $(3t + 1, 5 - t)$, $-5 \le t \le 5$

(b) $(\cos(t), \sin(t))$

(c) $(\sin(t), 3\cos(t))$

(d) $(\sin(2t), \cos(t))$

(e) $(t\cos(t), t\sin(t))$, $0 \le t \le 2\pi$

(f) $(3\sin(t), 2\sin(t))$

Problem 11.19. *Find the maximum and minimum of*
$$q(x, y) = \frac{1}{x^2 + y^2 + 4x - 12y + 45}$$
in the first quadrant: $0 < x, y$.

Problem 11.20. *Suppose that $P = f(x, y)$ is a plane and that we are considering the points where $x^2 + y^2 = r^2$ for some constant r. If $f(x, y)$ is not equal to a constant, explain why there is a unique minimum and a unique maximum value.*

Problem 11.21. *If*
$$f(x, y) = x^2 + y^2,$$
find the minimum of $f(x, y)$ for those points (x, y) on the line $y = mx + b$.

11.3. Lagrange Multipliers

This section introduces **constrained optimization** using a technique called **Lagrange Multipliers**. The basic idea is that we want to optimize a function $f(x, y)$ at those points where $g(x, y) = c$. The function $g(x, y)$ is called the **constraint**.

The proof that Lagrange multipliers work is beyond the scope of this text, so we will begin by just stating the technique.

Knowledge Box 11.7.

The method of LaGrange Multipliers with two variables

Suppose that $z = f(x, y)$ defines a surface and that $g(x, y) = c$ specifies points of interest. Then the optima of $f(x, y)$, subject to the constraint that $g(x, y) = c$, occur at solutions to the system of equations

$$f_x(x, y) = \lambda \cdot g_x(x, y)$$
$$f_y(x, y) = \lambda \cdot g_y(x, y)$$
$$g(x, y) = c$$

where λ is an **auxiliary variable**.

The variable λ is new and strange – it is the "multiplier" – and, as we will see, correct solutions to the system of equations that arise from Lagrange multipliers typically use λ in a fashion that causes it to drop out.

If the constraint $g(x, y) = c$ is thought of as the boundary of the domain of optimization, then using Lagrange multipliers gives us a tool for resolving the boundary as a source of optima as per the extreme value theorem.

With that context, let's practice our Lagrange multipliers.

Example 11.16. Find the maxima and minima of

$$f(x, y) = x + 4y - 2$$

on the ellipse $2x^2 + 3y^2 = 36$.

Solution:

The equations arising from the Lagrange multiplier technique are:

$$1 = \lambda \cdot 4x$$
$$4 = \lambda \cdot 6y$$
$$2x^2 + 3y^2 = 36$$
$$\text{Solving:}$$
$$x = \frac{1}{4\lambda}$$
$$y = \frac{2}{3\lambda}$$

Plug into the constraint and we get:

$$\frac{2}{16\lambda^2} + \frac{12}{9 \cdot \lambda^2} = 36$$
$$\frac{1}{8} + \frac{4}{3} = 36\lambda^2$$
$$35/24 = 36\lambda^2$$
$$35/864 = \lambda^2$$
$$\text{or}$$
$$\lambda = \pm\sqrt{35/864}$$

Which yields candidate points:

$$x = \pm\frac{1}{4}\sqrt{\frac{864}{35}} = \pm\sqrt{\frac{54}{35}}$$
$$y = \pm\frac{2}{3}\sqrt{\frac{864}{35}} = \pm\sqrt{\frac{384}{35}}$$

Since the equation of $f(x, y)$ is a plane that gets larger as x and y get larger, we see that the maximum is

$$f\left(\sqrt{54/35}, \sqrt{384/35}\right) \cong 12.5$$

and the minimum is

$$f\left(-\sqrt{54/35}, -\sqrt{384/35}\right) \cong -16.5$$

◊

Notice that in the calculations in Example 11.16, the auxiliary variable λ served as an informational conduit that let us discover the candidate points. The next example is much simpler.

Example 11.17. Find the point of closest approach of the line $2x + 5y = 3$ to the origin (0,0).

Solution:

This problem is like the closest approach of a plane to the origin problems, but in a lower dimension. The tricky part of this problem is phrasing it as a function to be minimized and a constraint. Since we are minimizing distance we get that the function is the distance of a point (x, y) from the origin:

$$d(x, y) = \sqrt{(x - 0)^2 - (y - 0)^2} = \sqrt{x^2 + y^2}$$

As per the monotone function shortcut, we can instead optimize the function

$$d^2(x, y) = x^2 + y^2.$$

The constraint function is the line. We can phrase the line as a constraint by saying:

$$g(x, y) = 2x + 5y = 3$$

With the parts in place, we can extract the Lagrange multiplier equations.

$$2x = \lambda 2$$
$$2y = \lambda 5$$
$$2x + 5y = 3$$
$$\text{Solve:}$$
$$x = \lambda$$
$$y = \frac{5}{2} \cdot \lambda$$
$$2(\lambda) + 5\left(\frac{5}{2} \cdot \lambda\right) = 3$$
$$\frac{29}{2} \cdot \lambda = 3$$
$$\lambda = \frac{6}{29}$$
$$\text{so}$$
$$x = \frac{6}{29}$$
$$y = \frac{15}{29}$$

Making the point on $2x + 5y = 3$ closest to the origin $\left(\dfrac{6}{29}, \dfrac{15}{29}\right)$.

◊

Example 11.18. Find the maximum value of

$$f(x, y) = x^2 + y^3$$

subject to the constraint $x^2 + y^2 = 9$.

Solution:

This problem is already in the correct form for Lagrange multipliers, so we may immediately derive the system of equations.

$$2x = 2x\lambda$$
$$2y = 3y^2\lambda$$
$$x^2 + y^2 = 9$$

The first equation yields the useful information that $\lambda = 1$ but that x may take on any value. Given that $\lambda = 1$, the second equation tells $y = 0$ or $y = 2/3$.

Plugging the values for y into the constraint that says the points lie on a circle, we can retrieve values for x: when $y = 0$, $x = \pm 3$; when $y = 2/3$, $x = \pm\sqrt{9 - 4/9} = \pm\sqrt{77/9}$.

This gives us the candidate points $(3, 0)$, $(-3, 0)$, $(2/3, \sqrt{77}/3)$, and $(2/3, -\sqrt{77}/3)$.

The sign of the x-coordinate is unimportant because $f(x, y)$ depends on x^2; since $f(x, y)$ depends on y^3, positive values yield larger values of f, and negative ones yield smaller values of f.

Since $f(3, 0) = 9$ and $f(2/3, \sqrt{77}/3) \cong 25.47$, we get that the maximum value of the function on the circle is:

$$f(2/3, \sqrt{77}/3) \cong 25.47$$

◊

In Section 11.2 we found the minimum of a line on a quadratic surface that opens upward (Example 11.12). The next example lets us try a problem like this using the formalism of Lagrange multipliers.

Example 11.19. Find the minimum of $f(x,y) = x^2 - xy + y^2$ on the line $5x + 7y = 18$.

Solution:

The constraint is the line – so $g(x,y) = 5x + 7y = 18$. With this detail we can apply the Lagrange multiplier technique:

$$2x - y = 5\lambda$$
$$2y - x = 7\lambda$$
$$5x + 7y = 18$$

Solve the constraint for y and we get $y = \dfrac{18 - 5x}{7}$ so

$$2x - \frac{18 - 5x}{7} = 5\lambda$$
$$14x - 18 + 5x = 35\lambda$$
$$19x - 35\lambda = 18$$
$$\text{and}$$
$$2\left(\frac{18 - 5x}{7}\right) - x = 7\lambda$$
$$36 - 10x - 7x = 49\lambda$$
$$17x + 49\lambda = -36$$
$$\lambda = -(17x + 36)/49$$
$$19x + 35 \cdot (17x + 36)/49 = 18$$
$$1526x = -378$$
$$x = -189/763 \cong -0.25$$
$$y = 2097/763 \cong 2.75$$

Making the minimum

$$\left(\frac{-189}{763}\right)^2 - \frac{-189 \cdot 2097}{74^2} + \left(\frac{2097}{763}\right)^2 \cong 8.30$$

$$\Diamond$$

This problem actually got harder when we used the Lagrange multiplier formalism to solve it. This is another example of how different tools are good for different problems.

At this point, we introduce the 3-space version of Lagrange multipliers. Or, we could say that this is the version of Lagrange multipliers that uses three independent variables. This version of the Lagrange multiplier technique widens the variety of problems we can work with.

Knowledge Box 11.8.

The method of LaGrange Multipliers with three variables

Suppose that $w = f(x, y, z)$ defines a surface and that $g(x, y, z) = c$ specifies points of interest. Then the optima of $f(x, y, z)$, subject to the constraint that $g(x, y, z) = c$, occur at solutions to the system of equations

$$f_x(x, y, z) = \lambda \cdot g_x(x, y)$$
$$f_y(x, y, z) = \lambda \cdot g_y(x, y)$$
$$f_z(x, y, z) = \lambda \cdot g_z(x, y)$$
$$g(x, y, z) = c$$

where λ is an *auxiliary variable*.

Example 11.20. Find the point of intersection of the function $w = 2x + y - z$ and the sphere $x^2 + y^2 + z^2 = 8$ that has the largest value for w.

Solution:
The problem is already in the correct form to apply Lagrange multipliers.

$$2x = 2\lambda$$
$$2y = \lambda$$
$$2z = -\lambda$$
$$x^2 + y^2 + z^2 = 8$$

Solving the first three equations tell us that

$$x = \lambda \qquad y = \frac{1}{2}\lambda \qquad z = -\frac{1}{2}\lambda$$

Plugging these into the last equation tells us that

$$\lambda^2 + \frac{1}{4}\lambda^2 + \frac{1}{4}\lambda^2 = 8$$
$$\frac{3}{2}\lambda^2 = 8$$
$$\lambda^2 = \frac{16}{3}$$
$$\lambda = \pm\frac{4}{\sqrt{3}}$$

So we see that: $x = \pm\dfrac{4}{\sqrt{3}} \qquad y = \pm\dfrac{2}{\sqrt{3}} \qquad z = \pm\dfrac{2}{\sqrt{3}}$

Looking at the formula for w, we see that w grows as x, y, and $-z$. So the point that maximizes w is: $(x, y, z) = \left(\dfrac{4}{\sqrt{3}}, \dfrac{2}{\sqrt{3}}, -\dfrac{2}{\sqrt{3}}\right)$.

◊

Example 11.21. Suppose we divide a rope of length 6 meters into three pieces. What size of pieces maximizes the product of the lengths?

Solution:

Name the length of the pieces x, y, z. That means the function we are maximizing is:

$$f(x, y, z) = xyz$$

and the constraint is $x + y + z = 6$.

Having put the problem into the form for Lagrange multipliers, we can move to the system of equations.

$$yz = \lambda$$

$$xz = \lambda$$

$$xy = \lambda$$

$$x + y + z = 6$$

So $x = \lambda/z = y$ and $z = \lambda/x = y$ making $x = y = z$. Since they sum to 6, we see $x = y = z = 2$ is the sole candidate point. Testing constrained points near $(2, 2, 2)$, like $(1.9, 2.1, 2)$ and $(2, 1.95, 2.05)$, shows that this point is a maximum.

Problems

Problem 11.22. *For each of the following sets of functions and constraints, write out but do not solve the Lagrange multiplier equations.*

(a) $f(x, y) = \cos(x)\sin(y)$
   constrained by $x^2 + y^2 = 8$

(b) $g(x, y) = xe^y$
   constrained by $2x - 3y = 12$

(c) $h(x, y) = \dfrac{x^2}{y^2 + 1}$
   constrained by $3x^2 + y^2 = 48$

(d) $r(x, y) = 2xy$
   constrained by $x^2 + y^2 = 8$

(e) $s(x, y) = \ln(x^2 + y^2 + 4)$
   constrained by $xy = 9$

(f) $q(x, y) = \tan^{-1}(xy + 1)$
   constrained by $x^2 - y^2 = 4$

Problem 11.23. *For each of the following lines, find the point on the line closest to the origin using the method of Lagrange multipliers.*

(a) $y = 2x + 1$

(b) $y = 4 - x$

(c) $2x + 3y = 17$

(d) $x + y = 12$

(e) $y = 5x + 25$

(f) $2x + 4y = 8$

Problem 11.24. *Using Lagrange multipliers, find the maximum and minimum values of*

$$f(x, y) = x^4 + y^4$$

subject to the constraint that

$$x^2 + y^2 = 1.$$

Problem 11.25. *Suppose that*

$$z = 3x - 5y + 2$$

For each of the following constraints, find the maximum and minimum values of z subject to the constraint, if any.

(a) $x^2 + y^2 = 4$

(b) $x^2 + y^2 = 0.25$

(c) $2x^2 + 6y^2 = 64$

(d) $x^2 - xy + y^2 = 1$

(e) $x^2 - y^2 = 4$

(f) $xy = 16$

Problem 11.26. *If*

$$f(x, y) = x^3 - y^2$$

and $x^2 + y^2 = 25$, find the maximum and minimum values $f(x, y)$ can take on.

Problem 11.27. *Use Lagrange multipliers to find the closest approach of the plane*

$$f(x, y) = 3x - y + 2$$

to the origin.

Problem 11.28. *If*

$$f(x, y, z) = x^2 + 4y^2 + 9z^2$$

and $x + y + z = 60$, find the values of x y, and z that maximize and minimize f.

Problem 11.29. *If*

$$h(x, y, z) = x^2 - y^2 + z^2$$

and $x + y + z = 6$, find the values of x y, and z that maximize and minimize h.

Problem 11.30. *If*

$$q(x, y, z) = x^2 - y^2 - 2z^2$$

and $x + y + z = 300$, find the values of x y, and z that maximize and minimize q.

Problem 11.31. *Find the largest point in intersection of*

$$x + 2y + 3z = 4$$

and a cylinder of radius 2 centered on the z-axis.

Chapter 12

Advanced Integration

This chapter covers various sorts of integration that compute volumes and areas. The first, volumes of revolution, is a small twist on the integration methods we have already studied. The other techniques involve multivariate integration, which is both newer and more difficult. This latter subject permits us to compute the volume under a surface as we computed the area under the curve in Chapter 4.

12.1. Volumes of Rotation

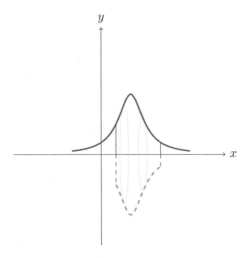

Figure 12.1. A portion of a curve rotated about the x-axis.

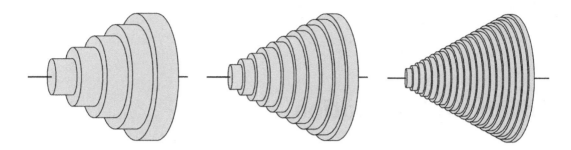

Figure 12.2. Approximations of a rotated volume using smaller and larger numbers of disks.

Figure 12.1 shows a portion of a curve and its shadow reflected about the x-axis. This is meant to evoke revolving the curve about the axis between the two vertical lines. When the curve is rotated, it encloses a volume – our goal is to compute that volume.

When integration was defined in Section 4.4, we first approximated the area represented by the integral with rectangles and then let the number of rectangles go to infinity, obtaining both the area and the integral as a limit. For volumes of rotation, we use a similar technique with *disks* instead of rectangles. Figure 12.2 shows how a conical volume is approximated using five, ten, and twenty disks. The more disks we use, the closer the sum of the volumes of the disks gets to the volume enclosed by rotating a curve about the x axis.

The final version of the area integral added up functional heights – rectangles of infinite thinness. Recall that the area under a curve from $x = a$ to $x = b$ was

$$\text{Area} = \int_a^b f(x) \cdot dx.$$

Given this, the analogous integral for volume of rotation is

$$\text{Volume} = \int_a^b \text{Area of Circles} \cdot dx = \int_a^b \pi r^2 \cdot dx$$

The radius of the circular slices of the volume we are trying to compute is given by the height of the function being rotated about the axis, giving us the final formula for *volume of rotation of a function about the x-axis.*

Knowledge Box 12.1.

Formula for volume of rotation about the x axis

The volume V enclosed by rotating the function $f(x)$ about the x-axis from $x = a$ to $x = b$ is:

$$V = \pi \int_a^b f(x)^2 \cdot dx$$

(Notice that we took the constant π out in front of the integral sign.)

Example 12.1. Find the volume enclosed by rotating $y = x^2$ about the x axis from $x = 0$ to $x = 2$.

Solution:

Start by sketching the situation:

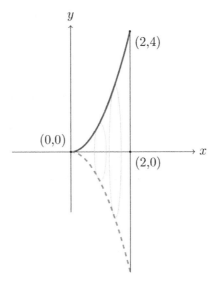

Using the equation in Knowledge Box 12.1 we get:

$$
\begin{aligned}
V &= \pi \int_0^2 f(x)^2 \cdot dx \\
&= \pi \int_0^2 \left(x^2\right)^2 \cdot dx \\
&= \pi \int_0^2 x^4 \cdot dx \\
&= \pi \frac{1}{5} x^5 \Big|_0^2 \\
&= \pi \left(\frac{32}{5} - 0 \right) \\
&= \frac{32\pi}{5} \text{units}^3
\end{aligned}
$$

And so we see the volume of the broad trumpet-shaped solid enclosed by rotating $y = x^2$ about the x-axis from $x = 0$ to $x = 2$ has a volume of $32\pi/5 \cong 503$ cubic units.

◇

The next example is similar except that it uses a more difficult integral. If you're not comfortable with integration by parts, please review Section 7.2.

Example 12.2. Find the volume enclosed by rotating $y = \ln(x)$ about the x axis from $x = 1$ to $x = 4$.

Solution:

Start again with a sketch.

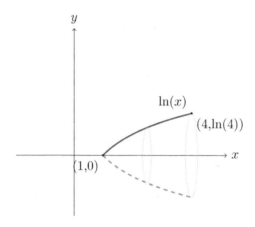

Applying the formula from Knowledge Box 12.1 we get:

$$\text{Volume} = \pi \int_1^4 \ln(x)^2 \cdot dx$$

$$\begin{array}{cc} U = \ln(x)^2 & V = x \\ dU = \dfrac{2\ln(x)}{x} \cdot dx & dV = dx \end{array}$$

$$= \pi \left(x\ln(x)^2 - \int_1^4 2\ln(x) dx \right)$$

$$\begin{array}{cc} U = \ln(x) & V = x \\ dU = \dfrac{dx}{x} & dV = dx \end{array}$$

$$= \pi \left(x\ln(x)^2 - 2x\ln(x) + 2\int_1^4 dx \right)$$

$$= \left. \left(\pi x \ln(x)^2 - 2\pi x \ln(x) + 2\pi x \right) \right|_1^4$$

$$= 4\pi \ln(4)^2 - 8\pi \ln(4) + 8\pi - 0 + 0 - 2\pi$$

$$\cong 8.16 \text{ units}^3$$

◊

Now that we have a formula for rotating objects about the x-axis, the next logical step is to rotate them about the y-axis. This turns out to be a little trickier. There are two basic techniques.

1 We can figure out a new function – the inverse of the original one – that gives us disks along the y-axis, and integrate with respect to dy, or

2 We can use a different type of slice, the *cylindrical shell*. Instead of slicing the shape into disks and integrating them, we slice it into cylinders.

The radius of the disks for rotation about the x-axis was simply the value of the function, $f(x)$, but the distance from a graph to the y axis is just x. We need to get the information about the y-distance in somehow. We will start with Method 2, the method that uses a different type of slice – cylinders centered on the y-axis. Figure 12.3 shows the result of rotating a line about the y axis.

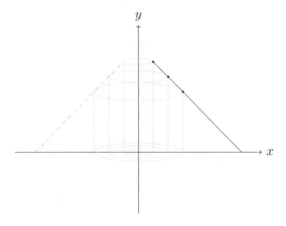

Figure 12.3. Cylindrical shells for computing a volume of revolution about the y-axis.

The area of a cylinder is $2\pi r h$ where r is its radius and h is its height. The radius, as already noted, is x, and the height of the cylinder is $y = f(x)$. Thus, we have the area of our cylindrical slices

$$\text{Area} = 2\pi \cdot r \cdot h = 2\pi \cdot x \cdot f(x)$$

Knowledge Box 12.2.

Volume of rotation with cylindrical shells about the y-axis

If we rotate a function $f(x)$ about the y-axis, then the volume enclosed by the curve between $x = a$ and $x = b$ is:

$$V = 2\pi \int_{x=a}^{x=b} x \cdot f(x) \cdot dx$$

(Notice that we took the constant 2π out in front of the integral sign.)

One tricky thing about this is that the limits of integration are the range of x-values that the radii of the cylindrical shells span. The y-distances only come in via the participation of $f(x)$. The formula in Knowledge Box 12.2 assumes we are *rotating the area below the curve* around the y axis. Finding other areas may require a more complicated setup where we need to figure out the height of the cylinders.

Example 12.3. Find the volume of rotation of the area below the curve $f(x) = 1/x$ about the y-axis from $x = 0.5$ to $x = 2.0$.

Solution:

Sketch the situation.

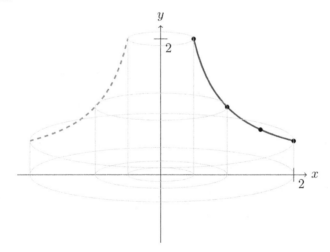

Applying the equation in Knowledge Box 12.2 we get:

$$
\begin{aligned}
\text{Volume} &= 2\pi \int_{0.5}^{2} x \cdot f(x) \cdot dx \\[2mm]
&= 2\pi \int_{0.5}^{2} x \cdot \frac{1}{x} \cdot dx \\[2mm]
&= 2\pi \int_{0.5}^{2} dx \\[2mm]
&= 2\pi x \Big|_{0.5}^{2} \\[2mm]
&= 2\pi (2 - 0.5) \\[2mm]
&= 3\pi \ \text{units}^3
\end{aligned}
$$

$\Diamond$

In the next example we will try and find the volume enclosed by rotating a curve about the y axis. This means we will need to be much more careful about the heights of the cylinders.

Example 12.4. Find the volume enclosed by rotating the area bounded by $f(x) = x^{2/3}$, $x = 1$, and $y = 3^{2/3}$ about the y-axis from $x = 1$ to $x = 3$.

Solution:

Start with the traditional sketch.

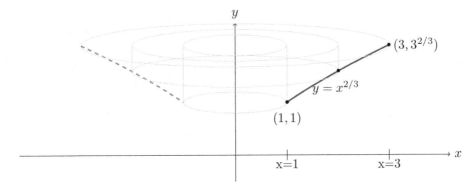

This example is one where the height of the cylinder is not just $f(x)$, so we need to find an expression for the height of the cylinder. This expression goes from $f(x)$ (at the bottom) to $y = 3^{2/3}$ (at the top). So, we get a formula of

$$h = 3^{2/3} - f(x) = 3^{2/3} - x^{2/3}$$

for the height of the cylinder. The area of a cylinder is thus $2\pi x \cdot \left(3^{2/3} - x^{2/3}\right)$ which makes the volume:

$$
\begin{aligned}
\text{Volume} &= 2\pi \int_1^3 x \cdot \left(3^{2/3} - x^{2/3}\right) dx \\
&= 2\pi \int_1^3 \left(x \cdot 3^{2/3} - x \cdot x^{2/3}\right) dx \\
&= 2\pi \int_1^3 \left(3^{2/3} x - x^{5/3}\right) dx \\
&= 2\pi \left(\frac{3^{2/3}}{2} x^2 - \frac{3}{8} x^{8/3}\right) \Bigg|_1^3 \\
&\cong 10.52 \text{ units}^3
\end{aligned}
$$

◇

Example 12.4 demonstrates that the formula in Knowledge Box 12.2 only covers one type of rotation about the y axis. In general, it is necessary to remember that you are adding up cylindrical shells and to carefully figure out the height and radius, plugging into $A = 2\pi r h$ to get the formula to integrate. Care is also needed in figuring out the limits of integration. Next we do an example of a problem where we use disks to rotate about the y axis.

In order to do this we need to use inverse functions. These were defined in Knowledge Box 2.9. Examples of inverse functions include:

- For $x \geq 0$, when $f(x) = x^2$ we have $f^{-1}(x) = \sqrt{x}$.

- If $g(x) = e^x$, we have $g^{-1}(x) = \ln(x)$.

- If $h(x) = \tan(x)$, we have $h^{-1}(x) = \tan^{-1}(x)$.

The notation for "inverse" and "negative first power" are identical and can only be told apart by examining context. Be careful!

Example 12.5. Rotation about the y-axis with disks: Find the volume of rotation when the area bounded by $y = x^2$, the y-axis, and $y = 4$ is rotated about the y axis.

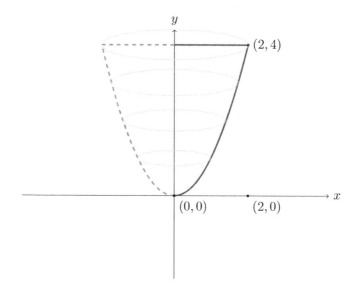

Solution:

For this problem, the radius of the disks is going from the y-axis to the curve $y = x^2$. We need to solve this equation for x which is not too difficult: $x = g(y) = \sqrt{y}$, ignoring the negative square root because we obviously need a positive radius. This means that we have a disk radius of $x = g(y)$. The volume formula becomes

$$V = \pi \int_{y_o}^{y_1} g(y)^2 \cdot dy$$

The function $x = g(y)$ is the *inverse function* of $y = f(x)$. We can now do the calculations for volume:

$$\text{Volume} = \pi \int_0^4 (\sqrt{y})^2 \cdot dy$$

$$= \pi \int_0^4 y \cdot dy$$

$$= \frac{\pi}{2} y^2 \Big|_0^4$$

$$= \frac{\pi}{2}(16 - 0)$$

$$= 8\pi \text{ units}^3$$

To see that is it possible, let's set this up (but not calculate the integral) with cylinders. The height of a cylinder is from $y = x^2$ to $y = 4$ meaning that $h = 4 - x^2$. So, the volume integral is

$$V = 2\pi \int_{x=0}^{x=2} x(4 - x^2) \cdot dx$$

Not a terribly hard integral, but one that is harder than the disk integral.

Knowledge Box 12.3.

Disks or Cylinders?

Both methods for finding volume of rotation involve adding up areas with integration to find a volume. The method of disks adds up circular disks and the method of cylinders adds up, well, cylinders. How do you tell which method to use?

You use whichever method you can set up and, if you can set up both, you use the one that yields the easier integral.

The next example is another one that lets us practice with the method of cylinders. It is an example where the calculations to find the radius values needed to use the method of disks is too hard.

Example 12.6. Compute the volume obtained by rotating the area bounded by the curve

$$y = 2x^2 - x^3$$

and the x-axis around the y-axis.

Solution:

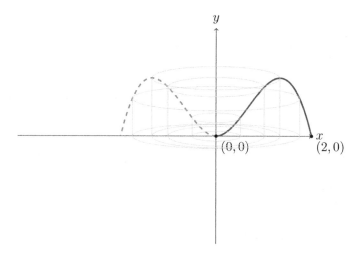

Let's do the calculations. The points where $y = 0$ are $x = 0$ and $x = 2$, giving us the limits of integration. In this case the radius of the cylinders is just x, and the height is just y. So, applying the formula from Knowledge Box 12.2 we get:

$$\text{Volume} = 2\pi \int_0^2 x(2x^2 - x^3) \cdot dx$$

$$= 2\pi \int_0^2 (2x^3 - x^4) \cdot dx$$

$$= 2\pi \left(\frac{1}{2}x^4 - \frac{1}{5}x^5 \Big|_0^2 \right)$$

$$= 2\pi \left(8 - \frac{32}{5} - 0 + 0 \right)$$

$$= 2\pi \cdot \frac{8}{5}$$

$$= \frac{16\pi}{5} \text{ units}^3$$

If we wanted to use the method of disks we would need to solve $y = 2x^2 - x^3$ for x to get the inverse function – a challenging piece of algebra.

◊

Suppose that we want to rotate the area between two curves about the x-axis. Then we get a large disk for the outer curve and a small disk for the inner curve – meaning that we get *washers*. A rendering of a washer is shown in Figure 12.4. The area of a washer with outer radius r_1 and inner radius r_2 is the difference of the area of the overall disk and the missing inner disk:

$$A = \pi r_1^2 - \pi r_2^2 = \pi(r_1^2 - r_2^2)$$

This area formula forms the basis of the integration performed with the **method of washers**.

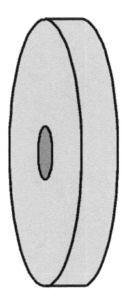

Figure 12.4. A washer or disk with a hole in the center. When we want to rotate the difference of two curves about the x-axis, this shape replaces the disks used when only one curve is rotated.

Knowledge Box 12.4.

Formula for finding volume with the method of washers

If $y = f_1(x)$ and $y = f_2(x)$ are functions for which $f_1(x) \geq f_2(x)$ on the interval $[a, b]$, then the volume obtained by rotating the area between the curves about the x-axis is:

$$V = \pi \int_a^b \left(f_1(x)^2 - f_2(x)^2 \right) \cdot dx$$

(Notice that we took the constant π out in front of the integral sign.)

Example 12.7. Find the volume resulting from rotating the area between $f_1(x) = x^2$ and $f_2(x) = \sqrt{x}$ about the x-axis.

Solution:

To apply the method of washers, we need to know where the curves intersect and which one is on the outside edge of the washers. It is easy to see that they intersect at $(0,0)$ and $(1,1)$. So the limits of integration will be from $x = 0$ to $x = 1$. On the range $0 \leq x \leq 1$, it's easy to see that $x^2 \leq \sqrt{x}$. So, the outer curve is $f_2(x)$. This gives us enough information to set up the integral.

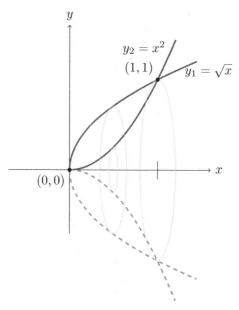

$$\text{Volume} = \pi \int_0^1 \left(f_1(x)^2 - f_2(x)^2 \right) \cdot dx$$

$$= \pi \int_0^1 \left(x - x^4 \right) dx$$

$$= \pi \left(\frac{1}{2}x^2 - \frac{1}{5}x^5 \right) \Big|_0^1$$

$$= \frac{3\pi}{10} \text{ units}^3$$

$\Diamond$

There are a large number of different ways to rotate objects about the x- and y-axis. While formulas are presented in this section, it is a really good idea to identify the radius and height of the disks, cylindrical shells, or washers you are integrating to create a volume of rotation. Without that understanding, it is easy to pick the wrong formula. For Example 12.4 *none* of the formulas apply!

Problems

Problem 12.1. *Find the volume of rotation about the x-axis for each of the functions below from $x = 0$ to $x = 1$.*

(a) $f(x) = 3x^5$

(b) $g(x) = 0.5x^6$

(c) $h(x) = 5x^7$

(d) $r(x) = 12x^5$

(e) $s(x) = 6x^{11}$

(f) $q(x) = e^{-x}$

(g) $a(x) = \cos(\pi x)$

(h) $b(x) = (e^x + e^{-x})$

Problem 12.2. *If we rotate $y = \sin(x)$ about the x-axis, the result is a string of beads. Find the volume of one bead.*

Problem 12.3. *If we rotate $f(x) = \sin(4x)$ about the x-axis, it makes a chain of beads. Find the volume of one bead.*

Problem 12.4. *Recalculate Example 12.5 using the method of cylinders.*

Problem 12.5. *Find the volume of rotation about the y axis of the area below each of the functions listed from $x = 0$ to $x = 1$.*

(a) $f(x) = 6x^9$

(b) $g(x) = 16x^8$

(c) $h(x) = 3x^{12}$

(d) $r(x) = 5x^8$

(e) $s(x) = 11x^8$

(f) $q(x) = e^x$

(g) $a(x) = \cos(\pi x)$

(h) $b(x) = \frac{x}{x^2+1}$

Problem 12.6. *We already know the area under $y = 1/x$ on the interval $[1, \infty)$ is infinite. Using the methhod of disks, find the volume of rotation of $y = 1/x$ on this interval.*

Problem 12.7. *For each of the following functions and starting and ending x values, find the volume of rotation of the function about the x-axis.*

(a) $f(x) = x^{-2/5}$ *between* $x = 3$ *and* $x = 7$

(b) $f(x) = x^{-2/5}$ *between* $x = 3$ *and* $x = 5$

(c) $f(x) = x^{2/7}$ *between* $x = 4$ *and* $x = 7$

(d) $f(x) = x^{3/7}$ *between* $x = 5$ *and* $x = 7$

(e) $f(x) = x^{-3}$ *between* $x = 5$ *and* $x = 9$

(f) $f(x) = x^{-2}$ *between* $x = 8$ *and* $x = 9$

Problem 12.8. *For each of the following pairs of functions, compute the volume obtained by rotating the area between the functions about the x-axis.*

(a) $f(x) = 3x^3$ *and* $g(x) = 9x^2$

(b) $f(x) = x^3$ *and* $g(x) = 4x^2$

(c) $f(x) = 3x^5$ *and* $g(x) = 48x$

(d) $f(x) = 3x^4$ *and* $g(x) = 3x^3$

(e) $f(x) = 2x^3$ *and* $g(x) = 6x^2$

(f) $f(x) = x^3$ *and* $g(x) = x^2$

Problem 12.9. *For each of the following pairs of functions, compute the volume obtained by rotating the area between the functions about the y-axis.*

(a) $f(x) = 7x^3$ *and* $g(x) = 28x^2$

(b) $f(x) = 4x^6$ *and* $g(x) = 64x^2$

(c) $f(x) = 2x^4$ *and* $g(x) = 18x^2$

(d) $f(x) = 3x^6$ *and* $g(x) = 3x^2$

(e) $f(x) = 7x^5$ *and* $g(x) = 21x^4$

(f) $f(x) = 2x^3$ *and* $g(x) = 8x$

Problem 12.10. *Using the method of disks, rotating about the x-axis, verify the formula for the volume of a cone of radius R and height H:*

$$V = \frac{1}{3}\pi \cdot R^2 H$$

12.2. Arc Length and Surface Area

In this section we will learn to compute the length of curves and, having done that, to find the surface area of figures of rotation. A piece of a curve is called an **arc**. The key to finding the length of an arc is the **differential of arc length**. In the past we have had quantities like dx and dy that measure infinitesimal changes in the directions of the variables x and y. The differential of arc length is different – it does not point in a consistent direction, rather it points along a curve and so, by integrating it, we can find the length of a curve.

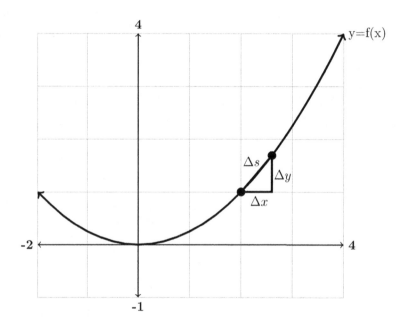

Figure 12.5. The triangle with sides Δx, Δy, and Δs shows how the change in the length of the curve is related to the changes in distance in the x and y directions.

Examine Figure 12.5. The relationship between the change in x and y and the change in the length of the curve is Pythagorean – based on a right triangle. If we take this relationship to the infinitesimal scale, we obtain a formula for the differential of arc length.

Knowledge Box 12.5.

The differential of arc length and arc length

If $y = f(x)$ is a continuous curve, then the rate at which the length of the graph of $f(x)$ changes is called the **differential of arc length**, denoted by ds. The value of ds is:

$$ds^2 = dy^2 + dx^2$$

$$ds = \sqrt{dy^2 + dx^2}$$

$$= \sqrt{\left(\frac{dy^2}{dx^2} + 1\right) \cdot dx^2}$$

$$= \sqrt{(y')^2 + 1} \cdot dx$$

The length S of a curve from $x = a$ to $x = b$ is:

$$S = \int_a^b ds$$

Example 12.8. Find the length of $y = 3x^{2/3}$ from $x = 1$ to $x = 8$.

Solution:

The first step in an arc length problem is to compute ds.

$$y = 3x^{2/3}$$
$$y' = 2x^{-1/3}$$
$$ds = \sqrt{4x^{-2/3} + 1} \cdot dx$$

This means that the desired length is

$$S = \int_1^8 ds$$
$$= \int_1^8 \sqrt{\frac{4}{x^{2/3}} + 1} \cdot dx$$
$$= \int_1^8 \sqrt{\frac{4 + x^{2/3}}{x^{2/3}}} \cdot dx$$
$$= \int_1^8 \sqrt{4 + x^{2/3}} \cdot \frac{dx}{x^{1/3}}$$

Let $u = 4 + x^{2/3}$, then $du = \frac{2}{3}x^{-1/3} \cdot dx = \frac{2}{3}\frac{dx}{x^{1/3}}$. So, $\frac{3}{2}du = \frac{dx}{x^{1/3}}$. Applying the substitution to the limits we see that the integral goes from $u = 5$ to $u = 8$. Transforming everything to u-space, the arc length is:

$$S = \int_5^8 \sqrt{u} \cdot \frac{3}{2}du$$
$$= \frac{3}{2} \int_5^8 u^{1/2} du$$
$$= \frac{3}{2} \left(\frac{2}{3}u^{3/2} \right) \Big|_5^8$$
$$= 8^{3/2} - 5^{3/2}$$
$$\cong 11.45 \text{ units}^2$$

$\Diamond$

Alert students will have noticed that the function chosen to demonstrate arc length is not one of our usual go-to functions for demonstration. This is because the formula for ds yields some very difficult integrals. The next example is one such, but yields a formula we already know how to integrate.

Example 12.9. Find the arc length of $y = x^2$ from $x = 0$ to $x = 2$.

Solution:
Since $y' = 2x$, it is easy to find that $ds = \sqrt{4x^2 + 1} \cdot dx$, meaning our integral is:

$$S = \int_0^2 \sqrt{4x^2 + 1} \cdot dx$$

This is a trig-substitution integral (Section 7.4). The triangle for this integral is

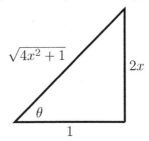

which means our substitutions are:

$$\sqrt{4x^2 + 1} = \sec(\theta)$$
$$2x = \tan(\theta)$$
$$x = \frac{1}{2} \tan(\theta)$$
$$dx = \frac{1}{2} \sec^2(\theta) \cdot d\theta$$

So

$$\int_0^2 \sqrt{4x^2 + 1} \cdot dx = \int_?^? \sec(\theta) \cdot \frac{1}{2} \sec^2(\theta) \cdot d\theta$$
$$= \frac{1}{2} \int_?^? \sec^3(\theta) \cdot d\theta$$

which is an integral we have done before (Example 7.25)

$$= \frac{1}{4} \left(\sec(\theta)\tan(\theta) + \ln|\sec(\theta) + \tan(\theta)| \right) \Big|_?^?$$
$$= \frac{1}{4} \left(\sqrt{4x^2 + 1} \cdot 2x + \ln|\sqrt{4x^2 + 1} + 2x| \right) \Big|_0^2$$

Now that we have performed the integral and transformed it back into x-space we can substitute in the limits and get the arc length.

$$S = \frac{1}{4} \left(\sqrt{17} \cdot 4 + \ln|\sqrt{17} + 4| - \sqrt{1} \cdot 0 + \ln|\sqrt{1} + 0| \right)$$
$$= \frac{1}{4} \left(4\sqrt{17} + \ln(\sqrt{17} + 4) \right) \cong 4.65 \text{ units}$$

$\Diamond$

Arc-lengths integrals are often challenging. Let's do one more example with a function chosen to keep the difficulty from getting out of hand. This example looks for a general formula for the arc length of a function.

Example 12.10. Find the length of $y = x^{3/2}$ from $x = 0$ to $x = a$.

Solution:

Since $y' = \dfrac{3}{2}x^{1/2}$ we see that

$$ds = \sqrt{\frac{9}{4}x + 1} \cdot dx = \frac{1}{2}\sqrt{9x + 4} \cdot dx$$

This means that the desired arc length is:

$$
\begin{aligned}
S &= \int_0^a \frac{1}{2}\sqrt{9x + 4} \cdot dx \\
&= \frac{1}{2}\int_0^a \sqrt{9x + 4} \cdot dx
\end{aligned}
$$

Let $u = 9x + 4$ so that $\dfrac{1}{9}du = dx$

$$
\begin{aligned}
&= \frac{1}{2}\int_?^? \sqrt{u} \cdot \frac{1}{9}du \\
&= \frac{1}{18}\int_?^? u^{1/2} \cdot du \\
&= \frac{1}{18}\left(\frac{2}{3}u^{3/2}\right)\Big|_?^? \qquad\qquad \text{Need to substitute back to } x \\
&= \frac{1}{27}\left(9x + 4\right)^{3/2}\Big|_0^a \\
&= \frac{(9a + 4)^{3/2} - 8}{27}
\end{aligned}
$$

which is the desired arc length formula.

We have already computed the volume of a solid that is enclosed by the graph of a function rotated about the x-axis. The solids defined in this fashion also have a surface area, the slices of which are circles. Figure 12.6 shows examples of the circles that appear in such a rotation.

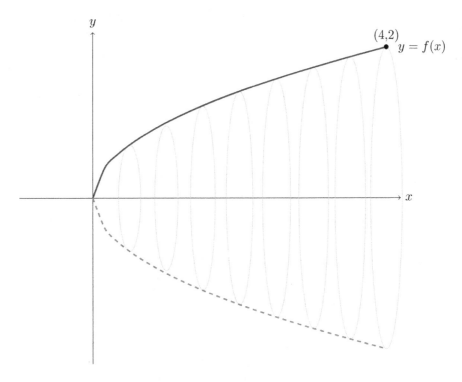

Figure 12.6. Shown are some circles that are the "slices" of the surface area obtained by rotating $f(x)$ about the x-axis.

Using Figure 12.6 as inspiration, we can compute surface area by integrating the circumference of circles with radius $f(x)$. Since these circle follow the arc of $f(x)$, the correct type of change is the differential of arc length, ds. This leads to the formula in Knowledge Box 12.6.

Knowledge Box 12.6.

Surface area of rotation

If $y = f(x)$ is a continuous curve, then the surface area A obtained by rotating $f(x)$ around the x-axis from $x = a$ to $x = b$ is:

$$A = 2\pi \int_a^b f(x) \cdot ds$$

Example 12.11. Find the surface area of rotation of $y = \sqrt{x}$ about the x-axis from $x = 0$ to $x = 4$.

Solution:

Since $y = x^{1/2}$, $y' = \dfrac{1}{2}x^{-1/2} = \dfrac{1}{2\sqrt{x}}$. So: $ds = \sqrt{\dfrac{1}{4x} + 1} \cdot dx$ Using the formula from Knowledge Box 12.6 we obtain the surface area integral,

$$A = 2\pi \int_0^4 \sqrt{x} \cdot \sqrt{\frac{1}{4x} + 1} \cdot dx$$

$$= 2\pi \int_0^4 \sqrt{x \cdot \left(\frac{1}{4x} + 1\right)} \, dx$$

$$= 2\pi \int_0^4 \sqrt{\frac{1 + 4x}{4}} \cdot dx$$

$$= 2\pi \int_0^4 \frac{1}{2}\sqrt{1 + 4x} \cdot dx$$

$$= \pi \int_0^4 \sqrt{1 + 4x} \cdot dx$$

Let $u = 4x + 1$, $\dfrac{1}{4}du = dx$

$$= \pi \int_1^{17} u^{1/2} \cdot \frac{1}{4} du$$

$$= \frac{\pi}{4} \int_1^{17} u^{1/2} \cdot du$$

$$= \frac{\pi}{4}\frac{2}{3} u^{3/2}\Big|_1^{17}$$

$$= \frac{\pi}{6}\left(17^{3/2} - 1\right) \cong 36.18 \text{ units}^2$$

$\Diamond$

Example 12.12. Find the surface area of rotation for $y = e^x$ from $x = 0$ to $x = 2$.

Solution:

Noting that $y' = e^x$, we see that
$$ds = \sqrt{e^{2x} + 1} \cdot dx.$$
This means the surface area is:
$$A = 2\pi \int_0^2 e^x \sqrt{e^{2x} + 1} \cdot dx$$

$$= 2\pi \int_0^2 \sqrt{e^{2x} + 1} \, (e^x \cdot dx)$$

Let $u = e^x$, then $du = e^x \cdot dx$

$$= 2\pi \int_?^? \sqrt{u^2 + 1} \cdot du \qquad\qquad \text{This is Example 7.33}$$

$$= 2\pi \left(u\sqrt{u^2 + 1} + \ln|u + \sqrt{u^2 + 1}| \right)\Big|_?^?$$

$$= 2\pi \left(e^x \sqrt{e^{2x} + 1} + \ln|e^x + \sqrt{e^{2x} + 1}| \right)\Big|_0^2$$

$$= 2\pi \left(e^2 \sqrt{e^4 + 1} + \ln(e^2 + \sqrt{e^4 + 1}) - \sqrt{2} - \ln|1 + \sqrt{2}| \right) \text{ units}^2$$

$$\Diamond$$

The types of integrals that arise from arc length and rotational surface area problems are often quite challenging. This justifies the large number of integration techniques we learned in Chapter 7, several of which came up in this section.

If you study multivariate calculus more deeply, the differential of arc length will appear again for tasks like computing the work done moving a particle along a path through a field. This section is a bare introduction to the power and applications of the differential of arc length.

Problems

Problem 12.11. *For each of the following functions, compute ds, the differential of arc length.*

(a) $f(x) = x^3$

(b) $g(x) = \sin(x)$

(c) $h(x) = \tan^{-1}(x)$

(d) $r(x) = \dfrac{x}{x+1}$

(e) $s(x) = e^{-x}$

(f) $q(x) = x^{3/4}$

(g) $a(x) = \dfrac{1}{3}$

(h) $b(x) = 2^x$

Problem 12.12. *For each of the following functions, compute the arc length of the graph of the function on the given interval.*

(a) $f(x) = 9x^{2/3}$; [0, 1]

(b) $g(x) = 2x + 1$; [0, 4]

(c) $h(x) = 2x^{3/2}$; [2, 5]

(d) $r(x) = x^2 + 4x + 4$; [0, 6]

(e) $s(x) = \sqrt{(x-2)^3}$; [4, 5]

(f) $q(x) = (x+1)^{2/3}$; [-1, 1]

Problem 12.13. *For each of the following functions, compute the surface area of rotation of the function for the given interval.*

(a) $f(x) = x$; [0, 3]

(b) $g(x) = \sqrt{x}$; [4, 9]

(c) $h(x) = e^x$; [0, 1]

(d) $r(x) = \sin(x)$; [0, π]

(e) $s(x) = \cos(3x)$; [0, $\pi/4$]

(f) $q(x) = \sin(x)\cos(x)$; [0, $\pi/2$]

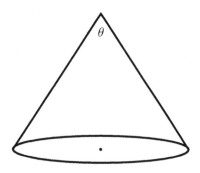

Problem 12.14. *Using the techniques for surface area of revolution, find the formula for the surface area of a cone with apex angle θ, as shown above. Don't forget the area of the bottom.*

Problem 12.15. *If*
$$y = x^{2/3},$$
find the formula for the arc length of the graph of this function on the interval $[a, b]$.

Problem 12.16. *Based on the material in this section, if*
$$(f(t), g(t))$$
is a parametric curve, what would the differential of arc length, ds, be?

Problem 12.17. *Derive the polar differential of arc length.*

Problem 12.18. *Derive the parametric differential of arc length for $(x(t), y(t))$.*

Problem 12.19. *Find the integral for computing the arc length of*
$$y = \sin(x).$$
Discuss: what techniques might work for this integral.

Problem 12.20. *We already know the area under $y = 1/x$ on the interval $[1, \infty)$ is infinite but that the enclosed volume of rotation is finite. Using comparison and cleverness, demonstrate the surface area of this shape is infinite.*

12.3. Multiple Integrals

In Chapter 4 we learned to use integrals to find the area under a curve. The analogous task in three dimensions is to find the volume under a surface over some domain in the x-y plane. As with partial derivatives we will find the idea of a *currently active* variable useful.

When we were integrating a single-variable function to obtain an area, we integrated over an interval on the x-axis. When we are finding volumes under surfaces, we will integrate over *regions* or subsets of the plane. An example of a relatively simple region in the plane is shown in Figure 12.7.

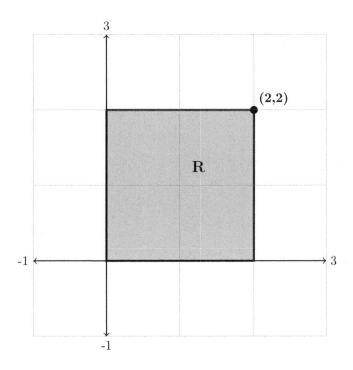

Figure 12.7. A rectangular region R, $0 \leq x, y \leq 2$ in the plane.

So far in this chapter we have found volumes by using integrals to add up slices. To perform volume integrals we will have to integrate slices that are, themselves, the result of integrating. This means we will use **multiple or iterated integrals**. Fortunately, these are done one at a time. We use differentials, like dy and dx, to cue which integral is performed next. Like forks at a fancy dinner party, the next integral is the one for the differential farthest to the left.

Example 12.13. Find the volume under

$$f(x, y) = x^2 + y^2$$

that is above R, $0 \leq x, y \leq 2$, as shown in Figure 12.7.

Solution:

The y and x variables are both in intervals of the form $[0, 2]$, so the limits of integration are 0 and 2 for both the integrals we must perform. The volume under the surface is:

$$V = \int_0^2 \int_0^2 \left(x^2 + y^2\right) \cdot dx \; dy \qquad\qquad \text{The variable } x \text{ is active.}$$

$$= \int_0^2 \left(\frac{1}{3}x^3 + xy^2\right)\bigg|_0^2 \cdot dy$$

$$= \int_0^2 \left(\frac{8}{3} + 2y^2 - 0 - 0\right) \cdot dy$$

$$= \int_0^2 \left(2y^2 + \frac{8}{3}\right) \cdot dy \qquad\qquad \text{Now } y \text{ is active.}$$

$$= \frac{2}{3}y^3 + \frac{8}{3}y \bigg|_0^2$$

$$= \left(\frac{16}{3} + \frac{16}{3} - 0 - 0\right)$$

$$= \frac{32}{3} \text{ units}^3$$

$$\diamond$$

Example 12.13 contained two integrals. During the first, x was the active variable, and y acted like a constant; when the limits of integration were substituted in they were substituted in for x *not* y. The second integral took place in an environment where x was gone and y was the active variable.

Integration with respect to a variable

When we are computing an integral

$$V = \int_a^b \int_c^d f(x,y) \cdot dx \, dy,$$

the first integral treats x as a variable and is said to be an integral **with respect to** x; the second integral treats y as a variable and is said to be an integral **with respect to** y.

When we are integrating with respect to one variable, other variables act as constants and so can pass through integral signs. This permits us to simplify some integrals.

Example 12.14. Use the constant status of variables to compute

$$\int_0^1 \int_0^1 \left(x^2 y^2 \right) \cdot dx \, dy$$

in a simple way.

Solution:

$$\int_0^1 \int_0^1 \left(x^2 y^2 \right) \cdot dx \, dy = \left(\int_0^1 y^2 \cdot dy \int_0^1 x^2 \cdot dx \right) \qquad \text{Pass } y \text{ through the } x \text{ integral}$$

$$= \left(\int_0^1 y^2 \cdot dy \right) \cdot \left(\int_0^1 x^2 \cdot dx \right)$$

$$= \left(\int_0^1 x^2 \cdot dx \right)^2 \qquad \text{Since the integrals are equal}$$

$$= \left(\frac{1}{3} x^3 \Big|_0^1 \right)^2$$

$$= \left(\frac{1}{3} - 0 \right)^2$$

$$= \frac{1}{9} \text{ units}^3$$

$\Diamond$

This sort of integral – that can be split up into two different integrals – is called a *decomposable integral* .

Knowledge Box 12.8.

Decoposable Integrals

The multiple integral of the product of a function of one variable by a function of the other variable can be factored into two single-variable integrals.

$$\int\int f(x)g(y) \cdot dx\, dy = \left(\int f(x)\, dx\right)\left(\int g(y)\, dy\right)$$

Example 12.15. Use the decomposition of integrals to perform the following:

$$\int_0^{\pi/2} \int_0^2 (x \cdot \cos(y)) \cdot dx\, dy$$

Solution:

$$\int_0^{\pi/2} \int_0^2 (x \cdot \cos(y)) \cdot dx\, dy = \left(\int_0^2 x\, dx\right) \times \left(\int_0^{\pi/2} \cos(y)\, dy\right)$$

$$= \left(\frac{x^2}{2}\Big|_0^2\right) \times \left(\sin(y)\Big|_0^{\pi/2}\right)$$

$$= (4/2 - 0/2) \times (\sin(\pi/2) - \sin(0))$$

$$= 2 \cdot 1 = 2$$

$$\Diamond$$

Volume integration becomes more difficult when the region R is not a rectangle. Over a rectangular region, the limits of integration are constants. If a region is not rectangular, then the curves that describe the boundaries of the region become involved in the limits of integration.

Example 12.16. Integrate the function $f(x, y) = 2x - y + 4$ over the region R given in Figure 12.8.

Solution:

The function is simple, but the limits of integration are tricky. In this case, $0 \leq x \leq 2$ and, *for a given value of* x, $0 \leq y \leq x$. This is because the upper edge of the region of integration is the line $y = x$.

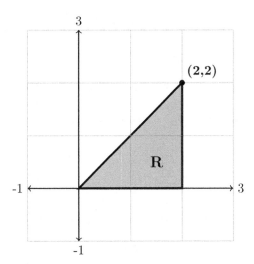

Figure 12.8. A non-rectangular region bounded by the x-axis, the line $x = 2$, and the line $y = x$.

Now, we can set up the integral, choosing the order of integration to agree with the limits.

$$= \int_0^2 \int_0^x (2x - y + 4) \, dy \, dx$$

$$= \int_0^2 \left(2xy - \frac{1}{2}y^2 + 4y \right) \Big|_0^x \, dx$$

$$= \int_0^2 \left(2x^2 - \frac{1}{2}x^2 + 4x - 0 - 0 - 0 \right) \cdot dx$$

$$= \int_0^2 \left(\frac{3}{2}x^2 + 4x \right) dx$$

$$= \frac{1}{2}x^3 + 2x^2 \Big|_0^2$$

$$= \frac{1}{2}8 + 8 - 0 - 0$$

$$= 12 \text{ units}^3$$

$\Diamond$

Example 12.17. Find the volume underneath $f(x, y) = 4 - x^2$ over the region bounded by $y = x$, $y = 2x$, and $x = 2$.

Solution:

Start by drawing the region of integration. This region has x bounds $0 \leq x \leq 2$. For a given value of x the region goes from the line $y = x$ to the line $y = 2x$.

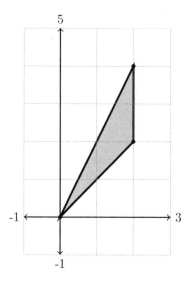

This gives us sufficient information to set up the integral.

$$V = \int_0^2 \int_x^{2x} \left(4 - x^2\right) dy \, dx$$

$$= \int_0^2 \left(4y - x^2 y\right) \bigg|_x^{2x} \cdot dx$$

$$= \int_0^2 \left(8x - 2x^3 - 4x + x^3\right) dx$$

$$= \int_0^2 \left(4x - x^3\right) dx$$

$$= 2x^2 - \frac{1}{4}x^4 \bigg|_0^2$$

$$= 8 - \frac{1}{4} \cdot 16 - 0 + 0$$

$$= 4 \text{ units}^3$$

◊

The regions we have used thus far have been based on functions that are easy to work with using Cartesian coordinates. To deal with other sorts of regions, we need to first develop a broader point of view.

Knowledge Box 12.9.

The differential of area

The **change of area** is

$$dA = dx\,dy = dy\,dx$$

This permits us to change our integral notation to the following for the integral of $f(x, y)$ over a region R

$$V = \int\int_R f(x, y)\,dA$$

In polar coordinates:

$$dA = r\,dr\,d\theta$$

Example 12.18. Find the area enclosed below the curve $f(x, y) = x^2 + y^2$ over a disk of radius 2 centered at the origin.

Solution:

This volume is below the function $f(x, y)$ for points no farther from the origin than 2. This means the region R is

$$x^2 + y^2 \leq 4.$$

In polar coordinates, this region is those points (r, θ) for which $0 \leq r \leq 2$ and $0 \leq \theta \leq 2\pi$. Since the polar/rectangular conversion equations tell us $r^2 = x^2 + y^2$, $f(r, \theta) = r^2$. Using polar coordinates the integral is:

$$V = \int \int_R f(r, \theta) \, dA$$

$$= \int_0^{2\pi} \int_0^2 r^2 \cdot r \cdot dr \, d\theta$$

$$= \int_0^{2\pi} \int r^3 \cdot dr \, d\theta$$

$$= \int_0^{2\pi} \frac{1}{4} r^4 \Big|_0^2 \, d\theta$$

$$= \int_0^{2\pi} (16/4 - 0) \, d\theta$$

$$= \int_0^{2\pi} 4 d\theta$$

$$= 4\theta \Big|_0^{2\pi}$$

$$= 8\pi - 0 = 8\pi \text{ units}^3$$

$$\diamondsuit$$

The next example is a very important one for the theory of statistics. As you know if you have studied statistics, the normal distribution has a probability distribution function of:

$$\frac{1}{\sqrt{2\pi}} \, e^{-x^2/2}$$

The area under the curve of a probability distribution function must be equal to one. Thus, if you have a function with an area greater than one, you must multiply it by a normalizing constant equal to one over the area. The following example shows where the normalizing constant $\frac{1}{\sqrt{2\pi}}$ in the normal distribution probabilty distribution function comes from.

The integral relies on a trick – squaring the integral and then shifting the squared integral to polar coordinates. This changes an impossible integral into one that can be done without difficulty by u-substitution. Sadly, this only permits the evaluation of the integral on the interval $[-\infty, \infty]$; the coordinate change is intractable *except* on the full interval where the function exists.

Example 12.19. Find: $A = \displaystyle\int_{-\infty}^{\infty} e^{-x^2/2} \cdot dx$

Solution:
The solution works if you compute the square of the integral.

$$A^2 = \left(\int_{-\infty}^{\infty} e^{-x^2/2} \cdot dx \right)^2$$

$$= \left(\int_{-\infty}^{\infty} e^{-x^2/2} \cdot dx \right) \left(\int_{-\infty}^{\infty} e^{-x^2/2} \cdot dx \right)$$

$$= \left(\int_{-\infty}^{\infty} e^{-x^2/2} \cdot dx \right) \left(\int_{-\infty}^{\infty} e^{-y^2/2} \cdot dy \right) \qquad \text{Rename}$$

$$= \int_{-\infty}^{\infty} \int_{-\infty}^{\infty} e^{-x^2/2} e^{-y^2/2} \cdot dy \, dx$$

$$= \int_{-\infty}^{\infty} \int_{-\infty}^{\infty} e^{-1/2(x^2+y^2)} \cdot dA$$

Change to polar coordinates

$$= \int \int_{R} e^{-1/2(r^2)} \cdot dA$$

$$= \int \int_{R} e^{-1/2(r^2)} \cdot r \cdot dr \, d\theta$$

The polar region in question is $0 \leq r < \infty$ and $0 \leq \theta < 2\pi$. Rebuild the integral with these limits and we get:

$$A^2 = \int_0^{2\pi} \int_0^\infty r \cdot \mathrm{e}^{-r^2/2} \cdot dr \; d\theta$$

$$= \left(\int_0^{2\pi} d\theta \right) \cdot \left(\int_0^\infty r \cdot \mathrm{e}^{-r^2/2} \cdot dr \right)$$

$$= \theta \Big|_0^{2\pi} \cdot \left(\int_0^\infty r \cdot \mathrm{e}^{-r^2/2} \cdot dr \right)$$

$$= 2\pi \int_0^\infty r \cdot \mathrm{e}^{-r^2/2} \cdot dr$$

$$= 2\pi \lim_{a \to \infty} \int_0^a r \cdot \mathrm{e}^{-r^2/2} \cdot dr$$

Let $u = -r^2/2$, then $-du = r \cdot dr$

$$= 2\pi \lim_{a \to \infty} \int_?^? \mathrm{e}^u \cdot -du$$

$$= -2\pi \lim_{a \to \infty} \int_?^? \mathrm{e}^u \cdot du$$

$$= -2\pi \lim_{a \to \infty} \mathrm{e}^{-r^2/2} \Big|_0^a$$

$$= -2\pi \lim_{a \to \infty} \left(\mathrm{e}^{-a^2/2} - 1 \right)$$

$$= -2\pi(0 - 1) = 2\pi$$

If $A^2 = 2\pi$ then $A = \sqrt{2\pi}$, which is the correct normalizing constant.

12.3.1. Mass and Center of Mass. The center of mass for an object is the average position of all the mass in an object. This section demonstrates techniques for computing the center of mass of flat plates with a density function $\rho(x, y)$. Density is the rate at which mass changes as you move through an object, which, in turn, means that the mass of an object is the integral of its density.

Knowledge Box 12.10.

Mass of a plate

Suppose that a flat plate occupies a region R with a density function $\rho(x, y)$ defined on R. Then the mass of the plate is

$$M = \int \int_R \rho(x, y) \cdot dA.$$

Remember that the function $\rho(x, y)$ is usually constant, or close enough to constant that we assume it to be constant, when we have a mass made of a relatively uniform material. The fairly high variation in the mass functions in the examples and homework problems is intended to give your integration skills a workout – not as a representation of situations encountered in physical reality.

Example 12.20. If a plate fills the triangular region R from Figure 12.8 with a density function $\rho(x, y) = x + 1$ grams/unit^2, find the mass of the plate.

Solution:
Using the mass formula, the integral is

$$\text{Mass} = \int_0^2 \int_0^x (x + 1) \, dy \, dx$$

$$= \int_0^2 (xy + y) \bigg|_0^x \cdot dx$$

$$= \int_0^2 \left(x^2 + x\right) \cdot dx$$

$$= \frac{x^3}{3} + \frac{x^2}{2} \bigg|_0^2$$

$$= \frac{8}{3} + 2 - 0 - 0$$

$$= \frac{14}{3} \text{grams}$$

◊

Once we have the ability to compute the mass of a plate from its dimensions and density, we can compute the coordinates of the center of mass of the plate using a type of averaging integral.

Knowledge Box 12.11.

Center of mass

Suppose that a flat plate occupies a region R with a density function $\rho(x, y)$ defined on R. Then if

$$M_x = \int\int_R y\rho(x, y) \cdot dA$$

and

$$M_y = \int\int_R x\rho(x, y) \cdot dA$$

the center of mass of the plate is

$$(\overline{x}, \overline{y}) = \left(\frac{M_y}{M}, \frac{M_x}{M}\right).$$

Example 12.21. Suppose that R is the square region $0 \leq x, y \leq 1$, and that $\rho(x, y) = 2y$ grams/unit^2. Find the center of mass.

Solution:

This problem requires three integrals.

$$M = \int_0^1 \int_0^1 2y \cdot dy \, dx \qquad M_x = \int_0^1 \int_0^1 y \cdot 2y \cdot dy \, dx \qquad M_y = \int_0^1 \int_0^1 x \cdot 2y \cdot dy \, dx$$

$$= \int_0^1 y^2 \Big|_0^1 dx \qquad\qquad = \int_0^1 \int_0^1 2y^2 \cdot dy \, dx \qquad\qquad = \int_0^1 xy^2 \Big|_0^1 \cdot dx$$

$$= \int_0^1 (1-0) dx \qquad\qquad = \int_0^1 \frac{2}{3} y^3 \Big|_0^1 \cdot dx \qquad\qquad = \int_0^1 (x - 0) \cdot dx$$

$$= \int_0^1 dx \qquad\qquad = \int_0^1 \left(\frac{2}{3} - 0 \right) \cdot dx \qquad\qquad = \frac{1}{2} x^2 \Big|_0^1 = \frac{1}{2}$$

$$= x \Big|_0^1 \qquad\qquad = \int_0^1 \frac{2}{3} \cdot dx$$

$$= (1-0) = 1 \text{ gram} \qquad = \frac{2}{3} x \Big|_0^1 = \frac{2}{3}$$

Now that we have the pieces we can use the formula for center of mass:

$$(\bar{x}, \bar{y}) = \left(\frac{1/2}{1}, \frac{2/3}{1} \right) = \left(\frac{1}{2}, \frac{2}{3} \right)$$

$\Diamond$

Problems

Problem 12.21. *Find the integral of each of the following functions over the specified region.*

(a) The function $f(x, y) = x + y^2$ on the strip

$$0 \leq x \leq 4 \quad 0 \leq y \leq 1$$

(b) The function $g(x, y) = xy$ on the rectangle

$$1 \leq x \leq 3 \quad 1 \leq y \leq 2$$

(c) The function $h(x, y) = x^2 y + x y^2$ on the square

$$0 \leq x \leq 2 \quad 0 \leq y \leq 2$$

(d) The function $r(x, y) = 2x + 3y + 1$ on the region bounded by $x = 0$, $y = 1$, and $y = x$.

(e) The function $s(x, y) = x^2 + y^2 + 1$ on the region bounded by the x axis and the function $y = 4 - x^2$.

(f) The function $q(x, y) = x + y$ on the region bounded by the curves $y = \sqrt{x}$ and $y = x^2$.

(g) The function $a(x, y) = x^2$ on the region bounded by the curves $y = 2x$ and $y = x^2$.

(h) The function $b(x, y) = y^2$ on the region bounded by the curves $y = \sqrt[3]{x}$ and $y = x$ for $x \geq 0$.

Problem 12.22. *Sketch the regions from Problem 12.21.*

Problem 12.23. *Explain why a density function $\rho(x, y)$ can never be negative.*

Problem 12.24. *Find a region R so that the integral over R of $f(x) = x^2 + y^2$ is 6 units^3.*

Problem 12.25. *Find the square region $0 \leq x, y \leq a$ so that*

$$\int \int_R \left(x^3 + y \right) \cdot dA$$

is 12 units^3.

Problem 12.26. *Find the mass of the plate $0 \leq x, y \leq 3$ if*

$$\rho(x, y) = y^2 + 1.$$

Problem 12.27. *Find the center of mass of the region bounded by the x-axis, the y-axis, and the line $x + y = 4$ if the density function is*

$$\rho(x, y) = y + 1.$$

Problem 12.28. *Find the center of mass of the region bounded by $y = \sqrt{x}$ and $y = x^2$ if the density function is*

$$\rho(x, y) = x + 2.$$

Problem 12.29. *Find the center of mass of the region $0 \leq x, y \leq 1$ if the density function is*

$$\rho(x, y) = (x + y)/2.$$

Problem 12.30. *Find the volume under*

$$f(x, y) = \sqrt{x^2 + y^2}$$

above the region $x^2 + y^2 \leq 16$.

Problem 12.31. *Find the volume under*

$$f(x, y) = x^2 + y^2$$

above the region bounded by the petal curve $r = 2\cos(3\theta)$.

Problem 12.32. *Find the volume under*

$$f(x, y) = (x^2 + y^2)^{3/2}$$

above the region bounded by the petal curve $r = \cos(2\theta)$.

Problem 12.33. *Find a plane $f(x, y)$ so that the area under the plane but over a circle of radius 2 centered at the origin is 16 units^3.*

Sequences, Series, and Function Approximation

This section is quite different from the earlier sections. Calculus is mostly part of the mathematics of continuous functions, while the sequences and series we study in this chapter are a part of **discrete mathematics** – math that is broken up into individual pieces. Discrete math is about things you can count, rather than things you can measure. Since this is a calculus book, we will apply what we learn to understanding and increasing the power of our calculus. The ultimate goal of the chapter is a much deeper understanding of transcendental (non-polynomial) functions like e^x or $\cos(x)$. We start, however, at the beginning.

13.1. Sequences and the Geometric Series

A sequence is an infinite list of numbers. Sometimes we give a sequence by listing an obvious pattern:

$$S = 1, \ \frac{1}{2}, \ \frac{1}{3}, \ \frac{1}{4}, \ \cdots$$

with the ellipsis meaning "and so on." We also can use more formal set notation to specify a series with a formula:

$$S = \left\{ \frac{1}{n} : n = 1, 2, \ldots \right\}$$

Like a function having a limit at infinity, there is a notion of a sequence **converging**.

Knowledge Box 13.1.

Definition of the convergence of a sequence to a limit

We call L the **limit of a sequence** $\{x_n : n = 1, 2, \ldots\}$ if, for each $\epsilon > 0$, there is a whole number N so that, whenever $n > N$, we have:

$$|x_n - L| < \epsilon$$

A sequence that has a limit is said to **converge**.

Example 13.1. Prove the the sequence

$$S = \left\{ \frac{1}{n} : n = 1, 2, \ldots \right\}$$

converges to zero.

Solution:

For $\epsilon > 0$ choose N to be the smallest whole number greater than $\dfrac{1}{\epsilon}$. This makes $\dfrac{1}{N} < \epsilon$. Then if $n > N$ we have:

$$n > N$$
$$\frac{1}{n} < \frac{1}{N} \qquad \text{Reciprocals reverse inequalities}$$
$$\frac{1}{n} < \epsilon \qquad \text{Since } 1/N < \epsilon$$
$$\left| \frac{1}{n} - 0 \right| < \epsilon \qquad \text{Value on the left did not change}$$

Which satisfies the definition of the limit of the sequence being $L = 0$.

$$\diamond$$

As always there are shortcuts that mean we only need to rely on the definition of the limits of sequences occasionally.

Knowledge Box 13.2.

Sequences drawn from functions

Suppose that we have a sequence

$$S = \{x_n\} = \{f(n) : n = 1, 2, \ldots\}$$

and that $\lim_{x \to \infty} f(x) = L$. Then we may conclude that $\lim_{n \to \infty} x_n = L$.

Example 13.2. Suppose that we have a sequence

$$\left\{ x_n = \frac{1}{1+n^2} : n = 0, 1, 2, \ldots \right\}$$

Find $\lim\limits_{n \to \infty} x_n$.

Solution:

We already know $\lim\limits_{x \to \infty} \dfrac{1}{1+x^2} = 0$. So, using Knowledge Box 13.2, we have that

$$\lim_{n \to \infty} x_n = 0$$

The rule in Knowledge Box 13.2 is *not reversible*.

Example 13.3. Determine if the sequence

$$S = \{\cos(2\pi n) : n = 0, 1, 2, \ldots\}$$

converges.

Solution:

Since $\lim\limits_{x \to \infty} \cos(2\pi x)$ jumps back and forth in the range $-1 \le y \le 1$, the function that we drew the sequence from does not have a limit. Leaving that aside, examine a listing of the first several terms of the sequence:

$$\{1, 1, 1, 1, 1, 1, \ldots\}$$

This sequence obviously converges to $L = 1$. Again – if the function has a limit, then so does the sequence. The reverse need not be true.

$\Diamond$

The next sequence resolution technique requires that we define several terms.

Definition 13.1. If a sequence $\{x_n\}$ has the property that $x_n \le x_{n+1}$, then we say that the sequence is **monotone increasing**.

Definition 13.2. If a sequence $\{x_n\}$ has the property that $x_n \ge x_{n+1}$, then we say that the sequence is **monotone decreasing**.

Definition 13.3. If a sequence $\{x_n\}$ is either monotone increasing or monotone decreasing, then we say the sequence is **monotone**.

Definition 13.4. If a sequence $\{x_n\}$ has the property that

$$x_n \le C$$

for all n and for some constant C, then we say the sequence is **bounded above**.

Definition 13.5. If a sequence $\{x_n\}$ has the property that

$$x_n \geq C$$

for all n and for some constant C, then we say the sequence is **bounded below**.

Definition 13.6. If a sequence $\{x_n\}$ is both bounded above and bounded below, then we say the sequence is **bounded**.

Knowledge Box 13.3.

Bounded monotone sequences

The following types of sequences all converge.

- A monotone increasing sequence that is bounded above

- A monotone decreasing sequence that is bounded below

- Any monotone bounded sequence

The reason that the sequences listed in Knowledge Box 13.3 converge is that they are required to move toward a bound of some sort without passing it. This is another useful shortcut for determining if a sequence converges – though this shortcut does not tell us the value of the limit.

Example 13.4. Demonstrate that the sequence

$$\left\{ x_n = \frac{n}{n+1} : n = 0, 1, 2, 3, \ldots \right\}$$

has a limit.

Solution:

While we could use Knowledge Box 13.2, let's use this example as a chance to demonstrate the technique from Knowledge Box 13.3. First notice that this sequence is monotone increasing. To see this, notice that:

$$n^2 + 2n < n^2 + 2n + 1$$
$$n(n+2) < (n+1)^2$$
$$\frac{n}{n+1} < \frac{n+1}{n+2}$$
$$x_n < x_{n+1}$$

Since $n < n+1$, we can deduce that $x_n = \dfrac{n}{n+1} < 1$. So the sequence is bounded above. This permits us to use Knowledge Box 13.3 to deduce that the sequence has a limit.

◊

It would have been easier to do Example 13.4 with Knowledge Box 13.2. But there will be times when the sequence is not drawn from a function when we have to use monotone sequence theory. The next Knowledge Box extends the reach of our ability to check sequences for convergence.

Knowledge Box 13.4.

Arithmetic combinations of sequences

If $\{x_n\}$ has a limit of L, and $\{y_n\}$ has a limit of M, and if a, b are constants, then:

- $\lim_{n \to \infty} ax_n \pm by_n = aL \pm bM$

- $\lim_{n \to \infty} x_n \cdot y_n = L \cdot M$

- $\lim_{n \to \infty} \dfrac{x_n}{y_n} = \dfrac{L}{M}$ if $M \neq 0$.

- $\lim_{n \to \infty} x_n^k = L^k$

Example 13.5. Find the limit of

$$S = \left\{ x_n = \frac{1}{n} + 3 \cdot \frac{n}{n+1} : n = 1, 2, 3, \ldots \right\}$$

Solution:

We already know the limit of $1/n$, as a series, is zero. Using Knowledge Box 13.2 it is easy to see that:

$$\lim_{n \to \infty} \frac{n}{n+1} = 1$$

Combining these results using the information in Knowledge Box 13.4 we get that the limit is:

$$L = 0 + 3 \cdot 1 = 3$$

$\Diamond$

This concludes our direct investigation of sequences. We now turn to using sequences as a tool to explore **series**. Where a sequence is an infinite list of numbers, a series is an infinite list of numbers *that you add up*. This may or may not result in a finite sum – and resolving that question requires sequence theory.

Knowledge Box 13.5.

Definition of series

If $\{x_n : n = 0, 1, \ldots\}$ is a sequence, then

$$\sum_{n=0}^{\infty} x_n = x_0 + x_1 + \cdots + x_k + \cdots$$

is the corresponding **infinite series**. If we sum only finitely many terms we have a **finite series**.

Example 13.6. Show that the infinite series:

$$\sum_{n=1}^{\infty} \frac{1}{2}^n = \frac{1}{2} + \frac{1}{4} + \frac{1}{8} + \cdots$$

sums to 1.

Solution:

Examine Figure 13.1. The figure divides a unit square – with area one – into rectangles of size 1/2, 1/4, 1/8, and every other number in the series we are trying to sum. This constitutes a geometric demonstration that the series sums to 1.

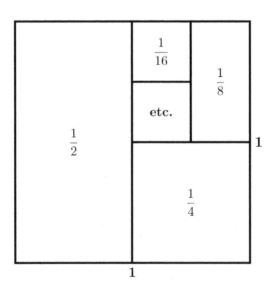

Figure 13.1. A unit square cut into pieces 1/2, 1/4, 1/8, ...

Clearly, finding a cool picture is not a general technique for demonstrating that a series has a sum. Proving no such picture exists when the series fails to have a sum is even more impossible – we need a more general theory.

Knowledge Box 13.6.

The sequence of partial sums of a series

Suppose that $S = \sum_{n=1}^{\infty} x_n$ is a series. If we set

$$p_n = \sum_{k=1}^{n} x_k,$$

then $\{p_n\}$ is called the **sequence of partial sums of** S. We say that a series **converges to a sum** L if and only if its sequence of partial sums has L as a limit.

Example 13.7. Find the sequence of partial sums of the series in Example 13.6 and compute its limit.

Solution:

Compute the first few members of the sequence of partial sums:

$$p_1 = 1/2$$
$$p_2 = 1/2 + 1/4 = 3/4$$
$$p_3 = 1/2 + 1/4 + 1/8 = 7/8$$
$$p_4 = 1/2 + 1/4 + 1/8 + 1/16 = 15/16$$

Which is a clear pattern, and it is easy to see

$$p_n = 1 - \frac{1}{2^n}$$

This gives us the sequence of partial sums. Computing the limit we get:

$$\lim_{n \to \infty} p_n = \lim_{n \to \infty} 1 - \frac{1}{2^n} = 1 - 0 = 1$$

So we get the same sum using this more formal approach.

◇

At this point we need to call forward an identity from Chapter 6, Knowledge Box 6.9:

$$1 + x + x^2 + \cdots x^n = \frac{x^n - 1}{x - 1} = \frac{1 - x^n}{1 - x}$$

This identity is one that is true for polynomials, but it also applies to summing a finite series. Additionally, this formula can be used as the partial sum of a particular type of infinite series – at least when its limit exists.

Finite and infinite geometric series

The polynomial identity in the text tells us, for a constant $a \neq 1$, that:

$$\sum_{k=0}^{n} a^k = \frac{a^{n+1} - 1}{a - 1}$$

This is the **finite geometric series formula.**

If $|a| < 1$, then the limit of the finite series gives us the **infinite geometric series formula**:

$$\sum_{n=0}^{\infty} a^n = \frac{1}{1 - a}$$

Applying Knowledge Box 13.4 we also get that:

$$\sum_{n=0}^{\infty} c \cdot a^n = \frac{c}{1 - a}$$

for a constant c. The number a is called the **ratio** of the geometric series.

Example 13.8. Compute

$$\sum_{n=0}^{\infty} \frac{3}{5^n}$$

Solution:

This is a geometric series with ratio $a = 1/5$. It has a leading constant $c = 3$. Applying the appropriate formula we see that:

$$\sum_{n=0}^{\infty} \frac{3}{5^n} = \frac{3}{1 - 1/5} = \frac{3}{4/5} = \frac{15}{4}$$

$\Diamond$

Problems

Problem 13.1. *Prove formally, using the definition of the limit of a sequence, that*

$$\{\cos(2\pi n) : n = 0, 1, 2, \ldots\}$$

converges to 1.

Problem 13.2. *Prove formally, using the definition of the limit of a sequence, that*

$$\left\{\frac{1}{n^2} : n = 1, 2, 3, \ldots\right\}$$

converges to 0.

Problem 13.3. *Prove formally, using the definition of the limit of a sequence, that*

$$\left\{\frac{n}{n+1} : n = 0, 1, 2, \ldots\right\}$$

converges to 1.

Problem 13.4. *Compute the limit of each of the following sequences or give a reason why the limit does not exist. Assume $n = 1, 2, \ldots$*

(a) $\left\{x_n = \tan^{-1}(n)\right\}$

(b) $\left\{y_n = \sin\left(\frac{\pi}{2}n\right)\right\}$

(c) $\left\{z_n = \dfrac{n^2}{n+1}\right\}$

(d) $\{y_n = \sin(\pi n)\}$

(e) $\left\{z_n = \dfrac{3n^2}{n^2+1}\right\}$

(f) $\left\{z_n = \dfrac{\cos(n)}{n+1}\right\}$

Problem 13.5. *Do the calculation to prove the infinite geometric series formula from the finite one: see Knowledge Box 13.7.*

Problem 13.6. *Compute the following sums or give a reason they fail to exist.*

(a) $\displaystyle\sum_{k=0}^{20} 1.2^k$

(b) $\displaystyle\sum_{n=1}^{\infty} \left(\frac{1}{3}\right)^n$

(c) $\displaystyle\sum_{n=1}^{\infty} 2 \cdot \left(\frac{1}{7}\right)^n$

(d) $\displaystyle\sum_{n=1}^{\infty} \left(\frac{-1}{4}\right)^n$

(e) $\displaystyle\sum_{n=1}^{\infty} 3 \left(\frac{3}{2}\right)^n$

(f) $\displaystyle\sum_{n=1}^{\infty} 0.05^n$

(g) $\displaystyle\sum_{n=1}^{\infty} 112 \, (0.065)^n$

(h) $\displaystyle\sum_{n=1}^{\infty} 2 \cdot (-1)^n$

Problem 13.7. *Compute* $\displaystyle\sum_{k=12}^{24} 3^n$

Problem 13.8. *A swinging pendulum is losing energy. Its first swing is 2m long, and each after that is 0.9985 times as long as the one before it. Estimate the total distance traveled by the pendulum.*

Problem 13.9. *A ball is dropped from a height of 8m. If each bounce is 3/4 the height of the one before it, estimate the total vertical distance traveled by the ball.*

Problem 13.10. *A rod is initially displaced 2.1mm from equilibrium and undergoes damped vibration with a decay in the length of each subsequent swing of 0.937. Find the total vertical distance traveled by the end of the rod.*

13.2. Series convergence tests

In this section we will develop a number of tests to determine if a series converges. At present, we know that an infinite geometric series with a ratio of a with $|a| < 1$ will converge, but not much else. We begin with a motivating example.

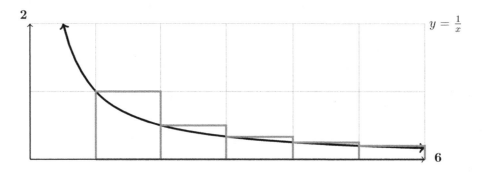

Figure 13.2. Shown is a portion of the graph of $y = 1/x$ and a sequence of rectangles of width 1 and height $1/n$ for $n = 1, 2, 3, \ldots$.

Example 13.9. Determine if the series $\displaystyle\sum_{n=1}^{\infty} \frac{1}{n}$ has a finite or infinite sum.

Solution:

Examine Figure 13.2. This shows that a series of rectangles with area $1, 1/2, 1/3, \ldots$ have a larger area than $\displaystyle\int_{1}^{\infty} \frac{dx}{x}$ because the area under the curve is strictly smaller than the sum of the areas of the rectangles. Computing the integral:

$$
\begin{aligned}
\int_{1}^{\infty} \frac{dx}{x} &= \lim_{a \to \infty} \int_{1}^{a} \frac{dx}{x} \\
&= \lim_{a \to \infty} \ln(x) \Big|_{1}^{a} \\
&= \lim_{a \to \infty} \ln(a) - \ln(1) \\
&= \infty
\end{aligned}
$$

shows us that:

$$
\sum_{n=1}^{\infty} \frac{1}{n} \geq \infty
$$

Thus, we conclude that this series has an infinite sum.

$\Diamond$

The series

$$\sum_{n=1}^{\infty} \frac{1}{n}$$

is sufficiently important that it has its own name: the **harmonic series**.

Definition 13.7. When considering

$$\sum_{n=1}^{\infty} x_n$$

we call x_n the **general term** of the series.

Example 13.9 is another example of proving something about a series sum by drawing a clever picture. An interesting technique, but, as before, we need more general tools. One such tool involves taking the limit of the sequence generating the series. If it does not approach zero, there is no hope that the infinite sum converges.

Knowledge Box 13.8.

The divergence test

If $\lim_{n \to \infty} x_n \neq 0$ then

$$\sum_{n=0}^{\infty} x_n$$

does not have a finite value.

It is important to note that the divergence test is uni-directional. If the general term of a series *does* go to zero, that tells you exactly nothing about the behavior of the sum of the series.

Example 13.10. Show that

$$\sum_{n=0}^{\infty} \frac{n}{n+1}$$

diverges (has an infinite sum).

Solution:

Since

$$\lim_{n \to \infty} \frac{n}{n+1} = 1,$$

the series in question diverges by the divergence test. Colloquially, we are adding up an infinite number of terms that are approaching one – so the resulting sum is infinite.

$\Diamond$

When you can use it, the divergence test is often short and sweet. The next test is the formal version of the test we used in Example 13.9.

Knowledge Box 13.9.

The integral test

Suppose that $f(x)$ is a positive, decreasing function on $[0, \infty)$ and that $x_n = f(n)$. Then,

$$\sum_{n=1}^{\infty} x_n \text{ and } \int_a^{\infty} f(x)dx$$

both converge or both diverge for any finite $a \geq 1$.

Example 13.11. Show that

$$\sum_{n=1}^{\infty} \frac{1}{n^2}$$

is finite.

Solution:

Use the integral test.

$$\int_1^{\infty} \frac{dx}{x^2} = \lim_{a \to \infty} \int_1^a \frac{dx}{x^2}$$

$$= \lim_{a \to \infty} \left. \frac{-1}{x} \right|_1^a$$

$$= \lim_{a \to \infty} \frac{-1}{a} - (-1)$$

$$= 0 + 1 = 1$$

Since the improper integral is finite, so is the sum. Note that this *does not* tell us the value of the sum.

$$\Diamond$$

The fact that the series $\sum_{n=1}^{\infty} \frac{1}{n}$ diverges but $\sum_{n=1}^{\infty} \frac{1}{n^2}$ converges motivates our next test.

Definition 13.8. A *p-series* is series of the form

$$\sum_{n=1}^{\infty} \frac{1}{n^p}$$

where p is a constant.

Knowledge Box 13.10.

The p-series test

The p-series

$$\sum_{n=1}^{\infty} \frac{1}{n^p}$$

converges if $p > 1$ and diverges if $p \leq 1$.

The series convergence test in Knowledge Box 13.10 follows directly from the integral test – something you are asked to verify in the homework.

Example 13.12. Determine if

$$\sum_{n=1}^{\infty} \frac{1}{n^{\pi}}$$

converges to a finite number or diverges.

Solution:

This series has the form of a p-series with $p = \pi$. Since $\pi > 1$, we conclude the series converges.

$\Diamond$

Example 13.13. Determine if

$$\sum_{n=1}^{\infty} \frac{1}{\sqrt{n}}$$

converges to a finite number or diverges.

Solution:

This series has the form of a p-series with $p = 1/2$. Since $1/2 \leq 1$, we conclude the series diverges.

$\Diamond$

The next two tests leverage series we already understand to resolve even more series. The tests depend on various forms of comparison of a series under test to a series with known convergence or divergence behavior.

Knowledge Box 13.11.

The comparison tests

- If $\displaystyle\sum_{n=0}^{\infty} x_n$ converges and $0 \leq y_n \leq x_n$ for all n, we have that $\displaystyle\sum_{n=0}^{\infty} y_n$ also converges.

- If $\displaystyle\sum_{n=0}^{\infty} x_n$ diverges and $y_n \geq x_n \geq 0$ for all n, we have that $\displaystyle\sum_{n=0}^{\infty} y_n$ also diverges.

The colloquial versions of the comparison tests are pretty easy to believe. The first says, of a series with positive general terms, that if it is smaller, term by term, than another series with a finite sum, then it has a finite sum. The second reverses that: saying that, if a series is larger, term by term, than another series with an infinite sum, then it has an infinite sum.

Example 13.14. Determine if

$$\sum_{n=1}^{\infty} \frac{1}{n/3}$$

converges or diverges.

Solution:

Notice that, for $n \geq 1$,

$$n \geq n/3$$
$$\frac{1}{n} \leq \frac{1}{n/3}$$

So $0 \leq \dfrac{1}{n} \leq \dfrac{1}{n/3}$, which permits us to deduce that the series in the example diverges by comparison to the harmonic series, which is known to diverge.

◊

Example 13.15. Determine if

$$\sum_{n=0}^{\infty} \frac{1}{2^n + 5}$$

converges or diverges.

Solution:

The key to using the comparison test is to find a good known series to compare to. In this case the geometric series with ratio $1/2$ is natural.

$$2^n \leq 2^n + 5$$
$$\frac{1}{2^n} \geq \frac{1}{2^n + 5}$$
$$\left(\frac{1}{2}\right)^n \geq \frac{1}{2^n + 5}$$

Which means that the general term of the convergent geometric series

$$\sum_{n=0}^{\infty} \left(\frac{1}{2}\right)^n$$

is greater than the general term of our target series. Since all terms are positive, this tells us that we may conclude the target series converges, by comparison.

The next test is one of the most all-around useful tests. It permits us to resolve any series drawn from a rational function, for example, by checking to see which p-series a rational function is most like.

Knowledge Box 13.12.

The limit comparison test

Suppose that

$$\sum_{n=0}^{\infty} x_n \text{ and } \sum_{n=0}^{\infty} y_n$$

are series, and that

$$\lim_{n \to \infty} \frac{x_n}{y_n} = C$$

where $0 < C < \infty$ is a constant. Then, either both series converge or both series diverge.

Example 13.16. Determine if

$$\sum_{n=1}^{\infty} \frac{n+5}{n^3+1}$$

converges or diverges.

Solution:

Perform limit comparison to

$$\sum_{n=1}^{\infty} \frac{1}{n^2}$$

$$\lim \frac{x_n}{y_n} = \lim_{n \to \infty} \frac{\frac{n+5}{n^3+1}}{\frac{1}{n^2}} = \lim_{n \to \infty} \frac{n+5}{n^3+1} \cdot \frac{n^2}{1} = \lim_{n \to \infty} \frac{n^3+5n^2}{n^3+1} = 1$$

Since $0 < 1 < \infty$, we can conclude that $\sum_{n=1}^{\infty} \frac{n+5}{n^3+1}$ converges, because $\sum_{n=1}^{\infty} \frac{1}{n^2}$ is a convergent p-series.

$$\Diamond$$

Notice that choosing a p-series with $p = 2$ was exactly what was needed to make the ratio of the general terms of the series be something that had a finite, non-zero limit. The limit comparison test is useful for putting rigor into the intuition that one series is "like" another.

The next test uses the fact that if you have a sum of positive terms that is finite, making some or all of those terms negative cannot take the sum farther from zero than the original finite sum – leaving the sum finite.

Knowledge Box 13.13.

The absolute convergence test

If $\sum_{n=0}^{\infty} |x_n|$ converges, then so does $\sum_{n=0}^{\infty} x_n$.

Example 13.17. Prove that

$$1 - \frac{1}{4} + \frac{1}{9} - \frac{1}{16} + \frac{1}{25} - \cdots$$

converges.

Solution:

The series in question is given as an obvious pattern – begin by pulling it into summation notation:

$$\sum_{n=1}^{\infty} \frac{(-1)^{n+1}}{n^2}$$

If we take the absolute value of the general terms of this series, we obtain the series:

$$\sum_{n=1}^{\infty} \frac{1}{n^2}$$

We know this to be a convergent p-series. We may deduce that the series converges.

$\Diamond$

Definition 13.9. If $\displaystyle\sum_{n=0}^{\infty} |x_n|$ converges, then we say the series $\displaystyle\sum_{n=0}^{\infty} x_n$ **converges in absolute value.**

This means we can restate Knowledge Box 13.13 as: "A series that converges in absolute value, converges."

The next test uses the fact that if you go up, then down by less, then up by even less, and so on, you end up a finite distance from your starting point.

Knowledge Box 13.14.

The alternating series test

Suppose that x_n is a series such that $x_n > x_{n+1} \geq 0$, and suppose that $\displaystyle\lim_{n\to\infty} x_n = 0$. Then the series

$$\sum_{n=0}^{\infty} (-1)^n \cdot x_n$$

converges.

Example 13.18. Show that $\displaystyle\sum_{n=1}^{\infty} \frac{(-1)^n}{\sqrt{n}}$ converges.

Solution:

We already know that $\displaystyle\lim_{n\to\infty} \frac{1}{\sqrt{n}} = 0$, and the terms of the series clearly get smaller as n increases. So, we may conclude this series converges by the alternating series test.

$\Diamond$

The next two tests both check to see if a series is like a geometric series and deduce its convergence or divergence from that similarity.

Knowledge Box 13.15.

The ratio test

Suppose that $\displaystyle\sum_{n=0}^{\infty} x_n$ is a series. Compute

$$r = \lim_{n\to\infty} \left| \frac{x_{n+1}}{x_n} \right|$$

Then:

- if $r < 1$, then the series converges
- if $r > 1$, then the series diverges
- if $r = 1$, then the test is inconclusive

Definition 13.10. The quantity $n! = n \cdot (n-1) \cdot (n-2) \cdots 2 \cdot 1$ is called n **factorial**. We define $0! = 1$.

Example 13.19. Determine if $\displaystyle\sum_{n=0}^{\infty} \frac{1}{n!}$ converges or diverges.

Solution:

Use the ratio test.

$$\lim_{n\to\infty} \left| \frac{1/(n+1)!}{1/n!} \right| = \lim_{n\to\infty} \left| \frac{n!}{(n+1)!} \right| = \lim_{n\to\infty} \frac{n(n-1)\cdots 2 \cdot 1}{(n+1)n(n-1)\cdots 2 \cdot 1} = \lim_{n\to\infty} \frac{1}{n+1} = 0$$

Since $0 < 1$, we can deduce that the series converges by the ratio test.

$\Diamond$

Knowledge Box 13.16.

The root test

Suppose that $\sum_{n=0}^{\infty} x_n$ is a series. Compute

$$s = \lim_{n \to \infty} \sqrt[n]{|x_n|}$$

Then:

- if $s < 1$, then the series converges
- if $s > 1$, then the series diverges
- if $s = 1$, then the test is inconclusive

Example 13.20. Determine if $\sum_{n=0}^{\infty} \dfrac{1}{n^n}$ converges or diverges.

Solution:

Use the root test.

$$\lim_{n \to \infty} \sqrt[n]{\left| \frac{1}{n^n} \right|} = \lim_{n \to \infty} \frac{1}{n} = 0$$

Since $0 < 1$, we can deduce that the series converges by the root test.

The root test and, especially, the ratio test will get a big workout in the next section. This section contains nine tests for series convergence – many of which depend on knowing the convergence or divergence behavior of other series. This creates a mental space very similar to the "which integration method do I use?" issue that arose in Chapter 7. The method for dealing with this is the same here as it was there: practice, practice, practice.

It's also good to keep in mind, when you are searching for series to compare to, that the examples in this section may be used as examples for comparison. A list of series with known behaviors is an excellent resource and, given different learning styles, somewhat personal: you may want to maintain your own annotated list.

13.2.1. Tails of sequences. You may have noticed that we are a little careless with where we start the index of summation on our infinite series. This is because, while the value of a convergent series depends on every term, the convergence or divergence behavior does not.

Definition 13.11. If we take a sequence and make a new sequence by discarding a finite number of initial terms, the new sequence is a **tail** of the old sequence.

Knowledge Box 13.17.

Tail convergence

A sequence converges if and only if all its tails converge.

The practical effect of Knowledge Box 13.17 is that, if a few initial terms of a sequence are causing trouble, you may discard them and test the remainder of the sequence to determine convergence or divergence.

Example 13.21. Determine the convergence of the series:

$$\sum_{n=0}^{\infty} \frac{(-1)^n}{n^2 - 6n + 10}$$

If we look at the first several terms of this series we get:

$$\frac{1}{10} - \frac{1}{5} + \frac{1}{2} - 1 + \frac{1}{2} - \frac{1}{5} + \frac{1}{10} - \frac{1}{17} + \cdots$$

The series alternates signs, getting larger in absolute value for the first four terms, but then getting smaller in absolute value for the remaining terms. This means that the alternating series test works *for the tail of the sequence starting at the 4th term*. Remember that the alternating series test requires that the terms shrink in absolute value. We conclude that the series converges by the alternating series test applied to a tail of the sequence.

In Chapter 7 we encountered the annoying, natural, and difficulty question, "How do I choose a method of integration?" Tests for the convergence or divergence of a sequence have the same annoying character. As with integration, the only real way to get better at choosing the correct test is to work examples. Lots of examples.

Knowledge Box 13.18.

Choosing a convergence test

(a) If the general term of the series fails to converge to zero, then it diverges by the divergence test. Remember that this does not work in reverse – if the general term converges to zero, anything might happen.

(b) If the series *is* a geometric series or a p-series, use the tests for those series.

(c) If the general term of the series is $a_n = f(n)$ for a function $f(x)$ that you can integrate, try the integral test.

(d) A positive series that is, term-by-term, no larger than a series that converges, converges (comparison test) – look for this.

(e) Similarly, a positive series that is, term-by-term, no smaller than a divergent series diverges, again by the comparison test.

(f) If you have a series that looks like a series you know how to deal with, try the limit comparison test. This is useful for things like slightly modified p series or letting you use simpler integral tests.

(g) You can often set up a comparison test or limit comparison test by doing algebra or arithmetic to the general term of a series.

(h) Remember that if the term-by-term absolute value of a series converges, then the series converges. This is the absolute convergence test.

(i) If the terms of a series alternate in sign, look at the alternating series test.

(j) If you can take the limit of adjacent terms of a series in a reasonable way, then the ratio test is a possibility.

(k) If you can take the nth root of the general term of a series in a reasonable way, the root test is a possibility.

(l) If some finite number of initial terms of a series are preventing you from using a test, tail convergence says you can ignore them and then do your test.

(m) Unless you're taking a test or quiz, asking for advise and suggestions is not the worst possible option.

Problems

Problem 13.11. *For each of the following series, determine if the series converges or diverges. State the name of the test you are using.*

(a) $\sum_{n=1}^{\infty} \frac{1}{n^e}$

(b) $\sum_{n=0}^{\infty} \frac{n+1}{n^2+1}$

(c) $\sum_{n=0}^{\infty} \frac{n+1}{5n+4}$

(d) $\sum_{n=1}^{\infty} \ln\left(\frac{1}{n^2}\right)$

(e) $\sum_{n=0}^{\infty} \frac{2^n+1}{3^n+1}$

(f) $\sum_{n=0}^{\infty} \frac{\sqrt{n}}{n^2+1}$

Problem 13.12. *For each of the following series, determine if the series converges or diverges. State the name of the test you are using.*

(a) $\sum_{n=0}^{\infty} \frac{n^2}{2^n+1}$

(b) $\sum_{n=0}^{\infty} 0.0462^n$

(c) $\sum_{n=2}^{\infty} \frac{\sin(n)}{n^{\sqrt{2}}}$

(d) $\sum_{n=1}^{\infty} \frac{1}{\sqrt[3]{n}}$

(e) $\sum_{n=1}^{\infty} \frac{(-1)^n}{\sqrt[3]{n}}$

(f) $\sum_{n=0}^{\infty} \frac{1}{(2n)!}$

Problem 13.13. *For each of the following series, determine if the series converges or diverges. State the name of the test you are using.*

(a) $\sum_{n=1}^{\infty} \frac{1}{n^{n/2}}$

(b) $\sum_{n=1}^{\infty} \frac{5^n}{n^{n/2}}$

(c) $\sum_{n=1}^{\infty} e^{-n}$

(d) $\sum_{n=1}^{\infty} \frac{e^n}{\pi^n}$

(e) $\sum_{n=1}^{\infty} \frac{e^{2n}}{\pi^n}$

(f) $\sum_{n=1}^{\infty} e^{3+n-n^2}$

Problem 13.14. *Give an example to demonstrate that the divergence test does not work in reverse – i.e. a sequence whose general term goes to zero but whose sum is infinite.*

Problem 13.15. *Use the integral test to prove that the p-test works.*

Problem 13.16. *Suppose that $p(x)$ is a polynomial. Use the integral test to demonstrate that*

$$\sum_{n=0}^{\infty} p(n)e^{-n}$$

converges. Hint: look at some of the shortcuts in Chapter 7.

Problem 13.17. *Suppose that $q(x)$ is a polynomial with exactly three roots, all of which are negative real numbers. Demonstrate that*

$$\sum_{n=0}^{\infty} \frac{1}{q(n)}$$

converges.

Problem 13.18. *If x_n and y_n are the general terms of a convergent series, then $x_n + y_n$ are as well. This requires only simple algebra. What is startling is that the reverse is not true. Find an example of $a_n = x_n + y_n$ so that*

$$\sum_{n=1}^{\infty} a_n$$

converges, but neither of

$$\sum_{n=1}^{\infty} x_n \ \ or \ \ \sum_{n=1}^{\infty} y_n$$

converge.

Problem 13.19. *Show that, when you apply the ratio test to a geometric series, the limit that appears in the test is the ratio of the series.*

Problem 13.20. *Show that, when you apply the root test to a geometric series, the limit that appears in the test is the ratio of the series.*

Problem 13.21. *Suppose $r > 1$. Prove that*

$$\sum_{n=0}^{\infty} \frac{n^k}{r^n}$$

converges when k is an integer ≥ 1.

Problem 13.22. *Suppose that we are testing the convergence of a series*

$$\sum_{n=0}^{\infty} \frac{p(n)}{q(n)}$$

where $p(x)$ and $q(x)$ are polynomials. If we add the assumption that $q(x)$ has no roots at any of the values of n involved in the sum, explain in terms of the degrees of the polynomials when the series converges.

Problem 13.23. *For each of the following series, determine if the series converges or diverges. State the name of the test you are using.*

(a) $\displaystyle\sum_{n=0}^{\infty} \frac{n^3}{(1+n)(2+n)(3+n)(4+n)}$

(b) $\displaystyle\sum_{n=0}^{\infty} \frac{n^2}{(1+n)(2+n)(3+n)(4+n)}$

(c) $\displaystyle\sum_{n=0}^{\infty} \frac{n^5}{e^n}$

(d) $\displaystyle\sum_{n=0}^{\infty} \frac{\sqrt{n}}{n^2 + 1}$

(e) $\displaystyle\sum_{n=0}^{\infty} 1.25^{-n}$

(f) $\displaystyle\sum_{n=0}^{\infty} e^{\pi - n}$

Problem 13.24. *Compute exactly*

$$\sum_{n=1}^{\infty} \frac{1}{n^2 + n}$$

Problem 13.25. *Compute exactly*

$$\sum_{n=1}^{\infty} \frac{1}{n^2 + 2n}$$

Problem 13.26. *Compute exactly*

$$\sum_{n=1}^{\infty} \frac{1}{n^2 + 5n}$$

Problem 13.27. *Demonstrate convergence of the series:*

$$\sum_{n=1}^{\infty} \frac{\sin(n)}{n^2 + n}$$

Problem 13.28. *Compute exactly*

$$\sum_{n=0}^{\infty} \frac{1}{4n^2 + 8n + 3}$$

13.3. Power Series

In this section we study series again. The good news is that we do not have any additional convergence tests. The bad news is that these series will have variables in them.

Definition 13.12. A **power series** is a series of the form:

$$\sum_{n=0}^{\infty} a_n x^n$$

In a way, a power series is actually an infinite number of different ordinary series, one for each value of x you could substitute into it. The goal of this section will be: given a power series, find values of x which cause it to converge.

Knowledge Box 13.19.

The radius of convergence of a power series

The power series

$$\sum_{n=0}^{\infty} a_n x^n$$

converges in one of three ways:

   1 Only at $x = 0$
   2 For all $|x| < r$ and possibly at $x = \pm r$
   3 For all x

The number r is the **radius of convergence** of the power series. In the first case above, we say the radius of convergence is zero; in the third, we say the radius of convergence is infinite.

Definition 13.13. The **interval of convergence** of a power series is the set of all x where it converges.

Knowledge Box 13.19 implies that the interval of convergence of a power series is one of $[0,0]$, $(-r, r)$, $[-r, r)$, $(-r, r]$, $[-r, r]$ or $(-\infty, \infty)$. The results with an r in them occur in the case where $0 < r < \infty$. Once we have the radius of convergence, in the case where r is positive and finite, we determine the interval of convergence by checking the behavior of the series when we set $x = \pm r$.

Example 13.22. Find the radius and interval of convergence of:

$$\sum_{n=0}^{\infty} x^n$$

Solution:

We start by trying to determine the radius of convergence. Use the ratio test:

$$\lim_{n \to \infty} \left| \frac{x^{n+1}}{x^n} \right| = \lim_{n \to \infty} |x| = |x|$$

This is true because x does not depend on n. The series thus converges when $|x| < 1$, meaning we have convergence for sure when $-1 < x < 1$. This also means the radius of convergence is $r = 1$. We now need to check $x = \pm 1$ to determine the interval of convergence. These values both yield non-converging geometric series:

$$\sum_{n=0}^{\infty} (-1)^n \text{ and } \sum_{n=0}^{\infty} 1$$

So the potential endpoints of the interval of convergence are *not* part of the interval of convergence. This means that the interval of convergence is $(-1, 1)$.

$\Diamond$

Example 13.23. Find the radius of convergence of:

$$\sum_{n=0}^{\infty} \frac{x^n}{2^n}$$

Solution:

Again, use the ratio test.

$$\lim_{n \to \infty} \left| \frac{a_{n+1}}{a_n} \right| = \lim_{n \to \infty} \left| \frac{x^{n+1}/2^{n+1}}{x^n/2^n} \right|$$

$$= \lim_{n \to \infty} \left| \frac{x^{n+1}}{2^{n+1}} \cdot \frac{2^n}{x^n} \right|$$

$$= \lim_{n \to \infty} \left| \frac{x}{2} \right| = \frac{|x|}{2}$$

So the series converges when:

$$-1 < \frac{x}{2} < 1$$

$$-2 < x < 2$$

The radius of convergence is $r = 2$.

$\Diamond$

Example 13.24. Find the radius of convergence of the series:

$$\sum_{n=0}^{\infty} \frac{x^n}{n!}$$

Solution:

Like before:

$$\lim_{n\to\infty} \left| \frac{a_{n+1}}{a_n} \right| = \lim_{n\to\infty} \left| \frac{x^{n+1}/(n+1)!}{x^n/n!} \right|$$

$$= \lim_{n\to\infty} \left| \frac{x}{n+1} \right|$$

$$= |x| \lim_{n\to\infty} \left| \frac{1}{n+1} \right|$$

$$= |x| \cdot 0 = 0$$

Since $0 < 1$ for all values of x, this power series converges everywhere and the radius of convergence is infinite.

$$\Diamond$$

Example 13.25. Find the radius of convergence of the series:

$$\sum_{n=0}^{\infty} \frac{x^n}{n^n}$$

Solution:

This time use the root test.

$$\lim_{n\to\infty} \sqrt[n]{\left| \frac{x^n}{n^n} \right|} = \lim_{n\to\infty} \left| \frac{x}{n} \right|$$

$$= |x| \cdot \lim_{n\to\infty} \frac{1}{n} = 0$$

Which tells us that, as $0 < 1$ for all x, that this sequence converges everywhere.

$$\Diamond$$

Example 13.26. Find the interval of convergence of the series:

$$\sum_{n=1}^{\infty} \frac{x^n}{n^2}$$

Solution:

Another natural job for the ratio test.

$$\lim_{n\to\infty}\left|\frac{a_{n+1}}{a_n}\right| = \lim_{n\to\infty}\left|\frac{x^{n+1}/(n+1)^2}{x^n/n^2}\right|$$

$$= \lim_{n\to\infty}\left|\frac{n^2}{(n+1)^2}\cdot x\right|$$

$$= |x|\cdot\lim_{n\to\infty}\frac{n^2}{n^2+2n+1}$$

$$= |x|\cdot 1 = |x|$$

So $-1 < x < 1$, and the radius of convergence is $r = 1$. To find the interval of convergence we need to check the ends of the interval. These are of the form

$$\sum_{n=1}^{\infty} \frac{(\pm 1)^n}{n^2}$$

The absolute value of each of these is a p-series with $p = 2$. Both endpoints correspond to series that converge in absolute value – meaning they both converge. This makes the interval of convergence $[-1, 1]$.

13.3.1. Using calculus to find series. It is sometimes useful to represent functions as power series. The geometric series formula (Knowledge Box 13.7) does this, for example, for the function $f(x) = \dfrac{1}{1-x}$. We can use calculus to derive power series for other functions.

Example 13.27. If $|x| < 1$, then the geometric series formula tells us that:

$$\sum_{n=0}^{\infty} x^n = \frac{1}{1-x}$$

If we plug $-x$ into this identity we find that:

$$\sum_{n=0}^{\infty} (-x)^n = \frac{1}{1-(-x)}$$

$$\sum_{n=0}^{\infty} (-1)^n x^n = \frac{1}{1+x}$$

or

$$\frac{1}{x+1} = 1 - x + x^2 - x^3 + x^4 - \cdots$$

Integrate both sides and we get:

$$\ln(x+1) + C = x - \frac{1}{2}x^2 + \frac{1}{3}x^3 - \frac{1}{4}x^4 + \frac{1}{5}x^5 - \cdots$$

Plug in $x = 0$

$$\ln(1) + C = 0$$

$$C = 0$$

Which means, at least when $|x| < 1$,

$$\ln(x+1) = \sum_{n=1}^{\infty} \frac{(-1)^n x^n}{n}$$

$\Diamond$

This shows that we can find power series that, when they converge, are equal to familiar transcendental functions. The next section gives another technique for doing this, to be used when the sequences you already know don't give you enough power. Let's encode this as a Knowledge Box.

Knowledge Box 13.20.

Using calculus to modify power series

Suppose that $f(x) = \sum_{n=0}^{\infty} a_n x^n$. Then:

$$\int f(x) \cdot dx = \sum_{n=0}^{\infty} \frac{a_n}{n+1} x^{n+1} \text{ and } f'(x) = \sum_{n=1}^{\infty} n \cdot a_n \, x^{n-1}$$

Example 13.28. Find a power series for $\tan^{-1}(x)$.

Solution:

We already have seen that:

$$\frac{1}{u+1} = 1 - u + u^2 - u^3 + u^4 - \cdots$$

If we substitute $u = x^2$ we obtain:

$$\frac{1}{x^2+1} = 1 - x^2 + x^4 - x^6 + x^8 - \cdots$$

Integrate and we get:

$$\tan^{-1}(x) + C = x - \frac{1}{3}x^3 + \frac{1}{5}x^5 - \frac{1}{7}x^7 + \frac{1}{9}x^9 - \cdots$$

Substitute in $x = 0$ and we get $C = 0$. So we obtain the power series:

$$\tan^{-1}(x) = \sum_{n=0}^{\infty} \frac{(-1)^n x^{2n+1}}{2n+1}$$

$\Diamond$

Example 13.29. What is the interval of convergence for the series for $\tan^{-1}(x)$ found in Example 13.28?

Solution:

Apply the ratio test.

$$\lim_{n\to\infty} \left| \frac{a_{n+1}}{a_n} \right| = \lim_{n\to\infty} \left| \frac{x^{2n+3}/(2n+3)}{x^{2n+1}/(2n+1)} \right|$$

$$= \lim_{n\to\infty} \left| x^2 \frac{2n+1}{2n+3} \right|$$

$$= |x^2| \lim_{n\to\infty} \frac{2n+1}{2n+3}$$

$$= |x^2| \cdot 1 = x^2$$

So $x^2 < 1$ when $-1 < x < 1$, making the radius of convergence $r = 1$. Now check the endpoints $x = \pm 1$. If $x = -1$ the resulting series is:

$$\sum_{n=0}^{n} \frac{(-1)^n (-1)^{2n+1}}{2n+1} = \sum_{n=0}^{n} \frac{(-1)^{n+1}}{2n+1}$$

which converges by the alternating series test. If $x = 1$ we get:

$$\sum_{n=0}^{n} \frac{(-1)^n}{2n+1}$$

which also converges by the alternating series test. The interval of convergence is thus [-1,1].

$\Diamond$

It is possible to find a power series by just using algebra on a known series.

Example 13.30. Find a power series for:

$$f(x) = \frac{x^2}{1 - x^2}$$

Solution:
We start with the known form:

$$\frac{1}{1 - u} = 1 + u + u^2 + u^3 + u^4 + \cdots$$

Substitute in $u = x^2$ and we get:

$$\frac{1}{1 - x^2} = 1 + x^2 + x^4 + x^6 + x^8 + \cdots$$

Now multiply both sides by x^2 and we get:

$$\frac{x^2}{1 - x^2} = x^2 + x^4 + x^6 + x^8 + x^{10} + \cdots$$

So

$$f(x) = \sum_{n=0}^{\infty} x^{2n+2}$$

$\Diamond$

Example 13.31. Find a power series for:

$$g(x) = \ln(x^2 + 1)$$

Solution:
Start with the known result for $\dfrac{1}{1 + x^2}$.

$$\frac{1}{1 + x^2} = 1 - x^2 + x^4 - x^6 + x^8 - \cdots$$

$$\frac{2x}{1 + x^2} = 2x - 2x^3 + 2x^5 - 2x^7 + 2x^9 - \cdots$$

$$\int \frac{2x}{1 + x^2} \cdot dx = \int \left(2x - 2x^3 + 2x^5 - 2x^7 + 2x^9 - \cdots \right) dx$$

$$\ln(x^2 + 1) + C = 2 \left(\frac{1}{2}x^2 - \frac{1}{4}x^4 + \frac{1}{6}x^6 - \frac{1}{8}x^8 + \frac{1}{10}x^{10} - \cdots \right)$$

Set $x = 0$ and we get $C = 0$.

So: $\ln(x^2 + 1) = \sum_{n=0}^{\infty} \frac{2 \cdot (-1)^n x^{2n+2}}{2n + 2}$

$\Diamond$

An integral that is both important in statistics, because it is related to the normal distribution, and famous for not having a closed form solution is

$$\int e^{-x^2/2} \cdot dx$$

The next example shows why power series are useful – it is easy to get a power series for this integral.

Example 13.32. Find a power series for

$$F(x) = \int e^{-x^2/2} \cdot dx$$

Solution:

$$e^x = \sum_{n=0}^{\infty} \frac{x^n}{n!}$$

$$e^{-x^2/2} = \sum_{n=0}^{\infty} \frac{\left(-x^2/2\right)^n}{n!}$$

$$e^{-x^2/2} = \sum_{n=0}^{\infty} \frac{(-1)^n x^{2n}}{2^n \cdot n!}$$

$$\int e^{-x^2/2} \cdot dx = \int \sum_{n=0}^{\infty} \frac{(-1)^n x^{2n}}{2^n \cdot n!} \cdot dx$$

$$\int e^{-x^2/2} \cdot dx = \sum_{n=0}^{\infty} \frac{(-1)^n x^{2n+1}}{(2n+1) \cdot 2^n \cdot n!}$$

This power series is useful for computing probabilities connected with the normal distribution.

◇

Problems

Problem 13.29. *Find the radius of convergence for each of the following power series.*

(a) $\displaystyle\sum_{n=1}^{\infty} \frac{x^n}{n}$

(b) $\displaystyle\sum_{n=0}^{\infty} \frac{x^n}{5^n}$

(c) $\displaystyle\sum_{n=0}^{\infty} \frac{x^n}{(2n+1)!}$

(d) $\displaystyle\sum_{n=0}^{\infty} \frac{x^{2n+1}}{n!}$

(e) $\displaystyle\sum_{n=1}^{\infty} nx^n$

(f) $\displaystyle\sum_{n=0}^{\infty} \frac{nx^n}{3n+1}$

(g) $\displaystyle\sum_{n=0}^{\infty} \frac{x^n}{2n^2+4}$

(h) $\displaystyle\sum_{n=0}^{\infty} \frac{x^n}{n^3+1}$

Problem 13.30. *Demonstrate that the radius of convergence of*

$$\sum_{n=0}^{\infty} a_n x^n \ \text{ and } \ \sum_{n=0}^{\infty} c a_n x^n$$

are the same for any constant c.

Problem 13.31. *Compute the radius of convergence of*

$$\sum_{n=1}^{\infty} \frac{x^n}{c^n}$$

where $c > 0$ is a constant.

Problem 13.32. *If*

$$f(x) = \frac{1}{1-x} + \frac{1}{1-2x} + \frac{1}{1-3x}$$

find and simplify a power series for $f(x)$.

Problem 13.33. *If*

$$g(x) = \frac{x}{2-x} + \frac{x}{3-x}$$

find and simplify a power series for $g(x)$.

Problem 13.34. *If*

$$h(x) = \frac{x^2}{2-x} + \frac{x^2}{1-2x}$$

find and simplify a power series for $h(x)$.

Problem 13.35. *Find the interval of convergence for each of the following power series.*

(a) $\displaystyle\sum_{n=0}^{\infty} \frac{x^n}{4^n}$

(b) $\displaystyle\sum_{n=1}^{\infty} \frac{x^n}{(2n+1)!}$

(c) $\displaystyle\sum_{n=0}^{\infty} \frac{nx^n}{2n^3+7}$

(d) $\displaystyle\sum_{n=2}^{\infty} \frac{n^2 x^n}{n^4-1}$

(e) $\displaystyle\sum_{n=0}^{\infty} \frac{x^n}{n^n}$

(f) $\displaystyle\sum_{n=0}^{\infty} \frac{(-2x)^n}{n}$

Problem 13.36. *Using calculus, find a power series for:*

$$g(x) = \ln\left(\frac{1+x}{1-x}\right)$$

Problem 13.37. *Using only algebra, and known series, find a power series for:*

$$h(x) = \frac{3x^2}{1+x^3}$$

Problem 13.38. *Find a power series for:*

$$q(x) = \ln\left(1+x^3\right)$$

13.4. Taylor Series

In the last section, we managed to create power series for several of the standard transcendental functions. Notably absent were $\sin(x)$, $\cos(x)$, and e^x. The key to these is **Taylor series**. We need a little added notation to build Taylor series. We will denote the nth derivative of $f(x)$ by $f^{(n)}(x)$. Notice that this means that $f^{(0)}(x) = f(x)$.

Knowledge Box 13.21.

Taylor series

If $f(x)$ is a function that can be differentiated any number of times,

$$f(x) = \sum_{n=0}^{\infty} \frac{f^{(n)}(c)(x-c)^n}{n!}$$

This formula is called the **Taylor series expansion of $f(x)$ at c**. The constant c is called the **center** of the expansion.

Example 13.33. Use Taylor's formula to find a power series centered at $c = 0$ for $f(x) = e^x$ and find its radius of convergence.

Solution:

The function $f(x) = e^x$ is a very good choice for a first demonstration of the Taylor expansion. This is because *every* derivative of of e^x is e^x. In other words

$$f^{(n)}(x) = e^x \text{ and so } f^{(n)}(0) = 1$$

Applying the formula we get:

$$e^x = \sum_{n=0}^{\infty} \frac{f^{(n)}(0)(x-0)^n}{n!} = \sum_{n=0}^{\infty} \frac{1 \cdot x^n}{n!} = \sum_{n=0}^{\infty} \frac{x^n}{n!}$$

The radius of convergence of this series was computed in Example 13.24 – it is $r = \infty$. This means that the expansion of e^x converges everywhere.

$\Diamond$

Here is an interesting calculation. Recall Euler's identity from Knowledge Box 1.43: $e^{ix} = i\sin(x) + \cos(x)$. Let's look at the Taylor series for e^{ix}:

$$e^{ix} = \sum_{n=0}^{\infty} \frac{(ix)^n}{n!}$$

$$= \sum_{n=0}^{\infty} \frac{i^n x^n}{n!}$$

$$= 1 + ix - \frac{1}{2}x^2 - i\frac{1}{3!}x^3 + \frac{1}{4!}x^4 + i\frac{1}{5!}x^5 - \frac{1}{6!}x^6 - i\frac{1}{7!}x^7 + \cdots$$

$$= \sum_{n=0}^{\infty} \frac{(-1)^n x^{2n}}{(2n)!} + i\sum_{n=0}^{\infty} \frac{(-1)^n x^{2n+1}}{(2n+1)!}$$

From Euler's identity, we get that the real part of the expression above is cosine, and the imaginary part is sine. This gives us power series for $\sin(x)$ and $\cos(x)$.

Knowledge Box 13.22.

Taylor series for $\sin(x)$ and $\cos(x)$ and e^x

$$e^x = \sum_{n=0}^{\infty} \frac{x^n}{n!}$$

$$\sin(x) = \sum_{n=0}^{\infty} \frac{(-1)^n x^{2n+1}}{(2n+1)!}$$

$$\cos(x) = \sum_{n=0}^{\infty} \frac{(-1)^n x^{2n}}{(2n)!}$$

with all three expansions having an interval of convergence of $(-\infty, \infty)$.

The calculations above show that we can use algebraic manipulation to create power series based on the power series we get from Taylor expansions. Let's do a couple more examples.

Example 13.34. Find a power series for $h(x) = e^{2x}$.

Solution:

$$e^x = \sum_{n=0}^{\infty} \frac{x^n}{n!}$$

$$e^{2x} = \sum_{n=0}^{\infty} \frac{(2x)^n}{n!}$$

$$= \sum_{n=0}^{\infty} \frac{2^n x^n}{n!}$$

$$= \sum_{n=0}^{\infty} \frac{2^n}{n!} x^n$$

$\Diamond$

Example 13.35. Find the Taylor expansion for $s(x) = \cos(x)$ using $c = \dfrac{\pi}{2}$.

Solution:
The Taylor formula needs the nth derivatives at $\dfrac{\pi}{2}$. Let's start by computing these numbers.

n	$s^{(n)}(x)$	$s^{(n)}\left(\frac{\pi}{2}\right)$
0	$\cos(x)$	0
1	$-\sin(x)$	-1
2	$-\cos(x)$	0
3	$\sin(x)$	1
4	$\cos(x)$	0
5	$-\sin(x)$	-1

Which is enough to notice the values repeat every four steps. Plug these values into the Taylor expansion formula with $c = \pi/2$ and we get:

$$\cos(x) = -\frac{(x - \pi/2)}{1!} + \frac{(x - \pi/2)^3}{3!} - \frac{(x - \pi/2)^5}{5!} + \frac{(x - \pi/2)^7}{6!}$$

$$\cos(x) = \sum_{n=0}^{\infty} \frac{(-1)^{n+1}(x - \pi/2)^{2n+1}}{(2n + 1)!}$$

$\Diamond$

Notice that the expansion of $\cos(x)$ with a center of $c = \dfrac{\pi}{2}$ has the same form as the negative of the Taylor expansion of $-\sin(x)$ at $c = 0$. This is a fairly extreme way of proving the identity

$$\cos(x) = -\sin(x - \pi/2)$$

13.4.1. Taylor Polynomials. If we take the first n terms of a Taylor series for $f(x)$ the result is called the *Taylor polynomial of degree n* for $f(x)$. Taylor polynomials are approximations to the function they are derived from and, as we already know, polynomials are among the very nicest functions to work with.

Example 13.36. Find a fifth degree Taylor polynomial for $f(x) = \sin(x)$ centered at $c = 0$.

Solution:
Since the Taylor series for $\sin(x)$ centered at $c = 0$ is

$$\sum_{n=0}^{\infty} \frac{(-1)^n \, x^{2n+1}}{(2n+1)!}$$

we need only extract the terms of degree five or less from the infinite series. This gives us a solution of

$$p(x) = x - \frac{x^3}{6} + \frac{x^5}{120}$$

A natural question is "how good is this polynomial as an approximation to $\sin(x)$? Let's graph both functions on the same set of axes.

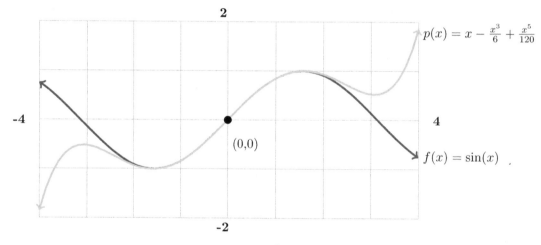

In the range $-2 \le x \le 2$ the polynomial and the sine function agree really well – after that they diverge from one another and the polynomial shoots off to positive and negative infinity. Not too surprisingly, the Taylor polynomial does a good job of approximating the function it was drawn from near the center c for expansion. How do we figure out where the polynomial is a good enough approximation?

If $T(x)$ is the Taylor series for a function $f(x)$ centered at c and $T_n(x)$ is the nth degree Taylor polynomial for that function, then we set $R_n(x) = T(x) - T_n(x)$ to get the *remainder* for the polynomial. Algebraically rearranging the terms:

$$T(x) = T_n(x) + R_n(x)$$

With this definition we can give Taylor's inequality for the remainder of a power series.

Knowledge Box 13.23.

Taylor's Inequality

Suppose we are looking at a Taylor polynomial for the function $f(x)$ in the interval $|x - a| \leq d$ and that $|f^{(n+1)}(x)| \leq M$ everywhere in this interval. Then for values of x in the interval we have that

$$|R_n(x)| \leq \frac{M}{(n+1)!} |x - a|^{n+1}$$

It is not immediately obvious how to use this result, so let's do a couple examples.

Example 13.37. Find a bound on the error of approximation of $T_5(x)$ for $f(x) = \sin(x)$ on the interval $-2 \leq x \leq 2$. Note that this is the interval that looked good in the graph associated with Example 13.36.

Solution:

We proceed by applying Taylor's inequality. The derivative of $\sin(x)$ are all sine and cosine functions. This means that $f^{(n+1)}(x)$ are always at most 1 so we may set $M = 1$ for Taylor's inequality. The value of a is zero so $|x - a| \leq 2$ on the interval we are using. Plugging these values into the formula for Taylor's inequality we get:

$$|R_5(x)| \leq \frac{1}{6!} \cdot 2^6 = \frac{64}{720} \cong 0.089$$

The error of $T_5(x)$ on $-2 \leq x \leq 2$ is *at most* 0.089. Not bad.

◇

Notice that Taylor's inequality has $(n + 1)!$ in the denominator. That means if we were to use

$$T_7(x) = x - \frac{x^3}{6} + \frac{x^5}{120} - \frac{x^7}{5040}$$

for $f(x) = \sin(x)$ on the interval $-2 \leq x \leq 2$ then the estimate of maximum error drops to

$$|R_7(x)| \leq \frac{1}{8!} \cdot 2^8 = \frac{256}{40320} \cong 0.0063$$

The factorial in the denominator lets the error drop really quickly.

◇

The hard part of using Taylor's inequality is finding the constant M. Standard practice is to simply find the largest value of $f^{n+1}(x)$ in the interval and live with it. The fact that both sine and cosine are bounded in absolute value by 1 make them especially easy functions to work with. Let's do an example with an exponential function.

Example 13.38. Suppose we are approximating $f(x) = e^x$ in the interval $-3 \le x \le 3$ with $T_n(x)$. What value of n makes the Taylor's inequality estimate of error at most 0.1?

Solution:

If $f(x) = e^x$ then $f^{(n+1)} = e^x$, which is very convenient. We again have that $a = 0$ so it is pretty easy to see that

$$|f^{(n+1)}(x)| \le e^3$$

everywhere on the interval. So, we set $M = e^3$. The largest value of $|x - a|$ on the interval is 3, so we need to find the smallest value of n that makes

$$\frac{e^3}{(n+1)!}3^{n+1} < 0.1$$

The simplest way to do this is to tabulate.

n	$|R_n(x)| \le \lvert \frac{e^3}{(n+1)!} \cdot 3^{n+1} \rvert$	n	$|R_n(x)| \le \lvert \frac{e^3}{(n+1)!} \cdot 3^{n+1} \rvert$
1	90.3849161543	6	8.7156883435
2	90.3849161543	7	3.2683831288
3	67.7886871158	8	1.0894610429
4	40.6732122695	9	0.3268383129
5	20.3366061347	10	0.0891377217

So the first value of n with an acceptable error is $n = 10$ and $T_{10}(x)$ is good enough to approximate $f(x) = e^x$ in the range $-3 \le x \le 3$.

The error estimates given by Taylor's inequality are not the best possible – they are actually fairly conservative. They usually over-estimate the error. If you take a course in numerical analysis later in your career you may study methods for building better error estimates. There is also a lot of room to be clever with how you use Taylor polynomials. The sine and cosine function are periodic, and so if you know their values even on a very small interval, like $[0, \pi/2]$, you can use those values to deduce any other value.

Problems

Problem 13.39. *Using the Taylor series formula, verify the formula for* $y = \sin(x)$.

Problem 13.40. *Using the Taylor series formula, verify the formula for* $y = \cos(x)$.

Problem 13.41. *Find the Taylor expansion of* $f(x) = \sin(x)$ *using a center of* $c = \pi$. *Use the formula for Taylor expansion.*

Problem 13.42. *Find the Taylor expansion of* $f(x) = \sin(x)$ *using a center of* $c = \pi/2$. *Use the formula for the Taylor expansion.*

Problem 13.43. *Find the Taylor expansion of* $f(x) = \cos(x)$ *using a center of* $c = \pi/4$. *Use the formula for the Taylor expansion. Warning: this is a little messy.*

Problem 13.44. *Using the Taylor series for* $\sin(x)$, $\cos(x)$ *and* e^x, *prove Euler's identity:*
$$e^{i\theta} = i\sin(x) + \cos(x)$$

Problem 13.45. *Using any method, find power series for the following functions.*

(a) $f(x) = e^{-x}$

(b) $g(x) = xe^{2x}$

(c) $h(x) = \cos 2x$

(d) $r(x) = \sin(x^2)$

(e) $s(x) = \ln(x^4)$

(f) $q(x) = (e^x + e^{-x})/2$

(g) $a(x) = \tan^{-1}(3x)$

(h) $b(x) = \dfrac{3x^2}{1 - x^4}$

Problem 13.46. *For each of the series you found in Problem 13.45, find the radius and interval of convergence.*

Problem 13.47. *Prove that the Taylor series for a polynomial function* $p(x)$ *is just the polynomial itself.*

Problem 13.48. *Find a power series expansion for*
$$f(x) = \frac{1}{x^2 - 3x + 2}$$

Problem 13.49. *Find a power series expansion for*
$$f(x) = \frac{1}{4 - 4x + x^2}$$

Problem 13.50. *Find a power series expansion for*
$$f(x) = \frac{1}{x^3 - 6x^2 + 11x + 6}$$

Problem 13.51. *Find a power series expansion for*
$$f(x) = \frac{1}{x^3 + x}$$

Problem 13.52. *If*
$$f(x) = p(x)e^x$$
where $p(x)$ *is a polynomial, demonstrate that* $f(x)$ *has a power series expansion with radius of convergence* $r = \infty$.

Problem 13.53. *Find the Taylor polynomial of degree n for the given function with the given center c.*

(a) $f(x) = \cos(x)$ *for* $n = 6$ *at* $c = 0$,

(b) $g(x) = \sin(2x)$ *for* $n = 7$ *at* $c = 0$,

(c) $h(x) = e^x$ *for* $n = 5$ *at* $c = 0$,

(d) $r(x) = \log(x)$ *for* $n = 3$ *at* $c = 1$,

(e) $s(x) = \tan^{-1}(x)$ *for* $n = 8$ *at* $c = 0$,

(f) $q(x) = x^2 + 3x + 5$ *for* $n = 2$ *at* $c = 1$.

Problem 13.54. *Suppose we have* $T_5(x)$ *for* $f(x) = e^x$ *at* $c = 0$. *Compute a bound on the size of* $R_5(x)$ *with Taylor's inequality.*

Problem 13.55. *Find the smallest n for which* $T_n(x)$ *on* $f(x) = \cos(x)$ *has* $|R_n(x)| < 0.01$ *on* $-3 \le x \le 3$ *with* $c = 0$.

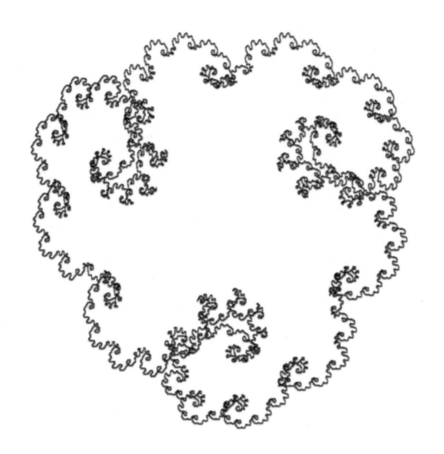

Useful Formulas

A.2. Trigonometric identites

Trig function definitions from sine and cosine.

- $\tan(\theta) = \dfrac{\sin(\theta)}{\cos(\theta)}$

- $\cot(\theta) = \dfrac{\cos(\theta)}{\sin(\theta)}$

- $\tan(\theta) = \dfrac{1}{\cot(\theta)}$

- $\sec(\theta) = \dfrac{1}{\cos(\theta)}$

- $\csc(\theta) = \dfrac{1}{\sin(\theta)}$

A.1. Powers, logs, and exponentials

Rules for Powers.

- $a^{-n} = \dfrac{1}{a^n}$

- $a^n \times a^m = a^{n+m}$

- $\dfrac{a^n}{a^m} = a^{n-m}$

- $(a^n)^m = a^{n \times m}$

Log and exponential algebra.

- $b^{\log_b(c)} = c$

- $\log_b(b^a) = a$

- $\log_b(xy) = \log_b(x) + \log_b(y)$

- $\log_b\left(\dfrac{x}{y}\right) = \log_b(x) - \log_b(y)$

- $\log_b(x^y) = y \cdot \log_b(x)$

- $\log_c(x) = \dfrac{\log_b(x)}{\log_b(c)}$

- If $\log_b(c) = a$, then $c = b^a$

Periodicity identities.

- $\sin(x + 2\pi) = \sin(x)$

- $\cos(x + 2\pi) = \cos(x)$

- $\sin(x) = \cos\left(x - \frac{\pi}{2}\right)$

- $\tan(x) = -\cot\left(x - \frac{\pi}{2}\right)$

- $\sec(x) = \csc\left(x + \frac{\pi}{2}\right)$

- $\cos(-x) = \cos(x)$

- $\sin(-x) = -\sin(x)$

- $\tan(x) = -\tan(x)$

- $\sin(x + \pi) = -\sin(x)$

- $\cos(x + \pi) = -\cos(x)$

- $\tan(x + \pi) = \tan(x)$

The Pythagorean identities.

- $\sin^2(\theta) + \cos^2(\theta) = 1$

- $\tan^2(\theta) + 1 = \sec^2(\theta)$

- $1 + \cot^2(\theta) = \csc^2(\theta)$

The law of sines, the law of cosines.

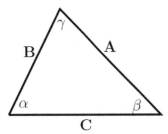

The law of sines

$$\frac{A}{\sin(\alpha)} = \frac{B}{\sin(\beta)} = \frac{C}{\sin(\gamma)}$$

The law of cosines

$$C^2 = A^2 + B^2 + 2AB \cdot \cos(\gamma)$$

The laws refer to the diagram.

Sum, difference, double angle.

- $\sin(\alpha+\beta) = \sin(\alpha)\cos(\beta)+\sin(\beta)\cos(\alpha)$

- $\cos(\alpha+\beta) = \cos(\alpha)\cos(\beta)-\sin(\alpha)\sin(\beta)$

- $\sin(\alpha-\beta) = \sin(\alpha)\cos(\beta)-\sin(\beta)\cos(\alpha)$

- $\cos(\alpha-\beta) = \sin(\alpha)\sin(\beta)+\cos(\alpha)\cos(\beta)$

- $\sin(2\theta) = 2\sin(\theta)\cos(\theta)$

- $\cos(2\theta) = \cos^2(\theta) - \sin^2(\theta)$

- $\cos^2(\theta/2) = \dfrac{1 + \cos(\theta)}{2}$

- $\sin^2(\theta/2) = \dfrac{1 - \cos(\theta)}{2}$

A.3. Speed of function growth

- Logarithms grow faster than constants

- Positive powers of x grow faster than logarithms

- Larger positive powers of x grow faster than smaller positive powers of x

- Exponentials (with positive exponents) grow faster than positive powers of x

- Exponentials with larger exponents grow faster than those with smaller exponents

A.4. Derivative rules

- If $f(x) = x^n$ then
$$f'(x) = nx^{n-1}$$

- $(f(x) + g(x))' = f'(x) + g'(x)$

- $(a \cdot f(x))' = a \cdot f'(x)$

- If $f(x) = \ln(x)$, then $f'(x) = \dfrac{1}{x}$

- If $f(x) = \log_b(x)$, then $f'(x) = \dfrac{1}{x\ln(b)}$

- If $f(x) = e^x$, then $f'(x) = e^x$

- If $f(x) = a^x$, then $f'(x) = \ln(a) \cdot a^x$

- $(\sin(x))' = \cos(x)$

- $(\cos(x))' = -\sin(x)$

- $(\tan(x))' = \sec^2(x)$

- $(\cot(x))' = -\csc^2(x)$

- $(\sec(x))' = \sec(x)\tan(x)$

- $(\csc(x))' = -\csc(x)\cot(x)$

- $\left(\sin^{-1}(x)\right)' = \dfrac{1}{\sqrt{1-x^2}}$

- $\left(\cos^{-1}(x)\right)' = \dfrac{-1}{\sqrt{1-x^2}}$

- $\left(\tan^{-1}(x)\right)' = \dfrac{1}{1+x^2}$

- $\left(\cot^{-1}(x)\right)' = \dfrac{-1}{1+x^2}$

- $\left(\sec^{-1}(x)\right)' = \dfrac{1}{|x|\sqrt{x^2-1}}$

- $\left(\csc^{-1}(x)\right)' = \dfrac{-1}{|x|\sqrt{x^2-1}}$

The product rule

$$(f(x) \cdot g(x))' = f(x)g'(x) + f'(x)g(x)$$

The quotient rule

$$\left(\frac{f(x)}{g(x)}\right)' = \frac{g(x)f'(x) - f(x)g'(x)}{g^2(x)}$$

The reciprocal rule

$$\left(\frac{1}{f(x)}\right)' = \frac{-f'(x)}{f^2(x)}$$

The chain rule

$$(f(g(x)))' = f'(g(x)) \cdot g'(x)$$

A.5. Sums and Factorization Rules

Factorizations of polynomials.

- $x^2 - a^2 = (x-a)(x+a)$

- $x^3 - a^3 = (x-a)(x^2 + ax + a^2)$

- $x^3 + a^3 = (x+a)(x^2 - ax + a^2)$

- $x^n - a^n = (x-a)(x^{n-1} + ax^{n-2} + \cdots a^{n-2}x + a^{n-1})$

Algebra of summation.

- $\sum_{i=a}^{b} f(i) + g(i) = \sum_{i=a}^{b} f(i) + \sum_{i=a}^{b} g(i)$

- $\sum_{i=a}^{b} c \cdot f(i) = c \cdot \sum_{i=a}^{b} f(i)$

Closed summation formulas.

- $\sum_{i=1}^{n} 1 = n$

- $\sum_{i=1}^{n} i = \dfrac{n(n+1)}{2}$

- $\sum_{i=1}^{n} i^2 = \dfrac{n(n+1)(2n+1)}{6}$

- $\sum_{i=1}^{n} i^3 = \dfrac{n^2(n+1)^2}{4}$

A.5.1. Geometric series.

- $\sum_{k=0}^{n} a^k = \dfrac{a^{n+1}-1}{a-1}$

- $\sum_{n=0}^{\infty} a^n = \dfrac{1}{1-a}$ if $|a| < 1$

- $\sum_{n=0}^{\infty} qa^n = \dfrac{q}{1-a}$ if $|a| < 1$

A.6. Vector Arithmetic

Vector arithmetic and algebra.

- $c \cdot \vec{v} = (cv_1, cv_2, \ldots, cv_n)$

- $\vec{v} + \vec{w} = (v_1 + w_1, v_2 + w_2, \ldots, v_n + w_n)$

- $\vec{v} - \vec{w} = (v_1 - w_1, v_2 - w_2, \ldots, v_n - w_n)$

- $\vec{v} \cdot \vec{w} = v_1 w_1 + v_2 w_2 + \ldots + v_n w_n$

- $c \cdot (\vec{v} + \vec{w}) = c \cdot \vec{v} + c \cdot \vec{w}$

- $c \cdot (d \cdot \vec{v}) = (cd) \cdot \vec{v}$

- $\vec{v} + \vec{w} = \vec{w} + \vec{v}$

- $\vec{u} \cdot (\vec{v} + \vec{w}) = \vec{u} \cdot \vec{v} + \vec{u} \cdot \vec{w}$

Cross product of vectors.

- $\vec{v} \times \vec{w} = (v_2 w_3 - v_3 v_2,\ v_3 w_1 - v_1 w_3,\ v_1 w_2 - v_2 w_1)$

Formula for the angle between vectors

$$\cos(\theta) = \frac{\vec{v} \cdot \vec{w}}{|v||w|}$$

A.7. Polar and Rectangular Conversion

- $x = r \cdot \cos(\theta)$

- $y = r \cdot \sin(\theta)$

- $r = \sqrt{x^2 + y^2}$

- $\theta = \tan^{-1}(y/x)$

A.8. Integral Rules

Basic Integration Rules.

- $\int x^n \cdot dx = \dfrac{1}{n+1} x^{n+1} + C$

- $\int a \cdot f(x) \cdot dx = a \cdot \int f(x) \cdot dx$

- $\int (f(x) + g(x)) \cdot dx =$
  $\int f(x) \cdot dx + \int g(x) \cdot dx$

Log and exponent.

- $\int \dfrac{1}{x} \cdot dx = \ln(x) + C$

- $\int e^x \cdot dx = e^x + C$

Trig and inverse trig.

- $\int \sin(x) \cdot dx = -\cos(x) + C$

- $\int \cos(x) \cdot dx = \sin(x) + C$

- $\int \sec^2(x) \cdot dx = \tan(x) + C$

- $\int \csc^2(x) \cdot dx = -\cot(x) + C$

- $\int \sec(x)\tan(x) \cdot dx = \sec(x) + C$

- $\int \csc(x)\cot(x) \cdot dx = -\cot(x) + C$

- $\int \dfrac{1}{\sqrt{1-x^2}} \cdot dx = \sin^{-1}(x) + C$

- $\int \dfrac{1}{1+x^2} \cdot dx = \tan^{-1}(x) + C$

- $\int \dfrac{1}{x\sqrt{x^2-1}} \cdot dx = \sec^{-1}(|x|) + C$

- $\int \tan(x) \cdot dx = \ln|\sec(x)| + C$

- $\int \sec(x) \cdot dx = \ln|\sec(x) + \tan(x)| + C$

- $\int \tan(x) \cdot dx = \ln|\sec(x)| + C$

- $\int \sec(x) \cdot dx = \ln|\tan(x) + \sec(x)| + C$

- $\int \sec^2(x) \cdot dx = \tan(x) + C$

- $\int \csc^2(x) \cdot dx = -\cot(x) + C$

- $\int \sec(x)\tan(x) \cdot dx = \sec(x) + C$

- $\int \csc(x)\cot(x) \cdot dx = -\csc(x) + C$

Integration by parts.

$$\int U \cdot dV = UV - \int V \cdot dU$$

Exponential/polynomial shortcut.

$$\int p(x)e^x \cdot dx =$$

$$(p(x) - p'(x) + p''(x) - p'''(x) + \cdots) e^x + C$$

Volume, surface, arc length.

Volume of rotation, disks.

$$V = \pi \int_a^b f(x)^2 \cdot dx$$

Volume of rotation, cylindrical shells.

$$V = 2\pi \int_{x=a}^{x=b} x \cdot f(x) \cdot dx$$

Volume of rotation with washers.

$$V = \pi \int_a^b \left(f_1(x)^2 - f_2(x)^2 \right) \cdot dx$$

Differential of arc length.

$$ds = \sqrt{(y')^2} \cdot dx = \sqrt{(f'(x))^2} \cdot dx$$

Arc length.

$$S = \int ds$$

Surface area of rotation.

$$A = 2\pi \int_a^b f(x) \cdot ds$$

A.9. Taylor Series

If $f(x)$ is a function that can be differentiated any number of times,

$$f(x) = \sum_{n=0}^{\infty} \frac{f^{(n)}(c)(x-c)^n}{n!}$$

Special Taylor Series.

$$e^x = \sum_{n=0}^{\infty} \frac{x^n}{n!}$$

$$\sin(x) = \sum_{n=0}^{\infty} \frac{(-1)^n x^{2n+1}}{(2n+1)!}$$

$$\cos(x) = \sum_{n=0}^{\infty} \frac{(-1)^n x^{2n}}{(2n)!}$$

Taylor's Inequality.

Suppose we are looking at a Taylor polynomial for the function $f(x)$ in the interval $|x - a| \leq d$ and that $|f^{(n+1)}(x)| \leq M$ everywhere in this interval. Then for values of x in the interval we have that

$$|R_n(x)| \leq \frac{M}{(n+1)!}|x - a|^{n+1}$$

Index

Made in the USA
Columbia, SC
13 August 2018